SOLUTIONS MANUAL TO ACCOMPANY
KINEMATICS AND DYNAMICS OF MACHINERY

Please return to 140 Reber Bldg. at end of semester

Charles E. Wilson
J. Peter Sadler

Second Edition

Prepared by

CHARLES E. WILSON

Department of Mechanical and Industrial Engineering
New Jersey Institute of Technology

and

J. PETER SADLER

Department of Mechanical Engineering
University of Kentucky

HarperCollins*CollegePublishers*

Solutions Manual to accompany Wilson/Sadler's KINEMATICS AND DYNAMICS OF MACHINERY, Second Edition

ISBN: 0-06-364592-0

92 93 94 95 96 9 8 7 6 5 4 3 2 1

This solutions manual has been prepared as a supplement to the second edition of KINEMATICS AND DYNAMICS OF MACHINERY. It is the authors' hope that it will be helpful to professors using the text. In solving the problems, the authors used various analytical and graphical methods, aided by various types of software. It is expected that professors assigning problems from the text will select solution methods and software in accordance with goals they have set for their students. The authors will be grateful for any corrections or comments related to the text or solutions manual.

Charles E. Wilson
J. Peter Sadler

CHAPTER ONE

1.1 a. The ball-joint (spherical pair) has three degrees-of-freedom, the prismatic pair one, and the cylindrical pair two. The number of degrees-of-freedom of the spatial linkage is given by

$$DF_{(spatial)} = 6(n_L - n_J - 1) + \Sigma f_i$$

$$= 6(4 - 3 - 1) + 3 + 1 + 2$$

$$= 6 \text{ degrees-of-freedom}$$

(We do not need the inequality sign for this open-loop chain).

b. Treating the construction equipment schmatic as a spatial linkage:

$$DF_{(spatial)} \geq 6(n_L - n_J - 1) + \Sigma f_i$$

$$\geq 6(9 - 11 - 1) + 9 + 2 \times 2$$

$$DF_{(spatial)} \geq -5$$

c. Treating the construction equipment schematic as a planar linkage, where a sliding pair has only one degree-of-freedom in plane motion:

$$DF_{(planar)} = 3(n_L - n_J - 1) + \Sigma f_i$$

$$DF_{(planar)} = 3(9 - 11 - 1) + 9 + 2$$

$$DF_{(planar)} = 2$$

d. The planar motion assumption applies if motion occurs in a plane or in a set of parallel planes. The planes of motion of the links must all be parallel. The revolute joint axes must be perpendicular to those planes. These conditions do apply to the construction machinery. The operator controls the linkage via the two hydraulic cylinders.

1.2(a) For the linkage as sketched originally,

$DF = 3(n_L-1) - 2n_{J'} = 3(9-1) - 2x12 = 0$

a) Removing any single link (one through seven) reduces n_L by one and $n_{J'}$ by two from which

$DF = 3(8-1) - 2x10 = 1.$

b) Also, we obtain DF = 1 by removing links 1, 2 and 5; or 3, 4 and 5; or 1, 2, 3, 4, and 5.

c) DF = 1 if all links are removed except link 1, or 6 or the slider.

d) DF = 1 if the slider is removed, but the joint at E retained.

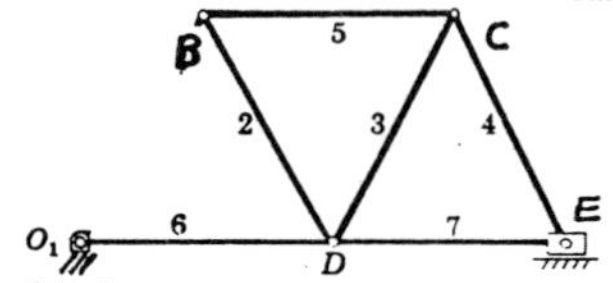

1.2(b) Add one link extending from O_1 to C, or B to D, or C to O_4, or B to O_4, or O_1 to D. Then, $DF = 3(n_L-1) - 2n_J' = 3(6-1) - 2x7 = 1.$

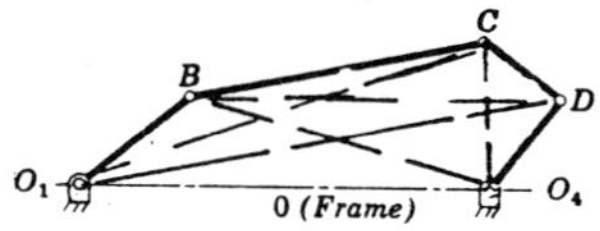

1.3 Stroke length S = 2R.

v(avg) = 2Sn = 2(4 in)(3000 rev/min) = 24,000 in/min or 400 in/sec).

1.4 R = 2, L = 4, E = 1 in. $\omega = \frac{2\pi}{60}(3000) = 314$ rad/sec

Referring to fig. 1.16

$\phi_1 = \arcsin E/(L-R) = 30^o.$

$\phi_2 = \arcsin E/(L+R) = 9.6^o.$

$\alpha = 180^o + \phi_1 - \phi_2 = 200.4^o$ or 3.50 rad.

$\beta = 180^o - \phi_1 + \phi_2 = 159.6^o$ or 2.79 rad.

Stroke $S = [(L+R)^2 - E^2]^{1/2} - [(L-R)^2 - E^2]^{1/2} = 4.184$ in.

"Forward" stroke time $t_1 = \alpha/\omega = 0.01114$ sec. $v_{1(avg)} = S/t_1 = 375.5$ in/sec.

"Return" stroke time $t_2 = \beta/\omega = 0.00889$ sec. $v_{2(avg)} = S/t_2 = 470.6$ in/sec.

1.5 (see 1.4) $\phi_1 = 48.5^o$, $\phi_2 = 14.5^o$, $\alpha = 3.73$ rad, $\beta = 2.55$ rad.

$S = 4.49$ in, $t_1 = 0.01187$ sec, $v_{1(avg)} = 378$ in/sec.

$t_2 = 0.00812$ sec, $v_{2(avg)} = 553$ in/sec.

1.6 $\omega = 3000\ \pi/30 = 314.16$ rad/s

$$\phi_1 = \arcsin (E/(L-R))$$
$$= \arcsin (50/(200-100)) = .5236 \text{ rad}$$
$$\phi_2 = \arcsin (E/(L+R))$$
$$= \arcsin (50/(200+100)) = .1674 \text{ rad}$$

$\alpha = \pi + \phi_1 - \phi_2 = 3.4977$ rad $\quad \beta = \pi - \phi_1 + \phi_2 = 2.7854$ rad

Forward stroke time $t_1 = \alpha/\omega = .01113$ s.

Return stroke time $t_2 = \beta/\omega = .008866$ s.

Stroke $S = \sqrt{(L+R)^2 - E^2} - \sqrt{(L-R)^2 - E^2} = 209.20$ mm

Avg. vel. forward $v_{1(avg)} = S/t_1 = 18790$ mm/s

Avg. vel. return $v_{2(avg)} = S/t_2 = 23595$ mm/s

1.7 The crank and connecting rod positions are determined as in the flow chart of section 1.12 The linkage skeleton diagram is shown on the sketch for L/R=1.5, E/R=0.2 where crank angle T_1 varies from 0° to 340° in 20° steps.

Angles T_1 and T_2 and slider location X_2 are tabulated below for R=1:

T1	T2	X2
0	-7.66225566077	2.48660687473
20	5.43290764513	2.43295424518
40	17.1690352815	2.19920149468
60	26.3634597467	1.84402758957
80	31.5474944957	1.45195830359
100	31.5474944957	1.10466195325
120	26.3634597467	0.844027589366
140	17.1690352815	0.667112607638
160	5.43290764513	0.553569083629
180	-7.66225566076	0.486606374732
200	-21.1829283574	0.438954641475
220	-34.1844166085	0.474805680816
240	-45.2985623587	0.555267662011
260	-52.1735359216	0.746259746265
280	-52.1735359216	1.09355610154
300	-4[illegible]	1.0000[illegible]
320	-34.1844166085	2.30669457325
340	-21.1829283574	2.39633968305
360	-7.66225566077	2.48660687473

```
120 R=1
130 L=1.5
140 E=0.2
150 WINDOW -1,2.5,-1.346,1.346
160 VIEWPORT 0,130,0,100
170 FOR T1=0 TO 2*PI STEP PI/9
180 X1=R*COS(T1)
190 Y1=R*SIN(T1)
200 T2=ASN((Y1-E)/L)
210 X2=X1+L*COS(T2)
220 Y2=E
230 MOVE @D:0,0
240 DRAW @D:X1,Y1
250 DRAW @D:X2,Y2
260 NEXT T1
270 AXIS @D:0.1,0.1,-1,-1
280 HOME
290 END
```

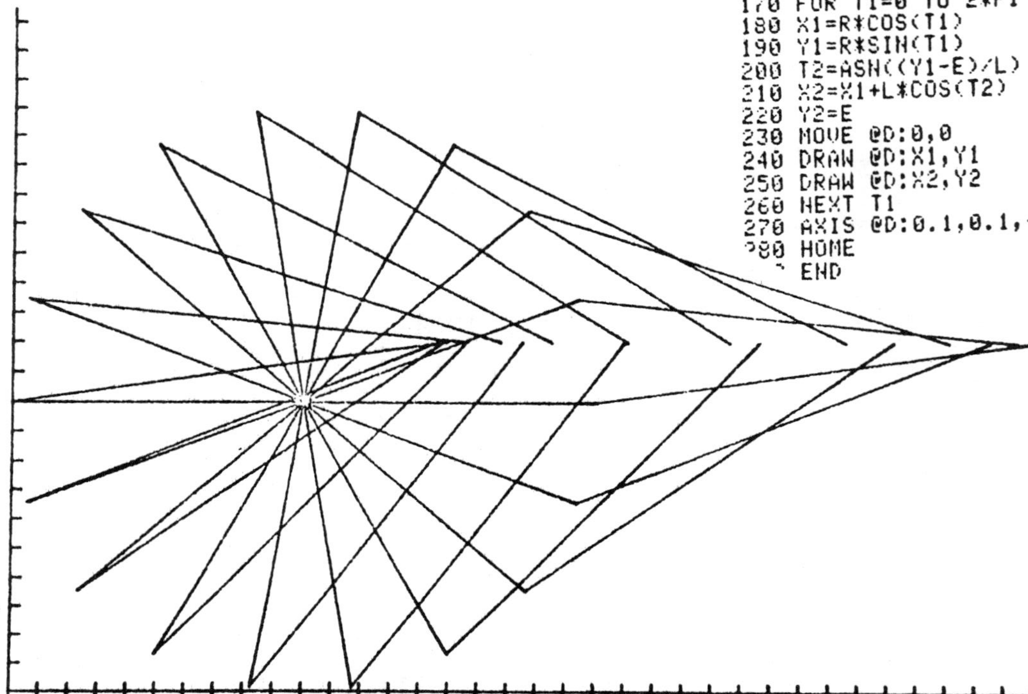

1.7

1.8 5000 rev/min x1min/60s x 360°/rev x 0.002s = 60°

From the tabulated data or the figure, the piston moves from

x_2 = 2.4866R to 1.8440R = -0.6426R

= 64.26 mm to the left as the crank rotates from T_1=0 to 60°.

Average velocity during the interval is

64.26 mm/0.002s =32,600 mm/s

= 32.6 m/s (ans)

to the left.

1.9 Angles T_1 and T_2 (degrees) and slider location X_2 are tabulated below for R=1:

T1	T2	X2
0	-12.8395884069	2.75499287748
20	-1.84587544996	2.73875858295
40	7.75179688929	2.54959544508
60	15.0049801317	2.2386259871
80	18.9591044355	1.87599943194
100	18.9591044355	1.52870310571
120	15.0049801317	1.2386259871
140	7.75179688929	1.01750655824
160	-1.84587544996	0.859373341279
180	-12.8395884069	0.754992877478
200	-24.3452117363	0.783248234398
220	-35.4031826419	0.731127687127
240	-44.6961909153	0.779523222522
260	-50.294446759	0.978263118083
280	-50.294446759	1.3235644734
300	-44.6961909153	1.77952322252
320	-35.4031826419	2.23321655337
340	-24.3452117363	2.57963350597
360	-12.8395884069	2.75499287748

1.10 a) The time interval is $\Delta t = (40^\circ \times \frac{\pi\,\text{rad}}{180^\circ})/\omega$ rad/s

The displacement, from the solution to the above problem, is $\Delta x = (2.5496 - 2.7550)R$

$$v_{avg} = \Delta x/\Delta t = \frac{(2.5496 - 2.7550)50}{(40\pi/180)/500}$$

v_{avg} = -7355 mm/s (7355 mm/s to the left). (ans)

1.10 b) $v_{avg} = \Delta x/\Delta t = \dfrac{(.7545 - 2.7545)R}{\pi/\omega}$

$= -2\ R\omega/\pi$

$= -2 \times 50 \times 500/\pi$

$v_{avg} = -15915$ mm/s (15.915 m/s) to the left. (ans)

c) $\Delta x = 0$ $v_{avg} = 0$ (ans)

1.11 a) Angles θ and T_2 and slider location x are tabulated and plotted below

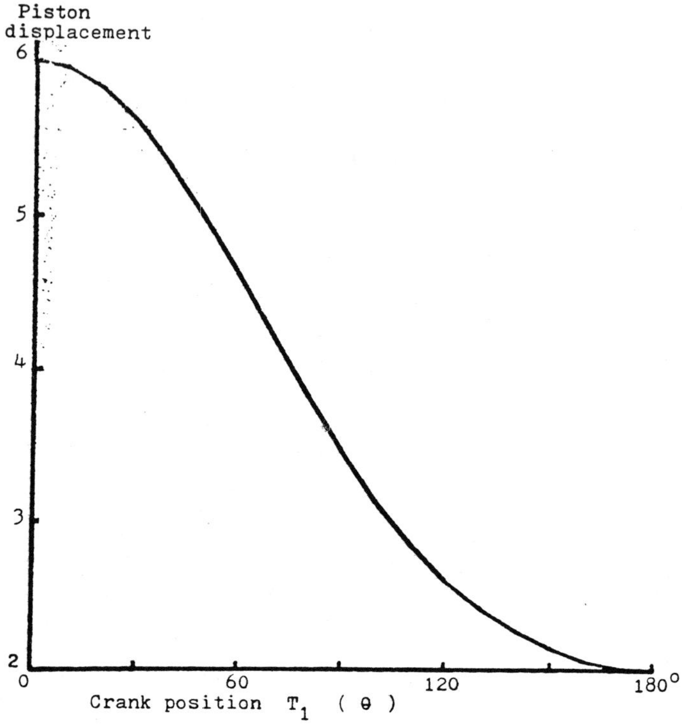

1.11 (cont'd.)

θ (T1°)	T2°	X2 (in)
0	0	6
30	14.4775121859	5.60503415378
60	25.6589062733	4.60555127546
90	30	3.46410161514
120	25.6589062733	2.60555127546
150	14.4775121859	2.14093253864
180	1.278976924E-12	2
210	-14.4775121859	2.14093253864
240	-25.6589062733	2.60555127546
270	-30	3.46410161514
300	-25.6589062733	4.60555127546
330	-14.4775121859	5.60503415378
360	-3.836930773E-12	6

b) For 30° crank rotation, the time interval is

(1 min/100 rev)x(1 rev/360°)x(60 s/min)x30° = 0.05 s/interval

For $30 \le \theta \le 60°$

v_{avg} = (4.60555 - 5.60503)/.05

= 19.99 in/s to the left (ans)

c) For the next interval

v_{avg} = (3.46410 - 4.60555)/.05

= 22.83 in/s to the left (ans)

1.12 through 1.20 The longest link is shorter than the sum of the other three, forming a mechanism in each case.

1.12 $L_{max} + L_{min} = 5 + 1.5 > L_a + L_b = 2 + 2$

The linkage is a non-Grashof (triple rocker) mechanism. See sketch.

1.13 Fixed link shortest. Try drag link.

$|L_1 - L_2| + L_0 < L_3 < L_1 + L_2 - L_0$ is not satisfied. Linkage is a triple rocker. $3.5 + 1 > 1.5 + 2$; $L_{max} + L_{min} > L_a + L_b$. (Non-Grashof). See sketch.

1.14 Fixed link shortest.

$|L_1 - L_2| + L_0 < L_3 < L_1 + L_2 - L_0$ is satisfied. Linkage is a drag link.

1.15 Fixed link is shortest, but drag link inequality is not satisfied since $|L_1 - L_2| + L_0 = L_3$. Motion is indefinite (a change point mechanism).

1.16 Fixed link is shortest, but drag link inequality is not satisfied since $L_3 = L_1 + L_2 - L_0$. Motion is indefinite (a change point mechanism).

1.12

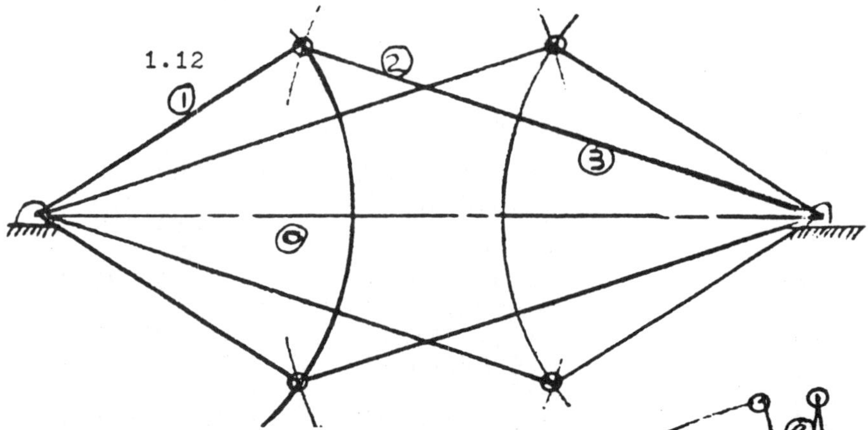

1.13

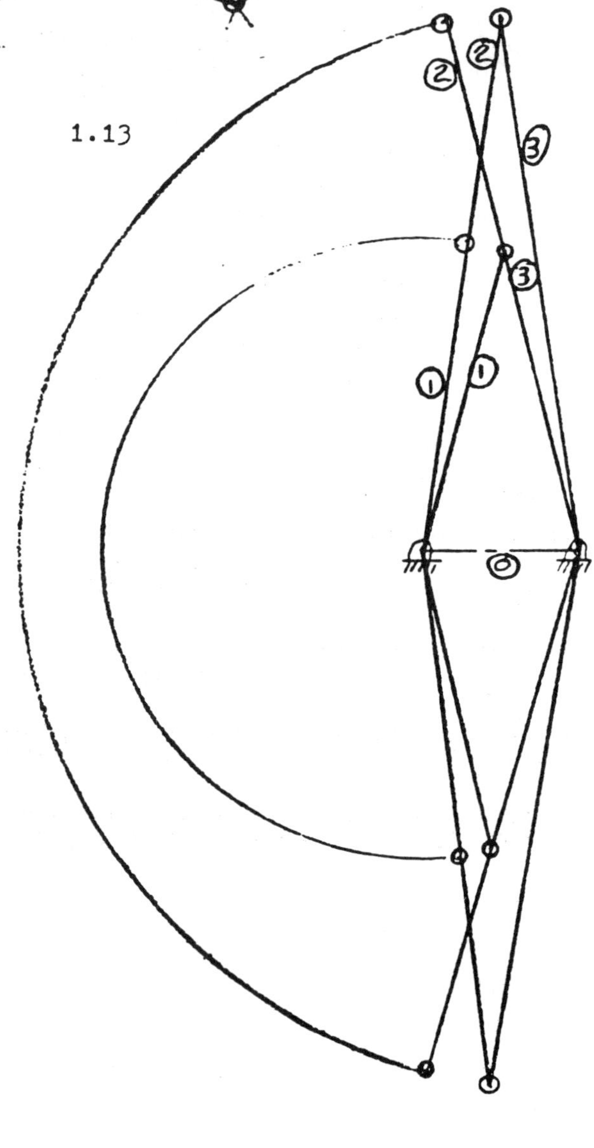

1.17

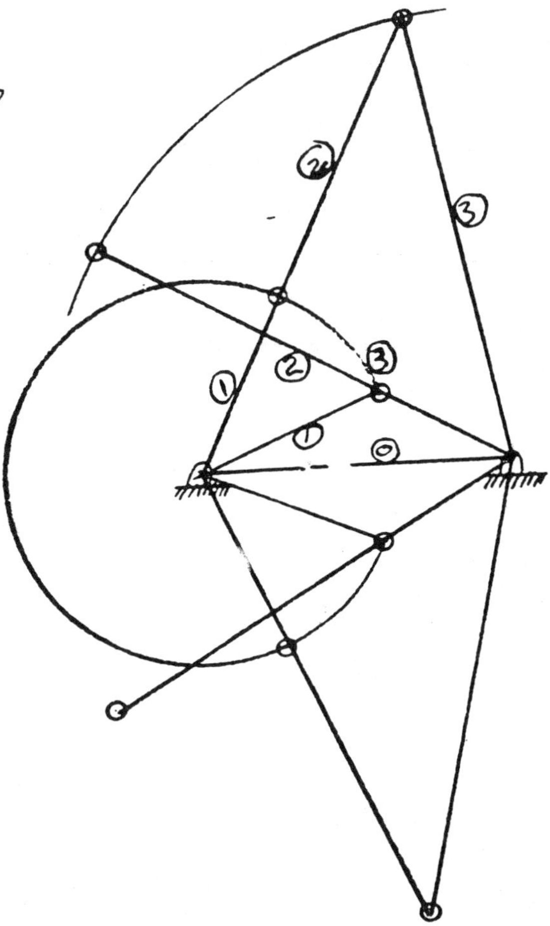

1.18

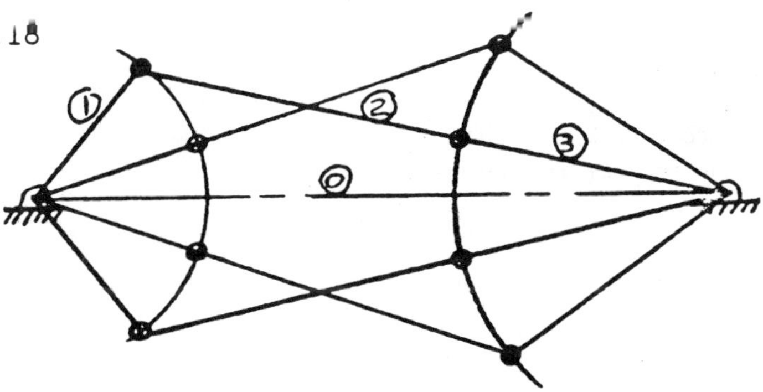

1.17 Crank is shortest, but crank-rocker inequality is not satisfied since $L_1 + |L_2 - L_3| > L_0$. Linkage is a triple rocker. See sketch. $3 + 1.25 > 2 + 2$; $L_{max} + L_{min} > L_a + L_b$ (Non-Grashof).

1.18 Crank is shortest, but crank rocker inequality is not satisfied since $L_0 > L_2 - L_1 + L_3$. Linkage is a triple rocker. $4 + 1 > 2 + 1.5$ $L_{max} + L_{min} > L_a + L_b$ (Non-Grashof). See sketch.

1.19 We may eliminate the drag link since the fixed link is not shortest. Trying the crank rocker criteria we find

$L_1 + |L_2 - L_3| = L_0 = L_2 - L_1 + L_3$. The motion is indefinite (a change point mechanism).

1.20 The crank is shortest. The inequality

$L_1 + |L_2 - L_3| < L_0 < L_2 - L_1 + L_3$ is satisfied. The linkage is a crank rocker. See sketch.

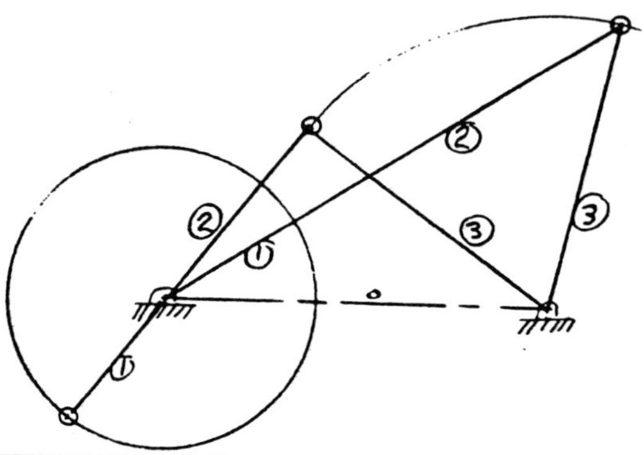

1.21 $L_{max} + L_{min} < L_a + L_b$ $15 + 5 < 15 + 10$ and the fixed link is shortest. The mechanism is a drag link.

1.22 $L_{max} + L_{min} < L_a + L_b$ $16 + 7 < 15 + 10$ and the crank is shortest. The mechanism is a crank rocker.

1.23 In the crank rocker, link 1 is shortest and $L_{max} + L_{min} < L_a + L_b$ from which $300 + 100 < 280 + L_2$ or $L_2 + 100 < 280 + 300$. Thus, $120 < L_2 < 480$.

1.24 Grashof linkages include the first four.

a) For the crank rocker

$L_1 + |L_2 - L_3| < L_0 < L_2 - L_1 + L_3$

$100 + |280 - 360| < L_0 < 280 - 100 + 360$

$180 < L_0 < 540$ mm. L_1 is shortest.

1.24 b) For the drag link

$|L_1 - L_2| + L_0 < L_3 < L_1 + L_2 - L_0$

$|100 - 280| + L_0 < 360 < 100 + 280 - L_0$

$L_0 < 180$ and $L_0 < 20$ $\quad L_0$ is shortest

$\underline{0 < L_0 < 20 \text{ mm.}}$

c) For the double rocker, L_2 is shortest. (none)

d) Change point mechanism

$L_{max} + L_{min} = L_a + L_b$

For L_0 largest:

$L_0 + 100 = 280 + 360 \quad \underline{L_0 = 540}$; or

for L_0 intermediate:

$360 + 100 = L_0 + 280 \quad \underline{L_0 = 180}$; or

for L_0 smallest

$360 + L_0 = 100 + 280 \quad \underline{L_0 = 20}$ mm.

e) Grashof linkages: A, B, C and D above

$\underline{0 < L_0 \le 20}$ and $\underline{180 \le L_0 \le 540}$ mm.

f) Triple rocker (non-Grashof)

$L_{max} + L_{min} > L_a + L_b$

For L_0 largest $L_0 + 100 > 280 + 360 \quad L_0 > 540$

For L_0 intermediate (i.e. $100 \le L_0 \le 360$)

$360 + 100 > L_0 + 280 \quad 180 > L_0 \ge 100$

For L_0 smallest (i.e. $0 < L_0 \le 100$)

$360 + L_0 > 100 + 280 \quad 100 \ge L_0 > 20$

Also, $L_{max} < L_{min} + L_a + L_b$

$L_0 < 100 + 280 + 360 \quad L_0 < 740$

Combining:

$\underline{20 < L_0 < 180}$ and $\underline{540 < L_0 < 740}$ mm

1.25 (see 1.24)

Grashof linkages: $L_{max} + L_{min} \le L_a + L_b$

If L_1 is shortest $0 < L_1 \le 40$)

$60 + L_1 \le 40 + 40$; $0 < L_1 < 20$: crank rocker

If $L_1 = 20$, the result is a change point linkage

If L_1 is intermediate

$60 + 40 \le L_1 + 40 \quad \underline{L_1 = 60}$ change point

If L_1 is largest; i.e. $60 \le L_1 < 40 + 40 + 60$

$L_1 + 40 \le 40 + 60 \quad L_1 = 60$ change point

1.25 (cont'd.)

Combining results:

a) CR: $0 < L_1 < 20$

b) DL: none

c) DR: L_2 shortest: none

d) $L_1 = 20$ and $L_1 = 60$ (CP)

e) Grashof: $0 < L_1 \leq 20$ and $L_1 = 60$

f) TR: $L_{max} + L_{min} > L_a + L_b$

For L_1 largest (i.e. $60 \leq L_1 < 60 + 40 + 40$)

$L_1 + 40 > 60 + 40$

$60 < L_1 < 140$

For L_1 intermediate (i.e. $40 \leq L_1 \leq 60$)

$60 + 40 > L_1 + 40$

$40 \leq L_1 < 60$

For L_1 smallest (i.e. $0 \leq L_1 \leq 40$)

$60 + L_1 > 40 + 40$

$20 < L_1 \leq 40$

Combining: $20 < L_1 < 60$ and $60 < L_1 < 140$ mm (ans)

1.26 a) A crank rocker is impossible since the crank is not shortest (ans).

b) Drag link: L_0 is shortest and $L_{max} + L_{min} < L_a + L_b$

If L_2 is intermediate:

$250 + 50 < L_2 + 200 \quad 100 < L_2 \leq 250$

If L_2 is largest:

$L_2 + 50 < 200 + 250 \quad 250 \leq L_2 < 400$

Combining: $100 < L_2 < 400$ (ans)

c) Double rocker L_2 is shortest and

$250 + L_2 < 50 + 200$: impossible (ans)

d) Change point: $L_{max} + L_{min} = L_a + L_b$

If L_2 is largest

$L_2 + 50 = 200 + 250 \quad L_2 = 400$

If L_2 is smallest

$250 + L_2 = 50 + 200$ impossible

If L_2 is intermediate

$250 + 50 = L_2 + 200 \quad L_2 = 100$

$L_2 = 100$ or 400 (ans)

1.26 (cont'd.)

e) Grashof: combining a through d:

$100 \le L_2 \le 400$ (ans)

f) Triple rocker:

$L_{max} + L_{min} > L_a + L_b$

If L_2 is largest

$L_2 + 50 > 200 + 250 \quad L_2 > 400$

If L_2 is intermediate

$250 + 50 > L_2 + 200 \quad 100 > L_2 \ge 50$ impossible

If L_2 is smallest

$250 + L_2 > 50 + 200 \quad 50 \ge L_2 > 0$

Also, $L_2 < L_0 + L_1 + L_3 \quad L_2 < 50 + 200 + 250$

Combining the results,

$0 < L_2 < 100$ and $400 < L_2 < 500$ (ans)

1.27 a) Crank rocker: L_1 is not shortest, but if L_3 shortest:

$L_{max} + L_{min} < L_a + L_b$

$220 + L_3 < 120 + 80$ impossible (ans)

b) Drag link: L_0 is not shortest impossible (ans)

c) Double rocker: L_2 shortest

$L_{max} + L_{min} < L_a + L_b$ If L_3 longest:

$L_3 + 80 < 120 + 220 \therefore 220 \le L_3 < 260$

If L_3 intermediate, $220 + 80 < 120 + L_3$

$180 < L_3 \le 220$. Thus:

$180 < L_3 < 260$ (ans)

d) Change point:

$L_{max} + L_{min} = L_a + L_b$

$L_3 + 80 = 120 + 220 \quad L_3 = 260$

$220 + L_3 = 80 + 120$ impossible

$220 + 80 = 120 + L_3 \quad L_3 = 180$

Thus, $L_3 = 180$ and 260 (ans)

e) Grashof linkage: Combining a through d:

$180 \le L_3 \le 260$ (ans)

1.27 (cont'd.)

f) Triple rocker:

$L_{max} + L_{min} > L_a + L_b$. and $L_{max} < L_{min} + L_a + L_b$ if $L_3 = L_{max}$:

$L_3 + 80 > 120 + 220 \Rightarrow L_3 > 260$ and $L_3 < 80 + 120 + 220 \Rightarrow L_3 < 420$

$220 + L_3 > 80 + 120 \Rightarrow L_3 > -20$ and $220 < L_3 + 120 + 80 \Rightarrow L_3 > 20$ if min

$220 + 80 > L_3 + 120 \Rightarrow 180 > L_3 \geq 80$ if L_3 intermediate

Combining: $20 < L_3 < 180$ and $260 < L_3 < 420$ (ans)

1.28 a) Crank rocker: L_1 shortest and $L_{max} + L_{min} < L_a + L_b$. If L_0 longest:

$L_0/L_1 \geq 1.5$ $\quad L_0/L_1 + 1 < 1.2 + 1.5$ $\quad 1.5 \leq L_0/L_1 < 1.7$

If L_0 intermediate:

$1.5 + 1 < L_0/L_1 + 1.2$ $\quad 1.3 < L_0/L_1 \leq 1.5$. Thus:

$1.3 < L_0/L_1 < 1.7$ (ans)

b) Drag link: L_0 is L_{min}

$1.5 + L_0/L_1 < 1 + 1.2$ $\quad 0 < L_0/L_1 < 0.7$ (ans)

c) Double rocker: L_2 is L_{min} $\quad$ none (ans)

d) Change point:

If $L_0 = L_{min}$: $1.5 + L_0/L_1 = 1 + 1.2$ $\quad L_0/L_1 = 0.7$

If L_0 intermediate: $1.5 + 1 = L_0/L_1 + 1.2$ $\quad L_0/L_1 = 1.3$

If $L_0 = L_{max}$: $L_0/L_1 + 1 = 1.5 + 1.2$ $\quad L_0/L_1 = 1.7$

$L_0/L_1 = 0.7$ or 1.3 or 1.7 (ans)

e) Grashof: combine above:

$0 < L_0/L_1 \leq 0.7$ and $1.3 \leq L_0/L_1 \leq 1.7$ (ans)

f) Triple rocker: If $L_0 = L_{max}$ $\quad L_0/L_1 < 1 + 1.2 + 1.5$

$L_0/L_1 < 3.7$ $\quad L_0/L_1 + 1 > 1.2 + 1.5$ $\quad L_0/L_1 > 1.7$

If $L_0 = L_{min}$ $\quad 1.5 + L_0/L_1 > 1 + 1.2$ $\quad 0.7 < L_0/L_1 \leq 1$

If L_0 intermediate: $1 + 1.5 > L_0/L_1 + 1.2$ $\quad 1.3 > L_0/L_1 \geq 1$

Combining: $0.7 < L_0/L_1 < 1.3$ and $1.7 < L_0/L_1 < 3.7$ (ans)

1.29 To form a mechanism, if $L_2 = L_{max}$

$L_2/L_1 < 1 + 2.2 + 1.7$ $L_2/L_1 < 4.9$, and if

$L_2/L_1 + 1 > 2.2 + 1.7$ (i.e. $L_2/L_1 > 2.9$) it is a triple rocker.

For a change point, $L_2/L_1 + 1 = 2.2 + 1.7$ $L_2/L_1 = 2.9$

For a crank rocker, $L_2/L_1 + 1 < 2.2 + 1.7$ $1.7 \leq L_2/L_1 < 2.9$

Suppose L_2/L_1 is intermediate. For a triple rocker:

$2.2 + 1 > L_2/L_1 + 1.7$ $1.5 > L_2/L_1 \geq 1$. For a change point,

$2.2 + 1 = L_2/L_1 + 1.7$ $L_2/L_1 = 1.5$. For a crank rocker,

$1 + 2.2 < L_2/L_1 + 1.7$ $1.5 < L_2/L_1 \leq 1.7$. Suppose $L_2/L_1 = L_{min}$

For a triple rocker, $2.2 + L_2/L_1 > 1 + 1.7$ $1 \geq L_2/L_1 > 0.5$.

For a change point, $L_2/L_1 = 0.5$

For a double rocker, $L_2/L_1 < 0.5$

Summary:

a) Crank rocker: $1.5 < L_2/L_1 < 2.9$

b) Drag link: none

c) Double rocker: $0 < L_2/L_1 < 0.5$

d) Change point: $L_2/L_1 = 0.5$ or 1.5 or 2.9

e) Grashof: $0 < L_2/L_1 \leq 0.5$ and $1.5 \leq L_2/L_1 \leq 2.9$

f) Triple rocker: $0.5 < L_2/L_1 < 1.5$ and $2.9 < L_2/L_1 < 4.9$

1.30 a) For a crank rocker, $L_1 + |L_2 - L_3| < L_0 < L_2 - L_1 + L_3$ and the driver L_1 is shortest.

Thus, $400 < L_0$ $400 + |600-750| < L_0 < 600 - 400 + 750$

$550 < L_0 < 950$ mm

b) For drag link, fixed link is shortest. Thus $L_0 < 400$ and

$|L_1-L_2| + L_0 < L_3 < L_1 + L_2 - L_0$

$|400-600| + L_0 < 750 < 400 + 600 - L_0$

$200 + L_0 < 750 < 1000 - L_0$

$L_0 < 550$ $L_0 < 250$

$0 < L_0 < 250$ mm

c) For double rocker, coupler shortest.   ans none

d) For change point $L_{max} + L_{min} = L_a + L_b$ $L_0 \geq 750$

If $L_0 = L_{max}$, $L_0 + 400 = 600 + 750$; $L_0 = 950$

If $L_0 \leq 400$ (min) $750 + L_0 = 600 + 400$ $L_0 = 250$

If $400 < L_0 < 750$ $400 + 750 = 600 + L_0$ $L_0 = 550$ mm

1.30 (cont'd.)

e) Grashof mechanism:

$L_{max} + L_{min} \le L_a + L_b$ includes all above

$0 < L_0 \le 250$ and $550 \le L_0 \le 950$ mm

f) Non-Grashof (triple rocker):

$L_{max} + L_{min} > L_a + L_b$

If $L_0 \ge 750$, L_0 is max

$L_0 + 400 > 600 + 750$

$L_0 > 950$

If L_0 intermediate, $400 < L_0 < 750$

$750 + 400 > L_0 + 600$

$550 > L_0 > 400$

If L_0 min, $0 < L_0 \le 400$

$750 + L_0 > 400 + 600$

$400 \ge L_0 > 250$

Also, $L_{max} < L_{min} + L_a + L_b$

$L_0 < 400 + 600 + 750$

$L_0 < 1750$

$\therefore$ $250 < L_0 < 550$ and $950 < L_0 < 1750$

1.31 Crank rocker: L_1 is shortest ($L_1 = L_{min}$)

$L_3/L_1 > 1$ and $L_2/L_1 > 1$ (1 and 2)

If $L_3 = L_{max}$,

$L_3/L_1 + 1 < L_2/L_1 + 2 \quad L_3/L_1 < L_2/L_1 + 1$ (3)

If L_3 and L_2 are intermediate:

$2 + 1 < L_3/L_1 + L_2/L_1$

$L_3/L_1 > 3 - L_2/L_1$ (4)

If L_3 is intermediate and $L_2 = L_{max}$:

$L_2/L_1 + 1 < L_3/L_1 + 2$

$L_3/L_1 > L_2/L_1 - 1$ (5)

Permissible values for the crank rocker lie in the region bounded by inequalities 3, 4 and 5 (1 and 2 are redundant). See figure.

1.31

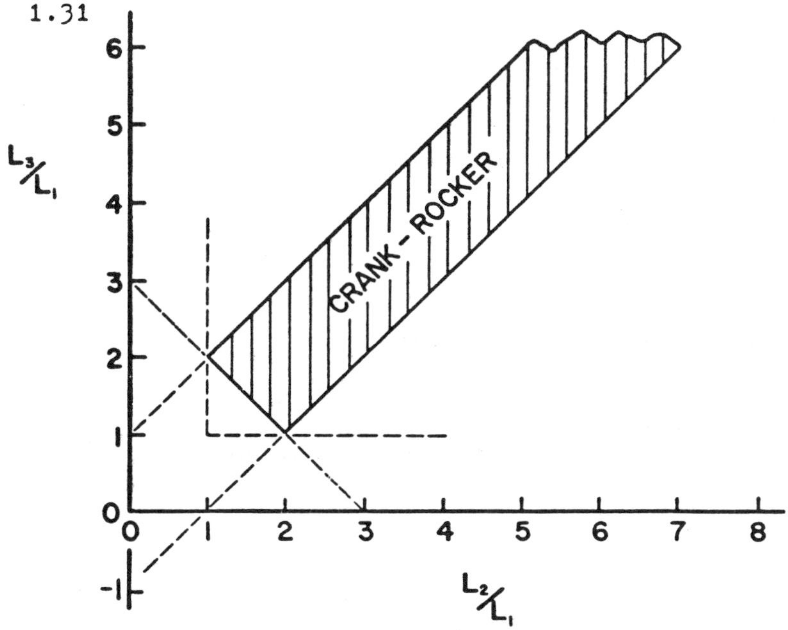

CRANK - ROCKER
L_3/L_1
L_2/L_1

1.32

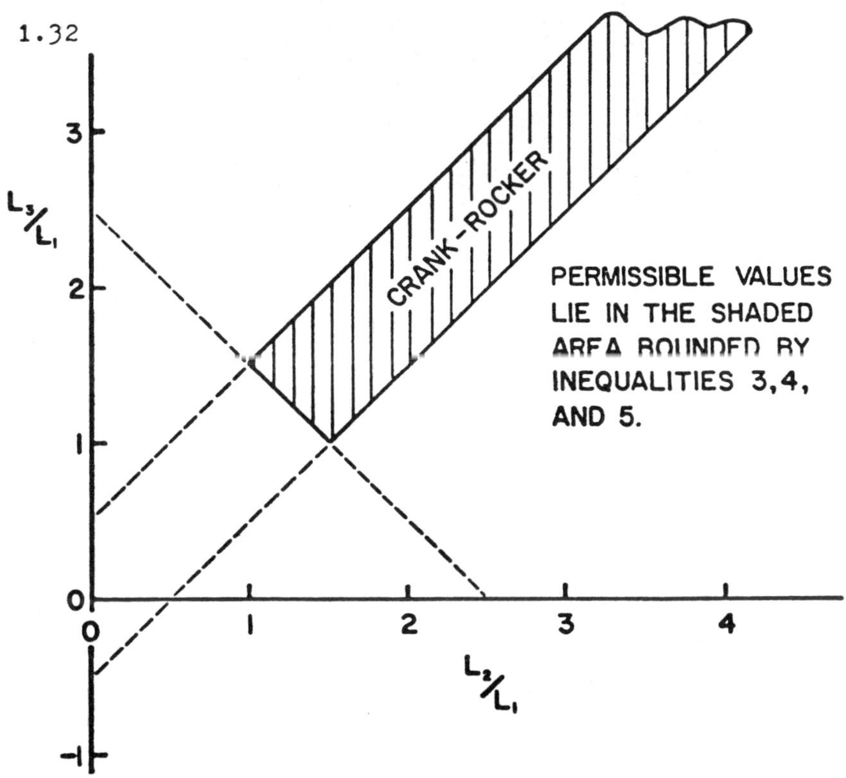

CRANK - ROCKER
PERMISSIBLE VALUES
LIE IN THE SHADED
AREA BOUNDED BY
INEQUALITIES 3,4,
AND 5.
L_3/L_1
L_2/L_1

1.32 L_1 is shortest for a crank rocker

$L_3/L_1 > 1$ (1) and $L_2/L_1 > 1$ (2)

If $L_3 = L_{max}$

$L_3/L_1 + 1 < L_2/L_1 + 1.5 \quad L_3/L_1 < L_2/L_1 + 0.5$ (3)

If $L_0 = L_{max}$

$1.5 + 1 < L_2/L_1 + L_3/L_1$

$L_3/L_1 > 2.5 - L_2/L_1$ (4)

If $L_2 = L_{max}$

$L_2/L_1 + 1 < L_3/L_1 + 1.5$

$L_3/L_1 > L_2/L_1 - 0.5$ (5) See figure.

1.33 See flow chart.

1.34 The program corresponds to the flow chart of Problem 1.33. It is tested for ten sets of values. Results are as follows:

FOUR BAR LINKAGE
IN MILLIMETERS

L0	L1	L2	L3	COMMENT
50	600	400	500	DRAG LINK
100	300	250	50	CHANGE POINT
300	100	150	200	TRIPLE ROCKER
150	100	250	500	NOT A MECHANISM
150	100	250	200	CHANGE POINT
500	900	300	400	TRIPLE ROCKER
700	350	560	600	CRANK ROCKER
300	100	400	800	NOT A MECHANISM
550	400	100	300	DOUBLE ROCKER
40	500	300	495	DRAG LINK

1.33

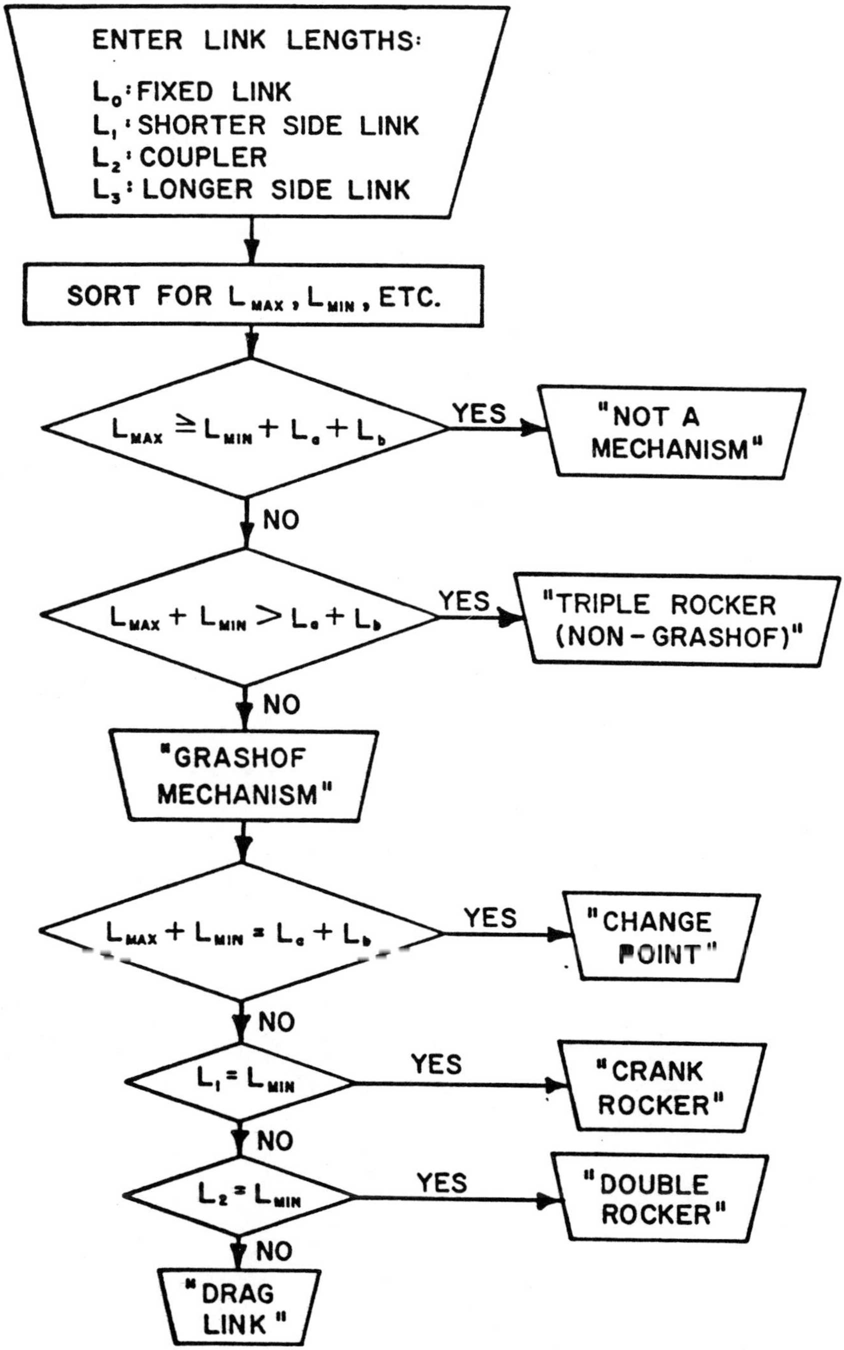
ENTER LINK LENGTHS:
L_0: FIXED LINK
L_1: SHORTER SIDE LINK
L_2: COUPLER
L_3: LONGER SIDE LINK
SORT FOR L_{MAX}, L_{MIN}, ETC.
$L_{MAX} \geq L_{MIN} + L_a + L_b$
YES
"NOT A MECHANISM"
NO
$L_{MAX} + L_{MIN} > L_a + L_b$
YES
"TRIPLE ROCKER (NON-GRASHOF)"
NO
"GRASHOF MECHANISM"
$L_{MAX} + L_{MIN} = L_a + L_b$
YES
"CHANGE POINT"
NO
$L_1 = L_{MIN}$
YES
"CRANK ROCKER"
NO
$L_2 = L_{MIN}$
YES
"DOUBLE ROCKER"
NO
"DRAG LINK"

1.35 From the Grashof criterion for a crank rocker linkage, if $L_2 = L_{max}$
$140 + 100 < 120 + L_3$ or $120 < L_3$

If $L_3 = L_{max}$

$L_3 + 100 < 120 + 140$ or $L_3 < 160$

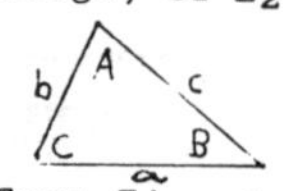

Thus, $120 < L_3 < 160$ mm without limiting ϕ. From Fig. 1.17, if ϕ_{min} or ϕ_{max} is specified, we may use the law of sines and the cosine law to find L_3. This may be programmed as a triangle solution.
$\phi_{min} = 45^o$, side $a = L_0 - L_1 = 20$, $b = L_2 = 140$. No triangle satisfies the input data. No value of L_3 satisfies the criteria.

1.36 See above problem.

From the Grashof criteria for a crank rocker, if $L_3 = L_{max}$,

$L_3 + 50 < 200 + 210 \quad L_3 < 360$

If $L_0 = L_{max} \quad 210 + 50 < 200 + L_3 \quad 60 < L_3$

$60 < L_3 < 360$ without limiting ϕ.

Let: $\phi_{min} = 45^o$; side $a = L_0 - L_1 = 210 - 50 = 160$

$b = L_2 = 200$ Triangle solution yields

$c = L_3 = 216.25$ mm (ans)

Let

$a = L_0 + L_1 = 210 + 50 = 260$, $b = L_2 = 200$

$c = L_3 = 216.25$ from which $77.20^o = \phi_{max}$ which is acceptable.
(Other solutions possible.)

1.37 If $L_3 = L_{max}$, $L_3 + 110 < 150 + 150$

$L_3 < 190$ If $L_{max} = 150$

$150 + 110 < L_3 + 150 \quad 110 < L_3$

Thus $110 < L_3 < 190$ to satisfy Grashof criteria (without limiting ϕ)

Let

$a = L_0 - L_1 = 150 - 110 = 40$; $b = L_2 = 150$ $\angle A = 45^o$.

No value of L_3 satisfies the criteria. (ans)

1.38 If $L_2 = L_{max}$

$1.5 + 1 < 1.2 + L_0/L_1 \quad 1.3 < L_0/L_1$

If $L_0 = L_{max} \quad L_0/L_1 + 1 < 1.2 + 1.5 \quad L_0/L_1 < 1.7$

Thus, $1.3 < L_0/L_1 < 1.7$ satisfies the Grashof criteria.

Let

$a = L_2/L_1 = 1.5$; $b = L_3/L_1 = 1.2$; $\angle C = 45^o$ from which
$c = 1.0698 = (L_0-L_1)/L_1 \quad L_0/L_1 = 2.0698$

which does not satisfy the crank rocker criteria.

No value of L_0/L_1 satisfies the criteria. (ans)

1.39 *Let*

$a = (L_0-L_1)/L_1 = 3.2 - 1 = 2.2$

$b = L_3/L_2 = 1.7$ $\quad <A = \phi_{min} = 45^o$ From which:

$c = L_2/L_1 = 3.045$ (ans)

Let

$a = (L_0+L_1)/L_1 = 3.2 + 1 = 4.2$

$b = L_3/L_2 = 1.7 \quad c = 3.045$

$<A = 121.95^o = \phi_{max}$ which is acceptable.

Grashof criteria L_1 is shortest;

$L_{max} + L_{min} < L_a + L_b \quad 3.2 + 1 < 3.045 + 1.7$ is satisfied for the crank rocker. Other solutions are possible.

1.40 Values must satisfy the Grashof criterion (see prob. 1.31).

Let $a = (L_0-L_1)/L_1 = 2 - 1 = 1$;

$b = L_2/L_1$; $LA = \phi_{min} = 45^o$

No combinations of L_3/L_1 and L_2/L_1 satisfy all criteria (without relaxing the transmission angle limitation). (ans)

1.41 First, checking the Grashof criteria for a crank rocker with $L_0/L_1 = 4$, if $L_0 = L_{max}$

$4 + 1 < (L_2+L_3)/L_1 \qquad 5 < (L_2+L_3)/L_1$

If $L_2 = L_{max}$

$L_2/L_1 + 1 < 4 + L_3/L_1 \qquad (L_2-L_3)/L_1 < 3$

If $L_3 = L_{max}$

$L_3/L_1 + 1 < 4 + L_2/L_1 \qquad (L_3-L_2)/L_1 < 3$

The following triangle solutions are based on the requirement that $\phi_{min} = 45^o$ (See Fig. 1.17A).

L_3/L_2 is determined from 2 sides and one angle (columns 1-4). Then, the configuration of Fig. 1.17B is checked for ϕ_{max} using the three known sides (columns 2, 4 and 5). Then, L_3/L_1 is computed based on $\phi_{max} = 135^o$ (columns 8-11) using the configuration of Fig. 1.17B. ϕ_{min} is checked (columns 9, 11, 12, 13 and 14) using the configuration of Fig. 1.17A. See figure.

1	2	3	4	5	6	7
$\frac{L_0-L_1}{L_1}$	$\frac{L_2}{L_1}$	$\phi_{min}{}^o$	$\frac{L_3}{L_1}$	$\frac{L_0+L_1}{L_1}$	$\phi_{max}{}^o$	Acceptable?
3	1.5	45°	3.867	5	132.22°	Yes
3	2	45°	4.060	5	106.15	Yes
3	3	45°	4.243	5	85.48	Yes
3	[illegible]	45°	3,823	5	79.36	Yes
3	4.5	45°	none	5	--	No
3	5	45°	none	5	--	No
3	4.22	45°	3.294	5	82.44	Yes

8	9	10	11	12	13	14
$\frac{L_0+L_1}{L_1}$	$\frac{L_2}{L_1}$	$\phi_{max}{}^o$	$\frac{L_3}{L_1}$	$\frac{L_0-L_1}{L_1}$	$\phi_{min}{}^o$	Acceptable?
5	1.5	135°	3.826	3	46.59°	Yes
5	2	135	3.382	3	61.58	Yes
5	3	135	2.406	3	66.36	Yes
5	3.5	135	1.870	3	58.97	Yes
5	4	135	1.295	3	33.11	No
5	3.7	135	1.645	3	52.59	Yes
5	4	108.2	2	3	46.56	Yes

1.41

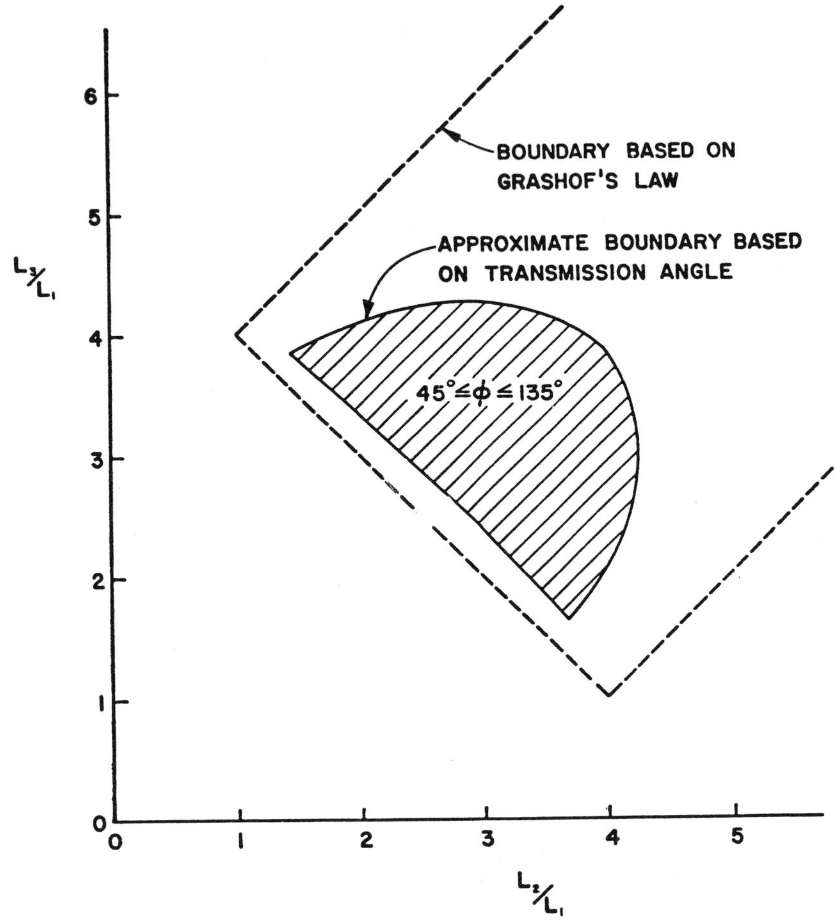
BOUNDARY BASED ON GRASHOF'S LAW
APPROXIMATE BOUNDARY BASED ON TRANSMISSION ANGLE
45° ≤ φ ≤ 135°
L3/L1
L2/L1
0
1
2
3
4
5
6

1.42 through 1.47: As shown in problem 1.51, extreme values of transmission angle occur when the crank is perpendicular to the slider path. Thus,

$$\sin \phi_{\substack{max \\ min}} = \frac{R \pm E}{L}$$

for crank length R, connecting rod length L and offset E.

1.42 $L_{min} = \dfrac{R}{\sin\phi_{max}} = \dfrac{100}{\sin 45^o} = \underline{141.42 \text{ mm}}$ (ans)

1.43 $L_{min} = \dfrac{R + E}{\sin\phi_{max}} = \dfrac{500 + 100}{\sin 45^o} = \underline{848.53 \text{ mm}}$ (ans)

1.44 $L_{min} = \dfrac{400 + 200}{\sin 45^o} = \underline{848.53 \text{ mm}}$ (ans)

1.45 $L_{min} = \dfrac{300 + 50}{\sin 45^o} = \underline{494.97 \text{ mm}}$ (ans)

1.46 $R + E_{max} = L \sin \phi_{max}$

$E_{max} = L \sin\phi_{max} - R = L(\sin 45^o - 0.5) = \underline{0.207 \quad L}$ (ans)

1.47 $E_{max} = L \sin\phi_{max} - R = L(\sin 45^o - 0.25) = \underline{0.4571 \text{ L}}$ (ans)

1.48 There are many possible solutions. Let the linkage have a forward-to-return-stroke ratio of 1:1. The limiting positions are shown in the figure (O_1BCO_3 and $O_1B'C'O_3$). $B'O_1BC'C$ is a straight line with $CO_1 = L_2 + L_1$ and $C'O_1 = L_2 - L_1$. Thus $CC' = 2L_1$ forming isosceles triangle O_3CC' with angles γ, $90^o - \gamma/2$ and $90^o - \gamma/2$. From the law of sines,

$$\frac{\sin \gamma}{2L_1} = \frac{\sin(90-\gamma/2)}{L_3} = \frac{\cos(\gamma/2)}{L_3} \qquad \text{(a)}$$

Since one link length may be selected arbitrarily, let $L_1 = 1$. Then, from (a),

$$L_3 = 2 \cos(\gamma/2)/\sin \gamma \qquad \text{(b)}$$

Referring to Fig. 1.17A, with $O_1O_3 = L_0$ and $L_1 = 1$, we have from the cosine law:

$$(L_0 - 1)^2 = L_2^2 + L_3^2 - 2L_2L_3 \cos \phi_{min} \qquad \text{(c)}$$

from which

$$\cos \phi_{min} = \frac{L_2^2 + L_3^2 - (L_0 - 1)^2}{2L_2L_3} \qquad \text{(d)}$$

1.48 (cont'd.)

Similarly, from Fig. 1.17B

$$\cos\phi_{max} = \frac{L_2^2 + L_3^2 - (L_0+1)^2}{2L_2L_3} \qquad \text{(e)}$$

If we specify that $\phi_{max} - 90^o = 90^o + \phi_{min}$, then

$$\phi_{max} = 180^o - \phi_{min} \text{ and } \cos\phi_{max} = -\cos\phi_{min} \qquad \text{(f)}$$

Using Eq. (f) in (d) and (e)

$$L_2^2 + L_3^2 - (L_0+1)^2 = -L_2^2 - L_3^2 + (L_0-1)^2$$

from which

$$L_2 = [L_0^2 + 1 - L_3^2]^{1/2} \qquad \text{(g)}$$

Substituting Eq. (g) in (d)

$$\cos\phi_{min} = \frac{L_0^2 + 1 - L_3^2 + L_3^2 - (L_0-1)^2}{2L_3\sqrt{L_0^2 + 1 - L_3^2}}$$

from which

$$L_0 = \left[\frac{L_3^2(1-L_3^2)\cos^2\phi_{min}}{1 - L_3^2\cos^2\phi_{min}}\right]^{1/2} \qquad \text{(h)}$$

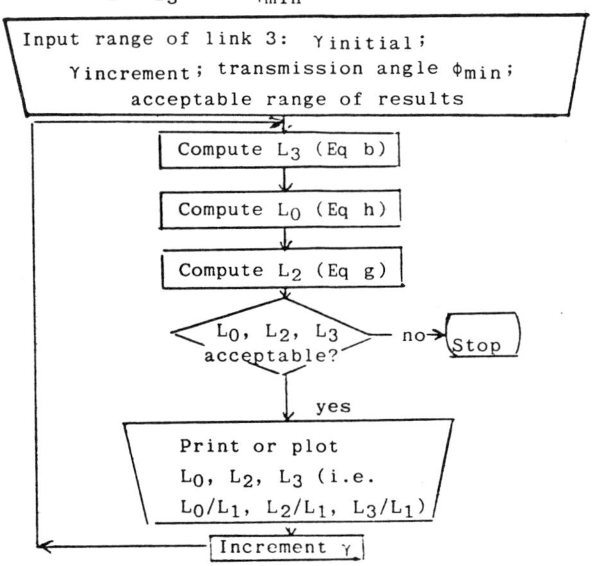

1.48

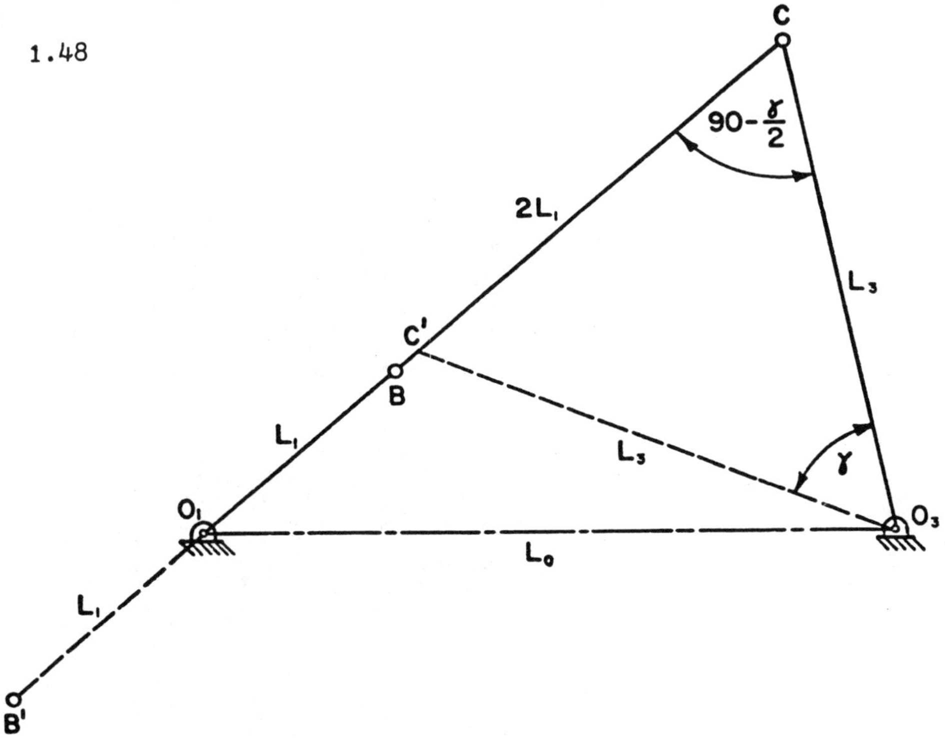

1.49

```
100 PRINT "OPTIMIZE FOLLOWER RANGE"
110 P=PI*5/18
112 P1=180*P/PI
115 PRINT "MINIMUM TRANSMISSION ANGLE (DEGREES) = ",P1
116 P1=180*P/PI
118 PRINT "RANGE(DEGREES)     L0/L1               L2/L1            L3/L1"
120 FOR G=PI*30/180 TO PI STEP PI*2.5/180
130 L3=2*COS(G/2)/SIN(G)
140 L0=(L3↑2*(1-L3↑2)*COS(P)↑2/(1-L3↑2*COS(P)↑2))↑0.5
150 L2=(L0↑2+1-L3↑2)↑0.5
160 G1=180*G/PI
170 P1=180*P/PI
220 PRINT G1,L0,L2,L3
230 NEXT G
240 END
```

```
OPTIMIZE FOLLOWER RANGE
MINIMUM TRANSMISSION ANGLE (DEGREES) =                      50
RANGE(DEGREES)     L0/L1              L2/L1              L3/L1
 30                4.07716844983      1.64167577129      3.86370330516
 32.5              3.81091585711      1.65903139821      3.57361084755
 35                3.58848196589      1.67874632662      3.32550952342
 37.5              3.40179920804      1.70114125105      3.11100567274
 40                3.24496838841      1.72661161566      2.92380440016
 42.5              3.11366278701      1.75564963511      2.75909229819
 45                3.00474636219      1.78887494709      2.61312592975
 47.5              2.91603439699      1.82707815646      2.48295026422
 50                2.8461578179       1.87128415594      2.36620158315
 52.5              2.79451426114      1.92284672843      2.26096669916
 55                2.76130750096      1.98359440983      2.1656805702
 57.5              2.74769908453      2.0560639006       2.07905062367
 60                2.75613255463      2.14389054261      2
 62.5              2.79096444356      2.25249786395      1.92762441832
 65                2.85970726987      2.39040014607      1.8611589967
 67.5              2.97563829177      2.5718853852       1.79995244627
 70                3.16388596242      2.8232193776       1.74344679562
 72.5              3.47809714994      3.19955203198      1.69116131077
 75                4.05934988833      3.84446682174      1.6426796317
 77.5              5.47536536389      5.33171400381      1.59763939913
 80                1.413811038E+7     1.413811038E+7     1.55572382686
```

1.49 An optional program and a plot of link length ratios for a minimum transmission angle of 40°.

```
LIST
80 D=32
90 PAGE
92 A=40
93 B=95
94 C=0
95 E=5
100 WINDOW A,B,C,E
101 VIEWPORT 0,80,0,90
102 AXIS 5,1,40,0
103 HOME
110 PRINT "OPTIMIZE FOLLOWER RANGE"
120 P=PI*40/180
130 P1=180*P/PI
140 PRINT "MINIMUM TRANSMISSION ANGLE (DEGREES) = ",P1
145 PRINT "FOLLOWER RANGE:",A,"TO",B,"L/L1:",C,"TO",E
150 P1=180*P/PI
180 X1=0
190 Y0=0
200 Y2=0
210 Y3=0
215 FOR G=PI*30/180 TO PI*95/180 STEP PI*2.5/180
220 L3=2*COS(G/2)/SIN(G)
230 L0=(L3↑2*(1-L3↑2)*COS(P)↑2/(1-L3↑2*COS(P)↑2))↑0.5
240 L2=(L0↑2+1-L3↑2)↑0.5
250 G1=180*G/PI
260 P1=180*P/PI
270 MOVE @D:X1,Y0
280 DRAW @D:G1,L0
290 MOVE @D:X1,Y2
300 DRAW @D:G1,L2
310 MOVE @D:X1,Y3
320 DRAW G1,L3
330 GOSUB 360
340 NEXT G
345 MOVE X1,Y0
346 PRINT "L0/L1"
347 MOVE X1,Y2
348 PRINT "L2/L1"
349 MOVE X1,Y3
350 PRINT "L3/L1"
351 END
360 X1=G1
370 Y0=L0
380 Y2=L2
390 Y3=L3
400 RETURN
```

1.49

OPTIMIZE FOLLOWER RANGE
MINIMUM TRANSMISSION ANGLE (DEGREES) = 40
FOLLOWER RANGE: 40 TO 95
L/L1: 0 TO 5

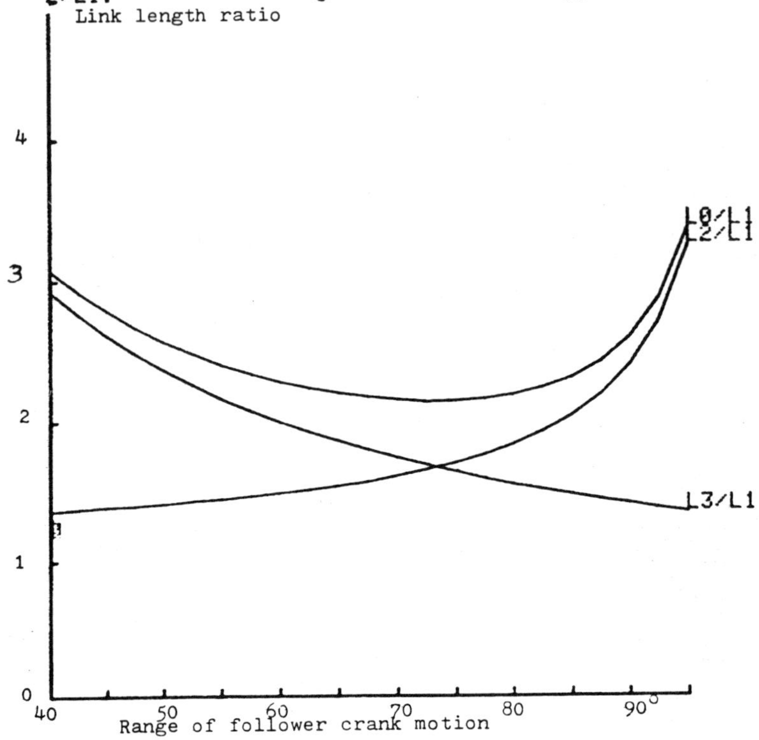

1.49 See listing and output. The program prints values of L_0, L_2 and L_3 where $L_1 = 1$ for a given value of ϕ_{min} (P) and a series of values of follower range γ (G). Computation continues until one of the variables is no longer real. The solution is the largest value of γ which produces a reasonable configuration (e.g., we could require $L_0 < 5$, etc.).

1.50

Using Eqs. 1.11 and 1.12, we have

$$L_0^2 + L_1^2 - 2L_0L_1 \cos\theta_1 = L_2^2 + L_3^2 - 2L_2L_3 \cos\phi$$

Differentiating with respect to θ_1,

$$2L_0L_1 \sin\theta_1 = 2L_2L_3 \sin\phi \frac{d\phi}{d\theta_1}$$

from which

$$\frac{d\phi}{d\theta_1} = \frac{L_0L_1 \sin\theta_1}{L_2L_3 \sin\phi}$$

$\frac{d\phi}{d\theta_1} = 0$ implies $\sin\theta_1 = 0$

or $\theta_1 = 0$ or 180° for extreme values of ϕ_1.

1.51

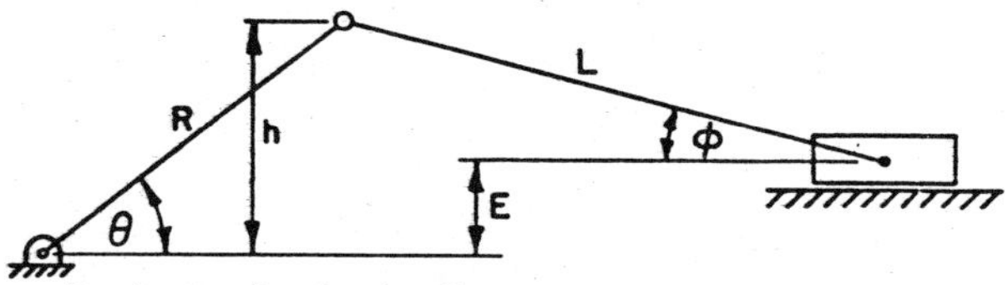

$$h = R \sin\theta = L \sin\phi + E$$

Differentiating with respect to θ:

$R \cos\theta = L \cos\phi \frac{d\phi}{d\theta}$ from which

$$\frac{d\phi}{d\theta} = \frac{R \cos\theta}{L \cos\phi} \qquad \frac{d\phi}{d\theta} = 0 \quad (\text{for } \phi \neq 0)$$

implies $\cos\theta = 0$ or $\theta = \pm 90^\circ$ for extreme values of ϕ.

1.52 a) $L_{max} + L_{min} < L_a + L_b \qquad 18 + 7 < 17 + 9$

The linkage theoretically acts as a crank rocker.

b) $\cos\phi_{min} = [L_2^2 + L_3^2 - (L_0 - L_1)^2]/(2L_2L_3) \qquad \phi_{min} =$ [illegible]

$\cos\phi_{max} = [L_2^2 + L_3^2 - (L_0 + L_1)^2]/(2L_2L_3) \qquad \phi_{max} = 146.4^\circ$

c) Values of transmission angle are not in the generally accepted range. The linkage will not operate properly unless special provisions are made to prevent jamming.

1.53 In a double rocker linkage (of the first kind) the coupler must be free to rotate through 360°. One possible configuration is shown.

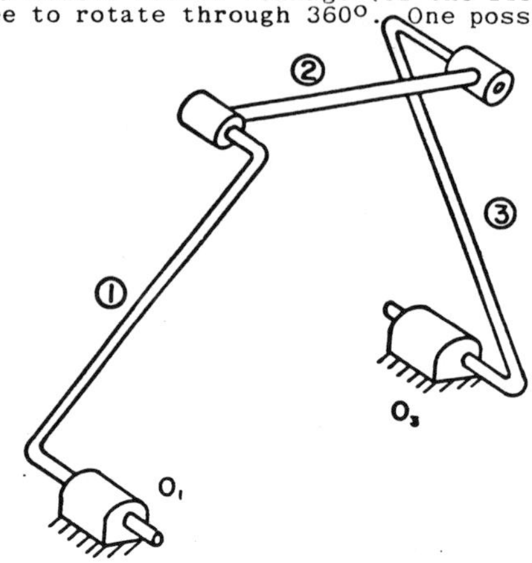

1.54 Referring to the figure, let the upper arm rotate through an angle θ_1 = 60° counterclockwise. This is equivalent to a 60° clockwise rotation of pulley 1 relative to the arm. Pulley 2 rotates

$$\theta 2 = \frac{r_1}{r_2} \theta 1 = \frac{50}{100} \times 60 = 30^\circ$$

clockwise relative to the arm. Thus, adding arm rotation, the total rotation of pulleys 2 and 3 is 30° counterclockwise, while the total rotation of pulley 1 is zero. Since pulleys 3 and 4 are equal, rotation of the lower arm is offset by equal relative motion. Thus, lower arm rotation causes no additional rotation, but the straight edges have rotated by an angle $\theta_1/2$. <u>The machine will not operate properly</u>. (ans)

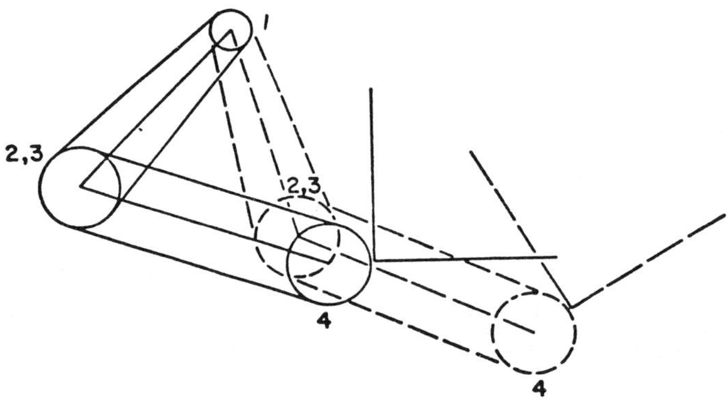

1.55 Use Fig. 1.25A. BC = DE. BE = CD. Size factor DA/DF = 2. Thus, tracing point F falls at midpoint of BE, and CA/CB = 2. Copy traced by A is twice the size of pattern traced by F. Lengths DE and DC are arbitrary, but large enough to allow F to trace the entire pattern.

1.56 Use Fig. 1.25B. Let BC = DE and BE = CD. Part to pattern size ratio = 1.10 = AF/DF where D traces the pattern and A reproduces it 10% larger. Thus, AB/BC = 1.10 and fixed point F is located by the proportion BF/EF = 1.10 since triangles ABF and DEF are similar. BE and DE are arbitrary, but must be large enough for D to trace the entire pattern.

1.57 Use Fig. 1.25A. Let BC = DE and BE = CD. Let AB/AC = 0.40. Point A traces the pattern and point F (on line AD) is the drawing tool. Lengths must be adequate to trace the entire pattern.

1.58 a) $Q = ANd \tan\phi$

$$Q = \frac{(120\ ft^3/hr)(1\ hr/60\ min)(1728\ in^3/ft^3)}{600\ rev/min}$$

$= 5.76\ in^3/rev$

where A, N, d and ϕ are arbitrary. E.g., let $A = .5\ in^2$, N = 6 cyl, $\phi = 30^o$.

Then, $d = \dfrac{Q}{AN \tan\phi} = 3.33$ in

b) Stroke length $L = d \tan\phi = 1.92$ in

Stroke time t = (1 min/600 rev)(60 sec/min)(1 rev/2 strokes)

= 0.05 sec/stroke.

$v(avg) = \dfrac{L}{t} = 38.5$ in/sec

1.59 a) $Q = ANd \tan\phi$

$$Q = \frac{0.01\ m^3}{s} \times \frac{1s}{300\ rad} \times \frac{2\ \pi rad}{rev} = 2.094\text{x}10^{-4}\ m^3/rev$$

Let $\phi = 30^o$, N = 6 cylinders, $A = (d/6)^2$

Then, $2.094\text{x}10^{-4} = (d/6)^2 \times 6 \times d \times \tan 30^o$

$d^3 = 2.094\text{x}10^{-4} \times 36/(6 \tan 30^o)$

d = 0.1296 m $A = 4.665\text{x}10^{-4}\ m^2$

d = 129.6 mm $A = 466.5\ mm^2$

b) Stroke length $L = d \tan\phi = 74.8$ mm

Stroke time $t = \dfrac{1s}{300\ rad} \times \dfrac{2\ \pi rad}{rev} = 2.094\text{x}10^{-2}$ s

$$V_{avg} = \frac{L}{t} = \frac{74.8\ mm}{2.094\text{x}10^{-2}\ s} = \underline{3573\ mm/s}\ \ (ans)$$

There are many possible solutions.

1.60 $v = (n/60)(L_2-L_1) = 2\pi\omega(L_2-L_1)$

from which

$L_2 - L_1 = v/(2\pi\omega) = 0.1/(2\pi \times 65) = .000245$

One solution would be two right hand single threads where the leads are $L_1 = 2$ mm, $L_2 = 2.000245$ mm.

1.61 $v = (n/60)(L_2-L_1)$

from which

$L_2 - L_1 = 60\ v/n = 60 \times 0.0005/60$

$= 0.0005$ in

One solution would be two right hand single threads where the leads are $L_1 = 0.05$ (20 threads/in) and $L_2 = 0.0505$ (19.8 threads/in would approximate this value).

1.62 There are many possible solutions, e.g. a linkage with the same dimensions as in the feed mechanism--example problem, but employing a single thread power screw with 10 threads per inch.

1.63 There are many possible solutions. Consider a linkage similar to Figs. 1.42 and 1.43, using a power screw with a lead $\underline{L = 15\text{mm}}$. Screw pitch would be 5mm if a triple thread screw is used. Feed per ratchet pitch is 0.100mm = L/N = 15/N from which $\underline{N \doteq 150}$ ratchet teeth. Let the fixed link have a radius r_o and let $r_3/r_o = 0.5$. See sketch. For a transmission angle of $\phi_{min} = 45^o$ (between links 2 and 3) the triangle solution yields

$c_1 = r_1 + r_2 = 1.289\ r_o$

$\theta_1 = 20.7^o$

$\angle C_1 = 114.3^o$ (extended position).

For the fixed position,

$\angle C_2$ is determined by the feed of 3mm divided by the lead:

$(3/15) \times 360 = 72^o = \Delta C$

from which

$\angle C_2 = 114.3 - 72^o = 42.3^o$

The triangle solution yields:

$c_2 = r_2 - r_1 = 0.7144\ r_o$; $\angle B = 28.1^o$

$\phi_{max} = 109.6^o$(which is acceptable).

Using $r_2 - r_1 + r_1 + r_2 = (0.7144 + 1.289)r_o = 2.0034\ r_o$

or $\underline{r_2 = 1.002\ r_o}$ and $\underline{r_1 = 0.287\ r_o}$ (maximum value)

These values would have to be changed slightly to account for tolerances in the actual linkage. Crank length r_1 must be adjustable to provide the range of feeds. If we try $r_1(\text{min}) = 0.090\ r_o$, then for the extended position,

$c_1 = r_1 + r_2 = 1.002\ r_o + 0.090\ r_o = 1.092\ r_o$

$\phi_{min} = 66.1^o$, $\theta_1 = 27.2^o$ and $\angle C_1 = 86.7^o$. For the flexed position,

$c_2 = r_2 - r_1 = 1.002\ r_o - 0.090\ r_o = 0.912\ r_o$

$\phi_{max} = 84.9^o$; $B = 29.9^o$; $\angle C_2 = 65.3^o$; $\Delta C = C_1 - C_2 = 21.4^o$

21.4 L/360 $\text{Feed}_{(min)} = 21.4\ L/360 = 21.4 \times 15/360 = 0.89$ mm

which is satisfactory (less than the required minimum).

1.63

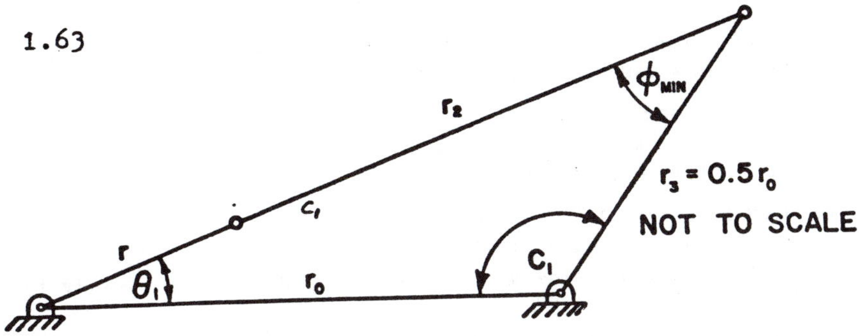

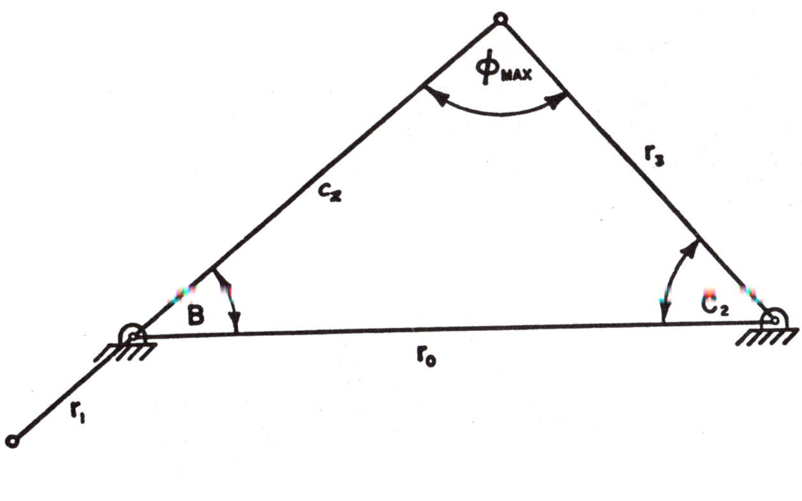

1.64 $n_2/n_1 = \omega_2/\omega_1 = \cos\phi$ to $1/\cos\phi$ where $\phi = 20^o$.

Thus, $n_2 = 939.7$ to 1064.2 RPM.

1.65 $\omega_2/\omega_1 = \cos\phi$ to $1/\cos\phi$

$n_2 = 1000 \cos 15^o$ to $1000/\cos 15^o$

$n_2 = 965.9$ to 1035.3 RPM

1.66 $0.98 \leq \omega_2/\omega_1 \leq 1.02$

$0.98 = \cos\phi$ yields $= 11^o28'$

$1.02 = 1/\cos\phi$ yields $\phi = 11^o22'$ or 11.36^o. Thus, permissible misalignment is 11.36^o.

1.67 $\pm 3\%$ variation requires $0.97 \leq \omega_2/\omega_1 \leq 1.03$

Also, $\cos\phi \leq \omega_2/\omega_1 \leq 1/\cos\phi$

$\arccos 0.97 = 14.07^o$; $\arccos (1/1.03) = 13.86^o$

Permissible misalignment is 13.86^o.

1.68 For 10 in. stroke, $L_3 = 10/2 = 5$ in. Other dimensions are arbitrary within certain ranges. Thus, there are many solutions. $L_4 > L_3$, say $L_4 = 8$ in. L_0 is shortest, say $L_0 = 2$ in. Let $L_1 = 4.5$ in. The drag link inequality requires that

$|L_1 - L_2| + L_0 < L_3 < L_1 + L_2 - L_0$ from which

$2.5 < L_2 < 7.5$.

Values of L_2 between 2.5 and 7.5 are tried as in the figure. See also Figs. 1.19 and 1.20. In trial 7, with $L_2 = 6.75$ in., we find $\angle BO_1B' = 121.5^o$ between limiting positions. For this value, the forward to return time ratio is

$$\frac{360^o - 121.5^o}{121.5^o} = 1.96$$

which is close enough to the required value.

1.69 See figure. There are many solutions. As an example, use the same dimensions as in Prob. 1.68 except for L_2. Trial I with $L_2 = 7.25$ produces $\angle BO_1B' = 106.5^o$. The forward to return ratio is

$$\frac{360^o - 106.5^o}{106.5^o} = 2.45, \text{ close enough to the required value.}$$

1.68 and 1.69

	Trial	L_2	BO_1B'	Ratio
1.68	1	3		
	2	4		
	3	5		
	4	6	135°	
	5	7	114°	
	6	6.5	126°	
	7	6.75	121.5^o	$\frac{238.5}{121.5} = 1.96$
1.69	I	7.25	106.5^o	$\frac{253.5}{106.5} = 2.45$

1.68 and 1.69

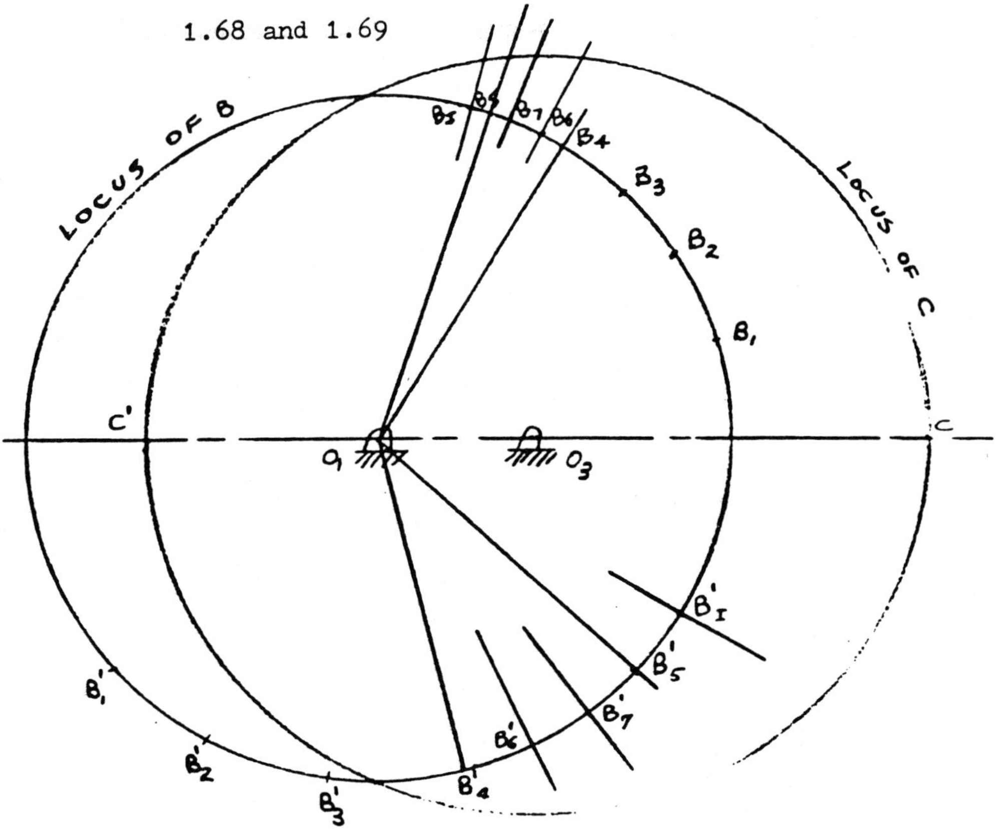
LOCUS OF B
LOCUS OF C
C'
C
O_1
O_3
B_1
B_2
B_3
B_4
B_5
B_6
B_7
B'_1
B'_2
B'_3
B'_4
B'_5
B'_6
B'_7

1.70 Refer to Figs. 1.19 and 1.20. L_3 = stroke/2 = 75mm. Other dimensions are arbitrary within limits. Referring to the solution to 1.68, we may use the same proportions, yielding $L_0 = 30$; $L_1 = 67.5$; $L_2 = 101.25$; and $L_4 = 120$mm.

1.71 Refer to Figs. 1.19 and 1.20. and the solution to Prob. 1.68. L_3 = stroke/2 = 100/2 = 50mm. Using the proportions of Prob. 1.68, $L_0 = 20$; $L_1 = 45$; $L_2 = 67.5$ and $L_4 = 80$ in.

1.72 For ratio of 1.5, set

$$\frac{360^\circ - \theta^\circ}{\theta^\circ} = 1.5 \text{ or } \theta = 144^\circ.$$

There are many solutions. For example, try a distance $O_1O_2 = 5$ in. Then, maximum drive crank length is $L_1(max) = O_1O_2 \cos\theta/2 = 1.55$ in. Then,

$$L_2 = O_2C = \frac{S(max)/2}{\sin(90^\circ - \frac{\theta}{2})} = 16.2 \text{ in.}$$

We will arbitrarily let $L_3 = 4$ in. The distance from O_1 to the path of D may be based on a horizontal position of CD at midstroke. Then, we have $L_2 - O_1O_2 = 11.2$ in., the distance from O_1 to the path of D. For minimum stroke, the crank must be adjustable to a length

$$L_1\ (min) = O_1O_2\{\frac{S(min)/2}{O_2C}\} = 0.75 \text{ in.}$$

The linkage must be checked to see that physical limits are not reached during operation.

1.73 See Prob. 1.72. Let $O_1O_2 = 7$ in and $L_3 = 4$ in.

$$\text{Ratio } 2.5 = \frac{360 - \theta}{\theta} \text{ whence } \theta = 103^\circ.$$

$$L_1\ (max) = O_1O_2 \cos\frac{\theta}{2} = 4.36. \text{ Then,}$$

$$L_2 = \frac{S(max)/2}{\sin(90 - \frac{\theta}{2})} = 8.03 \text{ in.}$$

If this design is used, link 2 must be estended beyond C and designed so that the slider clears pin C. (I.e., the actual configuration differs from the schematic.) For minimum stroke,

$$L_1\ (min) = O_1O_2\,\frac{S(min)/2}{O_2C} = 2.18 \text{ in.}$$

Let the distance from O_2 to the path of D equal O_2C or 8.03 in.

1.74

For 1.5 to 1 ratio, $\angle BO_1B' = 360 \times 1/(1.5+1) = 144^o$ $\angle BO_1O_2 = 72^o$

$\angle BO_2O_1 = 90 - 72 = 18^o$

For 200 mm stroke, $2O_2C \sin BO_2O_1 = 200$ mm

$O_2C = 100/\sin 18^o = \underline{323.6 \text{ mm}}$

$O_1B/O_2O_1 = \sin 18^o$

If we let $O_2O_1 = \underline{200 \text{ mm}}$, $O_1B_{(max)} = 200 \sin 18^o = \underline{61.8 \text{ mm}}$

For minimum stroke, $2\ O_2C \sin BO_2O_1 = 100$ mm

$\sin BO_2O_1 = 100/(2\text{x}323.6)$ $\angle BO_2O_1 = 8.888^o$

$O_1B_{(min)} = 200 \sin 8.888^o = \underline{30.9 \text{ mm}}$

1.75 See above problem. $\angle BO_1O_2 = 72^o$

$\angle BO_2O_1 = 18^o$

For 280 mm stroke, $2\ O_2C \sin BO_2O_1 = 280$ mm

$O_2C = 280/(2 \sin 18^o) = \underline{453 \text{ mm}}$

Selecting $O_1O_2 = \underline{250 \text{ mm}}$,

$O_1B_{max} = O_1O_2 \sin BO_2O_1 = 250 \sin 18^o = \underline{77.25 \text{ mm}}$

For 180 mm stroke, $2\ O_2C \sin BO_2O_1 = 180$ mm

$\sin BO_2O_1 = 180/(2\text{x}453)$ $\angle BO_1O_2 = 11.46^o$

$O_1B_{(min)} = 250 \sin BO_1O_2 = 250 \sin 11.46^o = \underline{43.67 \text{ mm}}$

1.76 As link 1 rotates about a fixed point, link 2 rotates (oscillates) about a fixed point, link 3 has combined rotation (oscillation) and translation and link 4 translates. As the slider moves from left to right, link 1 turns through $\beta = 257.4^o$. See sketch. Link 1 turns through $\theta = 102.6^o$ as the slider returns. The ratio of β to θ which corresponds to the time ratio of forward to return strokes is 2.5 to 1. The stroke, given by DD' or CC' is 5 in. Values may be computed or measured directly from the sketch.

1.77 As link 1 rotates about O_1, link 2 has combined rotation (oscillation) and translation and link 3 translates. As the slider moves from left to right, link 1 turns through = 159.6°; link 1 turns through $\theta = 200.4^o$ as the slider returns. See sketch. The ratio of to θ is 0.796 to 1. The stroke, CC', is 3.14 in. Values may be computed or measured directly from the sketch.

1.76

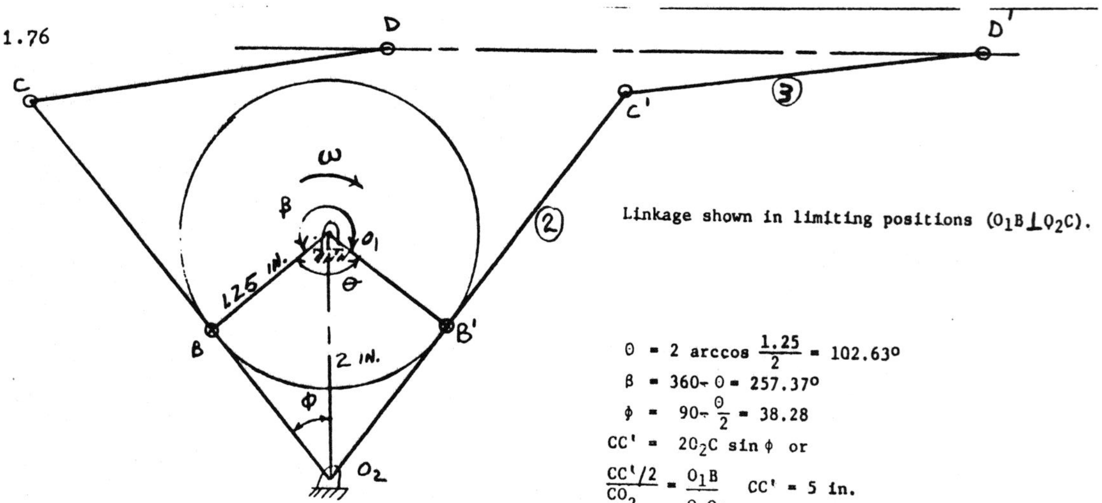

Linkage shown in limiting positions ($O_1B \perp O_2C$).

$$\Theta = 2 \arccos \frac{1.25}{2} = 102.63^\circ$$

$$\beta = 360 - \Theta = 257.37^\circ$$

$$\phi = 90 - \frac{\Theta}{2} = 38.28$$

$$CC' = 2O_2C \sin \phi \quad \text{or}$$

$$\frac{CC'/2}{CO_2} = \frac{O_1B}{O_1O_2} \qquad CC' = 5 \text{ in.}$$

1.77

$$\phi_1 = \arcsin \frac{E}{L-R} = 30^\circ$$

$$\phi_2 = \arcsin \frac{E}{L+R} = 9.6^\circ$$

$$\beta = 180 - \phi_1 + \phi_2 = 159.60^\circ$$

$$\Theta = 360 - \beta = 200.40^\circ$$

$$CC' = \sqrt{(L+R)^2 - E^2} - \sqrt{(L-R)^2 - E^2} = 3.137 \text{ in.}$$

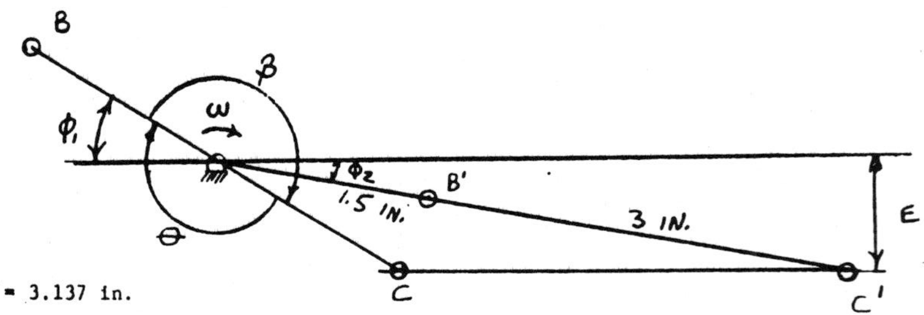

1.78 Link 1 rotates about O_1 , link 2 rotates and translates, and link 3 translates.

ϕ_1 = arcsin(E/(L-R)) = arcsin(20/(120-60))=19.47°

ϕ_2 = arcsin(E/(L+R)) = arcsin(20/(120+60))=6.38°

$\beta = 120^\circ - \phi_1 + \phi_2 = 180^\circ - 19.47^\circ + 6.38^\circ = 166.91^\circ$ (left to right)

$\alpha = 180^\circ + \phi_1 - \phi_2 = 180^\circ + 19.47^\circ - 6.38^\circ = 193.09^\circ$ (right to left)

Stroke $S = ((L+R)^2 - E^2)^{\frac{1}{2}} - ((L-R)^2 - E^2)^{\frac{1}{2}}$

$= ((120+60)^2 - 20^2)^{\frac{1}{2}} - ((120^2-60)^2 - 20^2)^{\frac{1}{2}} = 122.32\text{mm}.$

1.78

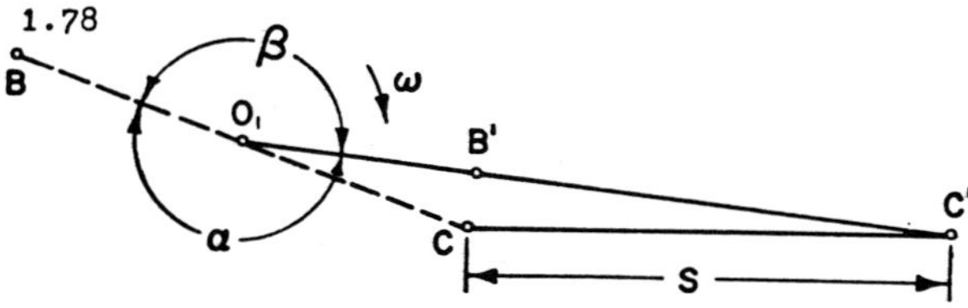

1.79 $\phi_1 = \arcsin \frac{E}{L-R} = 7.65^\circ$

$\phi_2 = \arcsin \frac{E}{L + R} = 3.28^\circ$

$\alpha = 180 + \phi_1 - \phi_2 = 184.37^\circ$

$\beta = 180 - \phi_1 + \phi_2 = 175.63^\circ$

a) $\frac{\alpha}{\beta} = 1.05$

b) $S = [(L + R)^2 - E^2]^{\frac{1}{2}} - [L-R)^2 - E^2]^{\frac{1}{2}} = 4.016$

1.80 $\phi_1 = \arcsin[E/(L-R)] = \arcsin[0.8/(5-2)] = 15.47^o$

$\phi_2 = \arcsin[E/(L+R)] = \arcsin[0.8/(5+2)] = 6.56^o$

$\alpha = 180^o + \phi_1 - \phi_2 = 180 + 15.47 - 6.56 = 188.9^o$

$\beta = 180^o - \phi_1 + \phi_2 = 180 - 15.47 + 6.56 = 171.1^o$

a) Time ratio $= \alpha/\beta = 188.9/171.1 = 1.104$

b) Stroke $S = [(L+R)^2 - E^2]^{1/2} - [(L-R)^2 - E^2]^{1/2}$

$= [(5+2)^2 - 0.8^2]^{1/2} - [5-2)^2 - 0.8^2]^{1/2} = 4.063$

1.81 $\phi_1 = \arcsin[E/(L-R)] = \arcsin[0.65/(5-2)] = 12.51^o$

$\phi_2 = \arcsin[E/(L+R)] = \arcsin[0.65/(5+2)] = 5.33^o$

$\alpha = 180^o = \phi_1 - \phi_2 = 180 + 12.51 - 5.33 = 187.19^o$

$\beta = 180^o - \phi_1 + \phi_2 = 180 - 12.51 + 5.33 = 172.81^o$

a) Time ratio $= \alpha/\beta = 187.19/172.81 = 1.083$

b) Stroke $S = [(L+R)^2 - E^2]^{1/2} - [(L-R)^2 - E^2]^{1/2}$

$= [(5+2)^2 - 0.65^2]^{1/2} - [(5-2)^2 - 0.65^2]^{1/2} = 4.041$

1.82 The linkage is plotted in various positions and the midpoint of the connecting rod is located as in the figure. Alternatively, a computer graphics program is used as in Fig. 1.44, plotting the linkage for increments of $T_1 = \pi/9$ radians (see second figure). Then, the midpoint of the connecting rod is located by $X3=(X_1+X_2)/2$, $Y3=(Y1+Y2)/2$. It is plotted to the same scale (third figure). (For this plot, the eccentricity is assumed to be positive.)

1.83 A computer graphics program is used as described in problem 1.82. First, the linkage is shown in 18 positions. See figure. The coupler curve is then plotted. The scale inside the coordinate axes applies to problem 1.82, while the outside scale applies to problem 1.83. The proportions of the linkage are the same in both problems.

1.84 As seen if Fig. 1.37B, when the one-way clutch is engaged, link 5 and the output shaft move as a unit. There are six links including the frame and seven one-degree-of-freedom pairs (point C represents two pairs). Degrees of freedom are given by

$$DF_{(planar)} = 3(n_L-1) - 2n_j$$
$$= 3(6-1) - 2(7) = 1$$

1.85 through 1.88

Driver L_1 shortest $(L_1<L_2)$ and $L_1 + |L_2-L_3| < L_0 < L_2 - L_1 + L_3$

or $L_{max} + L_1 < L_a + L_b$

1.83

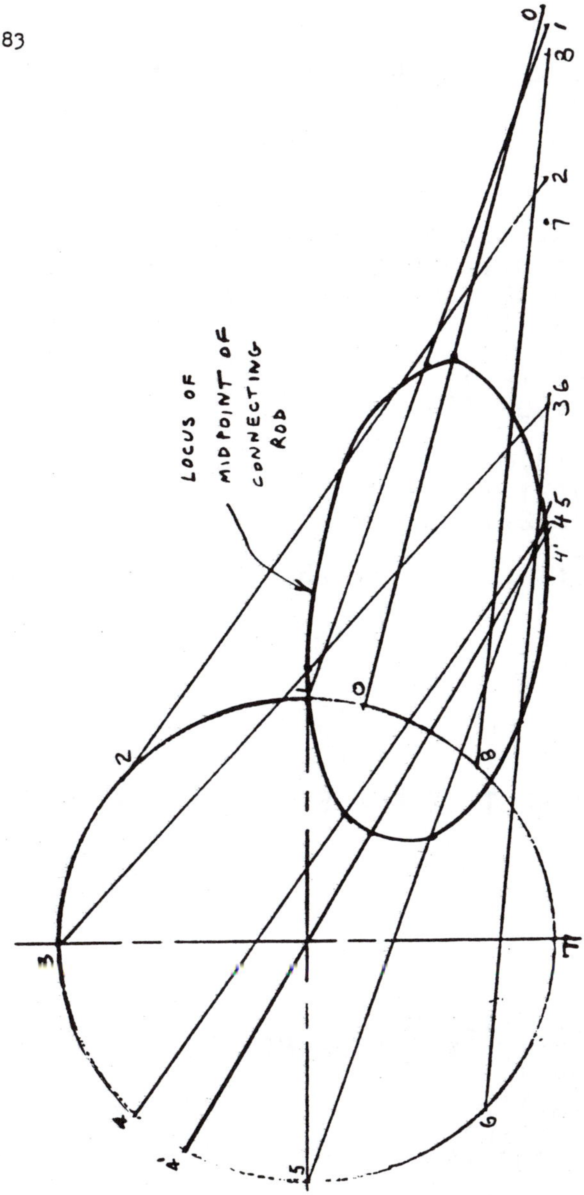

1.83 continued

```
LIST
100 PAGE
101 PRINT "INPUT 32 FOR SCREEN 1 FOR PLOTTER"
105 INPUT D
110 PAGE
120 R=150
130 L=450
140 E=150
150 WINDOW -150,600,-288.46,288.46
160 VIEWPORT 0,130,0,100
170 FOR T1=0 TO 2*PI STEP PI/9
180 X1=R*COS(T1)
190 Y1=R*SIN(T1)
200 T2=ASN((Y1-E)/L)
210 X2=X1+L*COS(T2)
220 Y2=E
230 MOVE @D:0,0
240 DRAW @D:X1,Y1
250 DRAW @D:X2,Y2
260 NEXT T1
270 AXIS @D:10,10,-150,-150
280 HOME
290 END
```

```
LIST
100 PAGE
101 PRINT "INPUT 32 FOR SCREEN 1 FOR PLOTTER"
105 INPUT D
110 PAGE
120 R=150
130 L=450
140 E=150
150 WINDOW -150,600,-288.46,288.46
160 VIEWPORT 0,130,0,100
165 MOVE @D:362.1,75
170 FOR T1=0 TO 2*PI STEP PI/30
180 X1=R*COS(T1)
190 Y1=R*SIN(T1)
200 T2=ASN((Y1-E)/L)
210 X2=X1+L*COS(T2)
220 Y2=E
230 X3=(X1+X2)/2
240 Y3=(Y1+Y2)/2
250 DRAW @D:X3,Y3
260 NEXT T1
270 AXIS @D:10,10,-150,-150
280 HOME
290 END
```

1.82 and 1.83

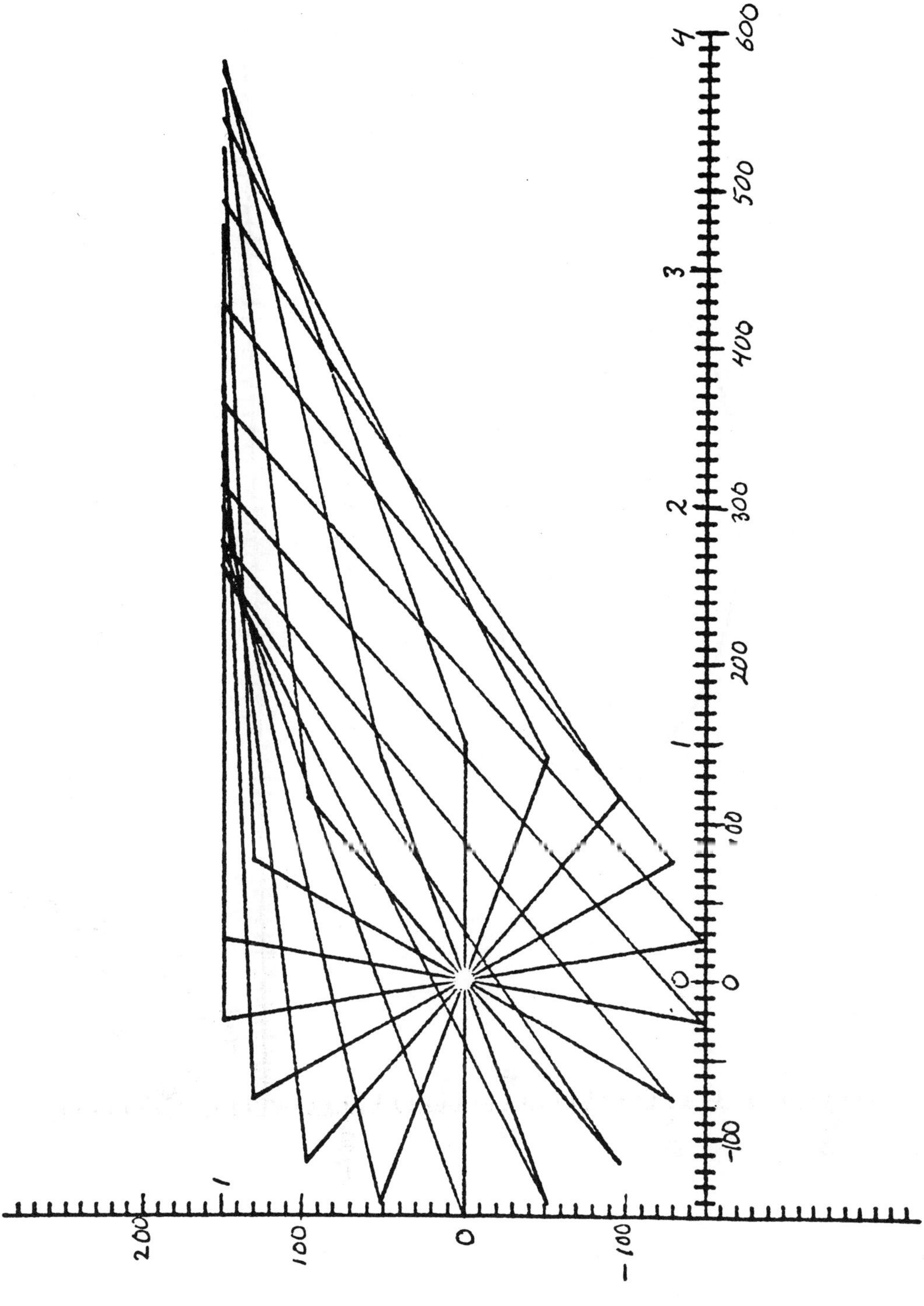

1.82 and 1.83

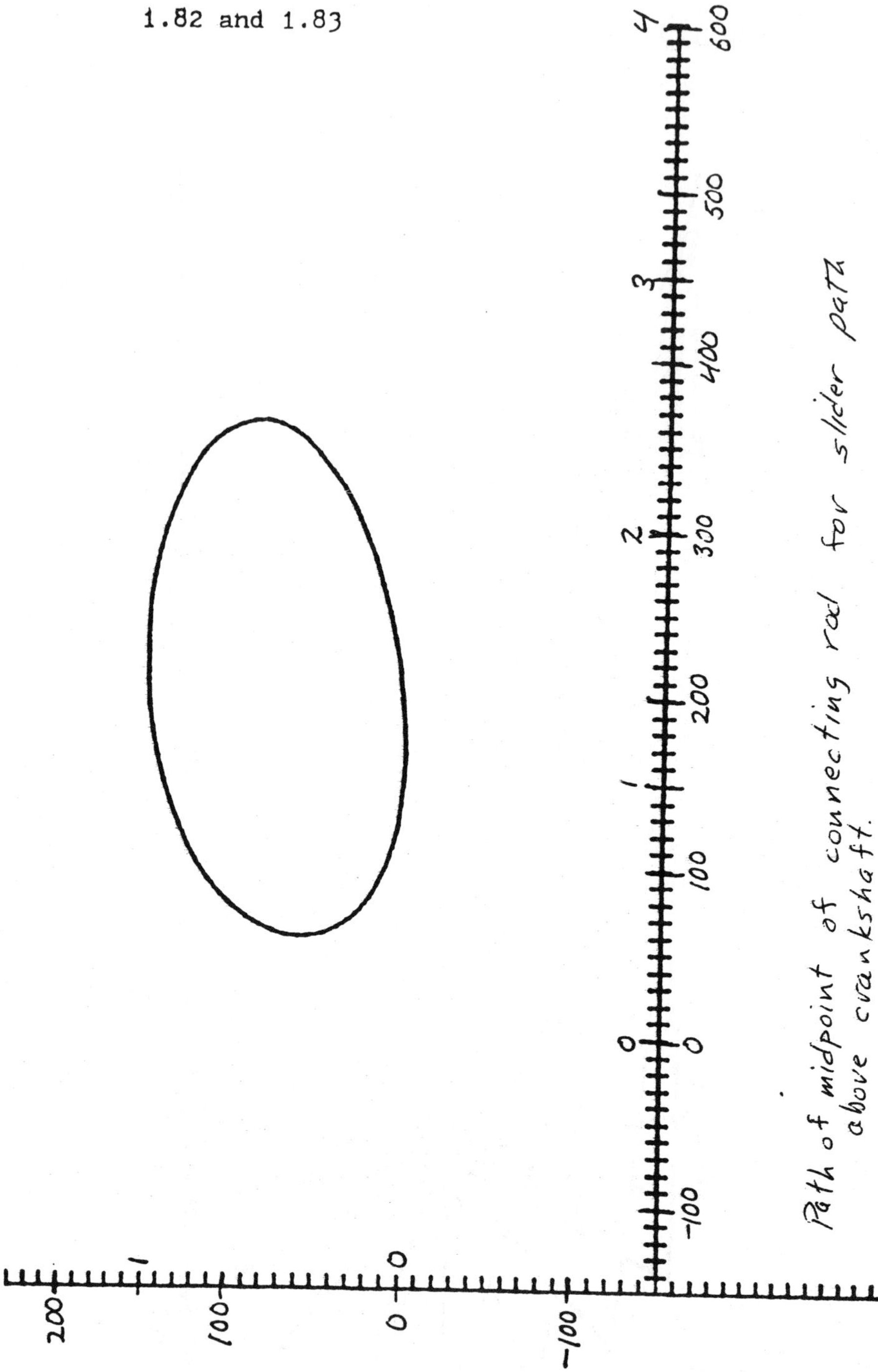

Path of midpoint of connecting rod for slider path above crankshaft.

1.85 $50 < L_2$ and $0 < L_2$

$50 + |L_2 - 150| < 200 < L_2 - 50 + 150 \therefore \underline{100 < L_2 < 300 \text{ mm}}$

.86 $50 < L_2$ and $50 + |L_2 - 150| < 210 < L_2 - 50 + 150$

$L_2 - 150 < 160 \qquad L_2 < 310 \qquad 110 < L_2$

$50 + 150 - L_2 < 210 \qquad -10 < L_2 \qquad \therefore \underline{110 < L_2 < 310 \text{ mm}}$

1.87 $45 < L_2$ and $45 + |L_2 - 150| < 200 < L_2 - 45 + 150$

$L_2 - 150 < 155 \qquad L_2 < 305 \qquad 45 + 150 - L_2 < 200 \quad -5 < L_2$

$200 + 45 - 150 < L_2 \qquad 95 < L_2 \qquad \therefore \underline{95 < L_2 < 305 \text{ mm}}$

1.88 $50 < L_2$ and $50 + |L_2 - 160| < 200 < L_2 - 50 + 160$

$L_2 - 160 < 150 \qquad L_2 < 310 \qquad 160 < L_2 < 150 \qquad 10 < L_2$

$200 < L_2 + 110 \qquad 90 < L_2 \qquad \therefore \underline{90 < L_2 < 310 \text{ mm}}$

1.89 through 1.92

$L_2^2 - 2L_2L_3 \cos 45^o + L_3^2 - (L_0 - L_1)^2 = 0$ for $\phi_{min} = 45^o$

$L_2^2 - 1.414\, L_3L_2 + L_3^2 - (L_0-L_1)^2 = 0$

$L_2 = (1/2)(1.414\, L_3 \pm \sqrt{2L_3^2 - 4[L_1^2 - (L_0-L_1)^2]}) =$

$0.707\, L_3 \pm \sqrt{-.5\, L_3^2 + (L_0-L_1)^2}$

$\cos \phi_{max} = (L_2^2 + L_3^2 - (L_0+L_1)^2)/(2L_2L_3)$

The root (L_2) which falls in the crank rocker range is selected.

1.89 $(200-50)^2 = L_2^2 + 150^2 - 2L_2 \times 150 \cos 45^o$

$L_2 = 0.707 \times 150 \pm \sqrt{-.5 \times 150^2 + (200-50)^2} = \underline{212.13 \text{ mm}}$ (or 0)

$\cos \phi_{max} = \dfrac{[212.13^2 + 150^2 - (200+50)^2)]}{2 \times 212.13 \times 150}$

$\underline{\phi_{max} = 85.5^o}$

1.90 $L_2 = .707 \times 150 \pm \sqrt{-.5 \times 150^2 + (210-50)^2} = \underline{225.84 \text{ mm}}$

(or -13.74 -- not possible)

$\cos \phi_{max} = \dfrac{225.84^2 + 150^2 - (210+50)^2}{2 \times 225.84 \times 150} \qquad \underline{\phi_{max} = 85^o}$

1.91 $L_2 = .707 \times 150 \pm \sqrt{-.5 \times 150^2 + (200-45)^2} = \underline{219.08 \text{ mm}}$

(or -6.98 -- not possible)

$\cos \phi_{max} = \dfrac{219.08^2 + 150^2 - (200+45)^2}{2 \times 219.08 \times 150} \qquad \underline{\phi_{max} = 80.83^o}$

1.92 $L_2 = .707 \times 160 \pm \sqrt{-.5 \times 160^2 + (200-50)^2} = \underline{211.6 \text{ mm}}$

(or 14.63 -- not a crank rocker)

$$\cos \phi_{max} = \frac{211.6^2 + 160^2 - (200+50)^2}{2 \times 211.6 \times 160} \qquad \underline{\phi_{max} = 83.32^\circ}$$

1.93 See figure.

$\phi_1 = \arcsin E/(R+L) = \arcsin(100/500) = 11.517^\circ$

$\phi_2 = \arcsin E(L-R) = \arcsin(100/200) = 30^\circ$

Stroke $S = \sqrt{(R+L)^2 - E^2} - \sqrt{(L-R)^2 - E^2} = 316.69$ mm

$\phi_3 = \phi_2 - \phi_1 = 18.463^\circ \qquad \alpha = 180^\circ - \phi_3 = 161.537^\circ = 2.8194$ rad

$\beta = 180^\circ + \phi_3 = 198.463^\circ = 3.4638$ rad

V_{av} away from crank $= S\omega/\alpha = 316.69 \times 240/2.8194 = \underline{26958 \text{ mm/s}}$

v_{av} toward crank $= S\omega/\beta = 316.69 \times 240/3.4638 = \underline{21942 \text{ mm/s}}$

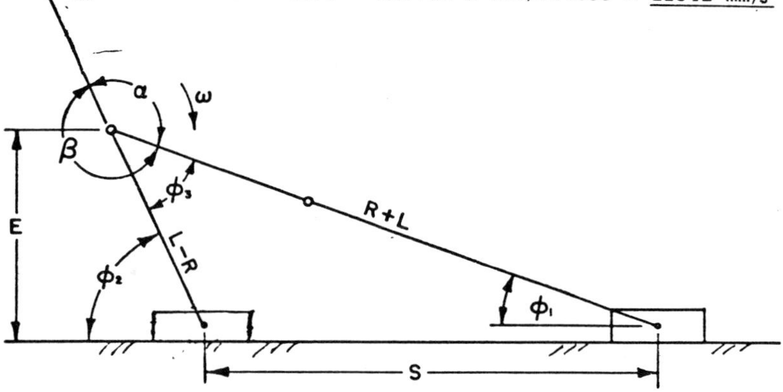

1.94 The RSSR mechanism becomes an RCSR mechanism.

$$DF_{(spatial)} \geq 6(n_L - n_J - 1) + \sum_{i=1}^{n_J} f_i$$

$$\geq 6(4-4-1) + 1 + 2 + 3 + 1 \geq 1$$

DF may exceed 1 for certain joint arrangements.

1.95 The mechanism becomes an SSSR mechanism.

$$DF_{(spatial)} \geq 6(n_L - n_J - 1) + \Sigma f_i$$

$$\geq 6(4-4-1) + 3 + 3 + 3 + 1 \qquad DF \geq 4$$

1.96 DF(spatial) $\geq 6(n_L - n_J - 1) + \Sigma f_i$

$$\geq 6(4-4-1) + 2 + 2 + 2 + 2 \qquad \text{DF} \geq 2$$

1.97 For an SSSS linkage,

$$\text{DF(spatial)} = 6(n_L - n_J - 1) + \Sigma f_i$$

$$= 6(4-4-1) + 3 + 3 + 3 + 3 \qquad \text{DF} = 6$$

1.98 For general link orientation, for a one-degree-of-freedom four-bar linkage

$$\text{DF(spatial)} = 6(n_L - n_J - 1) + \Sigma f_i$$

$$= 6(4-4-1) + \Sigma_{i=1}^{4} f_i \qquad \text{from which}$$

$\Sigma f_i = 7.$

The following linkages qualify: SCRR; SCRP; SCRH; SCPP; SCHH; SCHP; CCCR; CCCH and CCCP. In addition, others may be formed by interchanging joint positions. Others which do not satisfy the above criteria may have one-degree-of-freedom if the joints are oriented in a particular way.

1.99 One-degree-of-freedom linkages made up of lower pairs satisfy Grubler's criterion: $2n_J' - 3n_L + 4 = 0$. Since n_J' and n_L are positive integers, n_L must be even.

$n_J' = (3n_L - 4)/2$. Thus,

for $n_L = 2$, $n_J' = 1$

$n_L = 4$, $n_J' = 4$

$n_L = 6$, $n_J' = 7$

$n_L = 8$, $n_J' = 10$

Some examples are shown below:

1.99

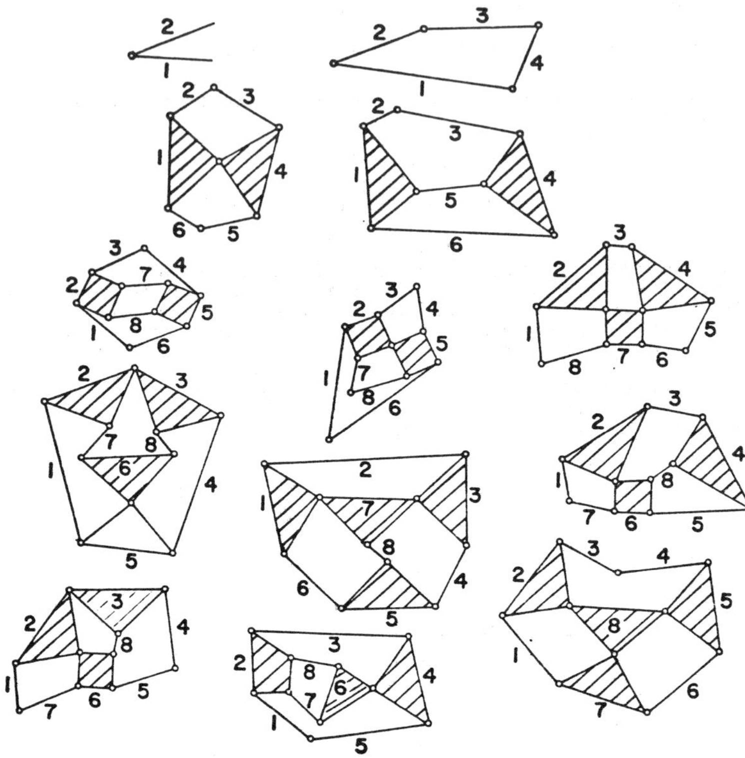

1.100 This problem uses methods discussed in section 1.13. See computer program S1.100a and iterative solution S1.100b for the given data. A first approximation of E/R = 1.1 and a first approximation of E/R = 1.9 both result in convergence to the same value. The required conditions are met with an eccentricity of 1.636 times the crank length.

1.101 This problem uses methods discussed in section 1.13. See computer program S1.101a and iterative solution S1.101b for the given data. The required conditions are met with an eccentricity of 1.163 times the crank length.

1.102 This problem uses methods discussed in section 1.13. See computer program S1.101a and iterative solution S1.102 for the given data. The required conditions are met with an eccentricity of 1.936 times the crank length.

51.100 a

```
10 REM Newton-Raphson method
20 LPRINT "Newton-Raphson method"
30 INPUT"max error, max iterations, 1st approx";TOLERANCE,N, X
35 LPRINT"max error, max iterations, 1st approx";TOLERANCE; N;
40 FOR I=1 TO N
45 F=SQR(16-X^2)-SQR(4-X^2)-2.5
55 LPRINT" x =";X;" F(x) =";F
65 G=-X/SQR(16-X^2)+X/SQR(4-X^2)
70 LPRINT"                      G(x) = ";G
75 XNEW=X-F/G
85 X=XNEW
95 IF ABS(F)<TOLERANCE GOTO 200
100 NEXT I
110 LPRINT "No root found after specified number of iterations
200 LPRINT "The root of F(x) = 0 is  x = ";X
```

51.100b for first approximation = 1.1

```
Newton-Raphson method
max error, max iterations, 1st approx .000001  20  1.1
 x = 1.1  F(x) =-.3245525
                    G(x) =  .3725247
 x = 1.971224  F(x) = .6425106
                    G(x) =  5.264875
 x = 1.849187  F(x) = .2849903
                    G(x) =  1.905686
 x = 1.69964  F(x) = 6.679583E-02
                    G(x) =  1.142945
 x = 1.641198  F(x) = 4.794121E-03
                    G(x) =  .9859428
 x = 1.636335  F(x) = 2.765656E-05
                    G(x) =  .9746375
 x = 1.636307  F(x) = 0
                    G(x) =  .9745721
The root of F(x) = 0 is  x =  1.636307
```

S1.100b contd.: For first approximation = 1.9

```
Newton-Raphson method
max error, max iterations, 1st approx .000001  20  1.9
 x = 1.9  F(x) = .3954435
                    G(x) =   2.502653
 x = 1.74199  F(x) = .1181762
                    G(x) =   1.289084
 x = 1.650316  F(x) = 1.388311E-02
                    G(x) =   1.007785
 x = 1.63654  F(x) = 2.272129E-04
                    G(x) =   .9751089
 x = 1.636307  F(x) = 2.384186E-07
                    G(x) =   .9745726
The root of F(x) = 0 is  x =  1.636307
```

S/1/01 a

```
10 REM Newton-Raphson method
20 LPRINT "Newton-Raphson method. L/R = 3  S/R = 2.2"
30 INPUT"max error, max iterations, 1st approx";TOLERANCE,N, X
35 LPRINT"max error, max iterations, 1st approx";TOLERANCE; N; X
40 FOR I=1 TO N
45 F=SQR(16-X^2)-SQR(4-X^2)-2.2
55 LPRINT" x =";X " F(x) =";F
65 G=-X/SQR(16-X^2)+X/SQR(4-X^2)
70 LPRINT"                          G(x) = ";G
75 XNEW=X-F/G
85 X=XNEW
95 IF ABS(F)<TOLERANCE GOTO 200
100 NEXT I
110 LPRINT "No root found after specified number of iterations"
200 LPRINT "The root of F(x) = 0 is  x = ";X
```

Sl.101 b

```
Newton-Raphson method. L/R = 3  S/R = 2.2
max error, max iterations, 1st approx .000001  20  1.5
 x = 1.5  F(x) = .1852233
                  G(x) =  .7293736
 x = 1.246052  F(x) = 3.656507E-02
                  G(x) =  .4686783
 x = 1.168034  F(x) = 2.179861E-03
                  G(x) =  .4141465
 x = 1.162771  F(x) = 9.298324E-06
                  G(x) =  .4107464
 x = 1.162748  F(x) =-2.384186E-07
                  G(x) =  .4107319
The root of F(x) = 0 is  x =  1.162749
```

S1.102

```
Newton-Raphson method. L/R = 3  S/R = 3
max error, max iterations, 1st approx .000001  20  1.8
 x = 1.8  F(x) =-.2996657
                   G(x) =  1.560838
 x = 1.99199  F(x) = .2898982
                   G(x) =  10.56564
 x = 1.964552  F(x) = .1094482
                   G(x) =  4.676678
 x = 1.94115  F(x) = .0158186
                   G(x) =  3.4756
 x = 1.936598  F(x) = 3.533363E-04
                   G(x) =  3.323069
 x = 1.936492  F(x) = 2.384186E-07
                   G(x) =  3.319702
The root of F(x) = 0 is  x =  1.936492
```

CHAPTER TWO

2.1 a) Piston; rack; reciprocating cam follower; straight edge on drafting machine; the straight section of a belt or chain which is in tension, etc.

b) Shaft; gear; pulley; crank; sprocket; cam, etc.

c) Connecting rod; planet gear; coupler of a four-bar linkage; screw; drill, etc. (Note: the motion of the first three above may be considered pure rotation at any instant.)

2.2 Stroke $2R = 0.060$ m $\qquad R = 0.030$ mm

$x = R(1-\cos\theta) \qquad v = dx/dt$

$v = \omega R \sin\theta$

$\omega = dR/dt = \pi n/30 = 1740\ \pi/30 = 182.2$ rad/s

$v_{max} = \omega R = \underline{5.466 \text{ m/s}}$ (ans a)

$a = dv/dt = \omega^2 R \cos\theta$

$a_{max} = \omega^2 R = \underline{996 \text{ m/s}^2}$ (ans b)

$j = da/dt = -\omega^3 R \sin\theta$

$j_{max} = \omega^3 R = \underline{181490 \text{ m/s}^3}$ (ans c)

2.3 See preceding problem. $\qquad R = .030$ m

$\omega = \pi n/30 = 3000\pi/30 = 100\pi$

$v_{max} = \omega R = \underline{9.425 \text{ m/s}}$ (ans a)

$a_{max} = \omega^2 R = \underline{2961 \text{ m/s}^2}$ (ans b)

$j_{max} = \omega^3 R = \underline{930{,}190 \text{ m/s}^3}$ (ans c)

2.4 See prob. 2.2. $\qquad 2R = 45$ mm $\qquad R = .0225$ m

$\omega = 1740\ \pi/30 = 182.2$ rad/s

$v_{max} = \omega R = \underline{4.100 \text{ m/s}}$ (ans a)

$a_{max} = \omega^2 R = \underline{747 \text{ m/s}^2}$ (ans b)

$j_{max} = \omega^3 R = \underline{136{,}120 \text{ m/s}^3}$ (ans c)

2.5 For stroke S, volumetric efficiency η and diameter D. the 2 cylinder pump delivers a flow rate (m^3/s)

$Q = 2\eta S(\pi D^2/4)n'$ at n' rev/sec

For $S = D$, $Q = \eta\pi D^3 n'/2$ or

$D^3 = 2Q/(\eta\pi n')$

$n' = \omega/2\pi$

$D^3 = 4Q/(\eta\omega) = 4 \times 0.5/(0.80 \times 90)$

$D = S = 0.303$ m $= \underline{303 \text{ mm}}$ (ans)

2.6 See preceding problem. $n' = n/60 = 880/60$

$D^3 = 2Q/(\eta\pi n') = 2Q/(\eta\pi n/60)$

$= 120 \times .25/(0.80\pi \times 880)$

$\underline{D = S = 0.2385 \text{ m} = 238.5 \text{ mm}}$ (ans)

2.7 See solution 2.5.

For inch units,

$Q(\text{in}^3/\text{min}) = 2\eta S(\pi D^2/4)n$

$= \eta\pi D^3 n/2$

$D^3 = 2Q/(\eta\pi n)$

1 US gallon = 231 in^3

$D^3 = 2 \times 231 \times 120/(0.80 \times \pi \times 1760)$

$\underline{D = S = 2.323 \text{ in}}$

2.8 See solution 2.5

$D^3 = 2Q/(\eta\pi n)$ $Q = 12 \text{ ft}^3/\text{min} \times 12^3 \text{ in}^3/\text{ft}^3$

$D^3 = 2 \times 12 \times 12^3/(0.80 \times \pi \times 700)$

$\underline{D = S = 2.867 \text{ in}}$

2.9 Let x(0) = 0 displacement:

$x = 20(1 - \cos 200t)$* for x(mm) and t(sec) displacement varies from 0 to 40 mm sinusoidally as in Fig. 2.2c, with a period τ where $\omega\tau = 2\pi$

$\tau = 2\pi/\omega = 2\pi/200 = \pi/100$ sec

$v = dx/dt = R\omega \sin \omega t = 20 \times 200 \sin 200t$

$v = 4000 \sin 200t$*

velocity varies from -4000 to 4000 mm/s

Acceleration:

$a = dv/dt = R\omega^2 \cos \omega t = 20 \times 200^2 \cos 200t$

$a = 800000 \cos 200t$*

Acceleration varies from -800,000 to 800,000 mm/s^2

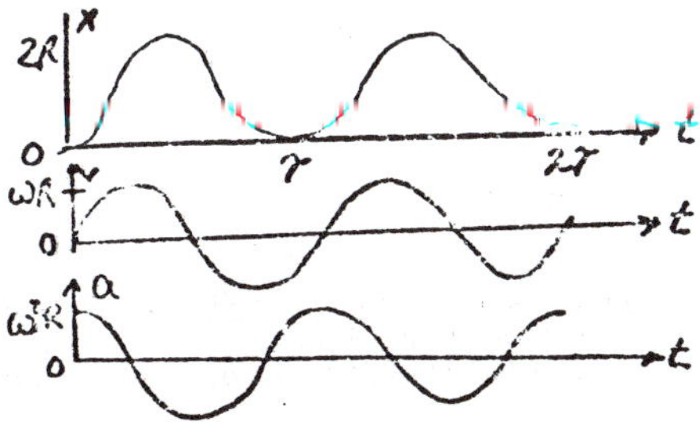

*the argument of the sine and cosine is in radians

2.10 For $x(0) = 0$, $\omega = 1500\pi/30 = 50\pi$ rad/s

$x = R(1-\cos\omega t) = 28(1 - \cos 50\pi t)$

for displacement x(mm) and

time t(sec).

The argument of the cosine is in radians.

$v = R\omega \sin \omega t = 28 \times 50\pi \sin 50\pi t$

$= 1400\pi \sin 50\pi t$

$a = R\omega^2 \cos \omega t = 28 \times (50\pi)^2 \cos 50\pi t$

$= 70{,}000\ \pi^2 \cos 50\pi t$

The plots are the same as in the preceding problem except that

$\tau = 2\pi/\omega = 2\pi/(50\pi) = 0.04$s.

x varies from 0 to 56 mm

v varies from -1400π to 1400π mm/s and

a varies from -690,870 to 690,870 mm/s^2

2.11

$x = R(1-\cos\omega t)$ $\quad v = R\omega \sin\omega t$

$a = R\omega^2 \cos \omega t$

$a_{max} = R\omega^2$ $\quad \omega^2 = a_{max}/R = 9800/5 = 1960$

$\omega_{max} = \underline{44.27 \text{ rad/s}}$ (ans)

2.12 Refer to Fig. 2.1.

$v = R\omega \sin \omega t$ $\quad a = dv/dt = R\omega^2 \cos \omega t$

$a_{max} = R\omega^2$ $\quad \omega = 4000\pi/30 = 418.9$ rad/s

$R = a_{max}/\omega^2 = 9800/(4000\pi/30)^2$

$\underline{R = 0.056 \text{ mm}}$ (ans)

2.13 Refer to Fig. 2.1.

$v = R\omega \sin \omega t$ $\quad \omega = 1200\pi/30 = 40\pi$

$v_{max} = R\omega = .25 \times 40\pi = \underline{31.42 \text{ in/s}}$ (ans)

$a = R\omega^2 \cos \omega t$

$a_{max} = R\omega^2 = 0.25 \times (40\pi)^2 = \underline{3948 \text{ in/s}^2}$ (ans)

2.14 See preceding solution. $\omega = 300$

$v_{max} = R\omega = 0.5 \times 300 = 150$ in/s

$a_{max} = R\omega^2 = 0.5 \times 300^2 = 45{,}000$ in/s^2

2.15 through 2.27
The vectors may be converted from polar form: $r\underline{/\theta}$ to rectangular form: $ir_x + jr_y$ where $r_x = r\cos\theta$ and $r_y = r\sin\theta$. Using a polar/rectangular conversion on a calculator and storing the results is most convenient.

2.15 through 2.27 (cont'd.)

$A = 1 \angle 30^\circ = i\ .866 + j\ .5$

$B = 2 \angle 60^\circ = i\ 1 + j\ 1.732$

$C = 1.5 \angle 180^\circ = -i\ 1.5 + j\ 0$

$D = 3 \angle 225^\circ = -i\ 2.121 - j\ 2.121$

$E = -2.5\ k$ using right hand coordinates.

2.15 $A + B = i(Ax + Bx) + j(Ay + By)$

$= i\ 1.866 + j\ 2.232$

$= \underline{2.909 \quad 50.10^\circ}$ (ans)

2.16 $A + B + D = -i\ 0.2553 + j\ 0.1107$

$= \underline{0.2783 \angle 156.55^\circ}$ (ans)

2.17 $A + B + C + D = -i\ 1.755 + j\ 0.1107$

$= \underline{1.759 \angle 176.39^\circ}$ (ans)

2.18 $A + B + C - D = i\ 2.487 + j\ 4.353$

$\underline{5.014 \angle 60.26^\circ}$ (ans)

2.19 $A \times B = k\ AB \sin(\theta_2 - \theta_1) = k\ 1 \times 2 \times \sin(60 - 30^\circ)$

$= \underline{k\ 1}$ (ans)

2.20 $C \times D = k\ CD \sin(\theta_2 - \theta_1) = k\ 1.5 \times 3 \sin(225 - 180^\circ)$

$= k\ 3.182$

2.21 CX(DXE)

$$= CX \begin{vmatrix} i & j & k \\ -2.121 & -2.121 & 0 \\ 0 & 0 & -2.5 \end{vmatrix}$$

$$= CX\ (i\ 5.3025 - j\ 5.3025)$$

$$= \begin{vmatrix} i & j & k \\ -1.5 & 0 & 0 \\ 5.3025 & -5.3025 & 0 \end{vmatrix}$$

$= k\ 7.954$ (ans)

2.22 DX(CXE) =

$$DX \begin{vmatrix} i & j & k \\ -1.5 & 0 & 0 \\ 0 & 0 & -2.5 \end{vmatrix}$$

$$= DX(-j\ 3.75) = \begin{vmatrix} i & j & k \\ -2.121 & -2.121 & 0 \\ 0 & -3.75 & 0 \end{vmatrix}$$

$= k\ 7.954$ (ans)

2.23 $\underset{\sim}{A} \cdot \underset{\sim}{B} = AB\cos(\theta_2 - \theta_1)$

$= 1 \times 2\cos(60 - 30^o)$

$= \underline{1.732}$ (ans)

2.24 $\underset{\sim}{A} \cdot (\underset{\sim}{B} + \underset{\sim}{C})$

$= (\underset{\sim}{i}\ 0.866 + \underset{\sim}{j}\ 0.5) \cdot (\underset{\sim}{i}\ 1 + \underset{\sim}{j}\ 1.732 - \underset{\sim}{i}\ 1.5)$

$= 0.433 + 0.866$

$= \underline{1.299}$ (ans)

2.25 $$\underset{\sim}{A} \cdot (\underset{\sim}{B} \times \underset{\sim}{C}) = \begin{vmatrix} A_x & A_y & A_z \\ B_x & B_y & B_z \\ C_x & C_y & C_z \end{vmatrix}$$

$$= \begin{vmatrix} 0.866 & 0.5 & 0 \\ 1 & 1.732 & 0 \\ -1.5 & 0 & 0 \end{vmatrix} = 0$$ (ans)

2.26 $\underset{\sim}{B} \cdot (\underset{\sim}{C} \times \underset{\sim}{A})$: $\underset{\sim}{B}$ lies in the xy plane and $\underset{\sim}{C \times A}$ in the z direction

$\underset{\sim}{B} \cdot (\underset{\sim}{C} \times \underset{\sim}{A}) = \underline{0}$ (ans)

2.27 $\underset{\sim}{C}$ lies in the xy plane

$(\underset{\sim}{A} \times \underset{\sim}{B})$ is a vector in the z direction

$\underset{\sim}{C} \cdot (\underset{\sim}{A} \times \underset{\sim}{B}) = \underline{0}$ (ans)

2.28 $\underset{\sim}{\omega} \times \underset{\sim}{r} = (5 + 3t)\underset{\sim}{k} \times (\underset{\sim}{i}\ r_x + \underset{\sim}{j}\ r_y)$

$$= \begin{vmatrix} \underset{\sim}{i} & \underset{\sim}{j} & \underset{\sim}{k} \\ 0 & C & 5 + 3t \\ r_x & r_y & 0 \end{vmatrix}$$

$= -\underset{\sim}{i}(5 + 3t)r_y + \underset{\sim}{j}(5 + 3t)r_x$ (ans)

2.29 $\frac{d}{dt}(\underset{\sim}{\omega \times r}) = -\underset{\sim}{i}[3r_y + (5 + 3t)\frac{dr_y}{dt}]$

$+ \underset{\sim}{j}[3r_x + (5 + 3t)\frac{dr_x}{dt}]$ (ans)

(See preceding problem.)

2.30 $\frac{d}{dt}(\omega \times r) = \omega \times \frac{dr}{dt} + \frac{d\omega}{dt} \times r$

$$\frac{dr}{dt} = \frac{di}{dt} r_x + i \frac{dr_x}{dt} + \frac{dj}{dt} r_y + j \frac{dr_y}{dt}$$

$$\frac{d\omega}{dt} = k3$$

If i and j rotate with angular velocity ω, then

$\frac{di}{dt} = \omega \times i = (5 + 3t)j$ and

$\frac{dj}{dt} = \omega \times j = -i(5 + 3t)$ from which

$$\frac{dr}{dt} = i\left[\frac{dr_x}{dt} - (5 + 3t)r_y\right] + j\left[(5 + 3t)r_x + \frac{dr_y}{dt}\right]$$

$$\omega \times \frac{dr}{dt} = \begin{vmatrix} i & j & k \\ 0 & 0 & \omega \\ \frac{dr_x}{dt} - \omega r_y & \omega r_x + \frac{dr_y}{dt} & 0 \end{vmatrix}$$

$$= i\left(-\omega^2 r_x - \omega\frac{dr_y}{dt}\right) + j\left(\omega \frac{dr_x}{dt} - \omega^2 r_y\right)$$

$$\frac{d\omega}{dt} \times r = \begin{vmatrix} i & j & k \\ 0 & 0 & 3 \\ r_x & r_y & 0 \end{vmatrix}$$

$$= i(-3r_y) + j(3r_x)$$

$$\frac{d}{dt}(\omega \times r) = i\left(-3r_y - \omega^2 r_x - \omega \frac{dr_y}{dt}\right)$$
$$+ j\left(3r_x - \omega^2 r_y + \omega\frac{dr_x}{dt}\right) \quad \text{(ans)}$$

where $\omega = 5 + 3t$

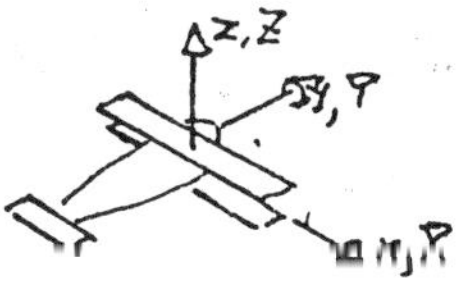

2.31 Consider a moving set of coordinates x, y and z for pitch, roll and yaw respectively. Fixed coordinates X, Y and Z are initially in the x, y and z directions, respectively.

In the Immelman turn, the aircraft goes from the +Y direction to the -Y direction (upside-down) by turning about the x axis. The aircraft is righted by a 180° turn about the y axis. If the order of operations is reversed, the aircraft turns upside down by a half roll about the y axis and then does a half loop (a 180° turn about the x axis), righting itself to fly in the -Y direction. The result is the same except that the reversed order results in an altitude loss. In general, the order of the finite rotations is critical. Consider, for example, a 90° rotation about the x axis followed by a 90° rotation about the y axis. The result is different if the

2.31 (Cont'd.)

order of the rotations is changed. The commutative law of vector addition does not apply to finite rotations about coordinates referred to the aircraft.

2.32 $\underset{\sim}{r}_3 = -\underset{\sim}{r}_1 - \underset{\sim}{r}_2 = -\underset{\sim}{i}\,10 - \underset{\sim}{j}\,25 - \underset{\sim}{i}\,40 + \underset{\sim}{j}\,20 - \underset{\sim}{k}\,10$

$\underset{\sim}{r}_3 = -\underset{\sim}{i}\,50 - \underset{\sim}{j}\,5 - \underset{\sim}{k}\,10$

2.33 $r_{1y}{}^{u}/r_{1x}{}^{u} = r_{1y}/r_{1x} = 2$ etc.

$r_{1x}{}^{u2} + r_{1y}{}^{u2} = 1 \qquad r_{1x}{}^{u2} + (2r_{1x}{}^{u})^2 = 1$

$r_{1x}{}^{u} = 1/\sqrt{5} \qquad r_{1y}{}^{u} = 2/\sqrt{5}$

$\underset{\sim}{r}_1{}^{u} = \underset{\sim}{i}/\sqrt{5} + \underset{\sim}{j}\,2/\sqrt{5}$

$r_{2x}{}^{u2} + (0.5\,r_{2x}{}^{u})^2 = 1$

$r_{2x}{}^{u} = \sqrt{4/5} \qquad r_{2y}{}^{u} = \sqrt{1/5}$

$\underset{\sim}{r}_2{}^{u} = \underset{\sim}{i}\,2/\sqrt{5} + \underset{\sim}{j}/\sqrt{5}$

$$\underset{\sim}{r}_2{}^{u} \times \underset{\sim}{k} = \begin{vmatrix} \underset{\sim}{i} & \underset{\sim}{j} & \underset{\sim}{k} \\ 2/\sqrt{5} & 1/\sqrt{5} & 0 \\ 0 & 0 & 1 \end{vmatrix} = \underset{\sim}{i}/\sqrt{5} - \underset{\sim}{j}2/\sqrt{5}$$

$$\underset{\sim}{r}_1{}^{u} \times \underset{\sim}{k} = \begin{vmatrix} \underset{\sim}{i} & \underset{\sim}{j} & \underset{\sim}{k} \\ 1/\sqrt{5} & 2/\sqrt{5} & 0 \\ 0 & 0 & 1 \end{vmatrix} = \underset{\sim}{i}\,2/\sqrt{5} - \underset{\sim}{j}/\sqrt{5}$$

$$r_1 = \frac{-\underset{\sim}{r}_3 \cdot (\underset{\sim}{r}_2{}^{u} \times \underset{\sim}{k})}{\underset{\sim}{r}_1{}^{u} \cdot (\underset{\sim}{r}_2{}^{u} \times \underset{\sim}{k})} = \frac{-(\underset{\sim}{i}50 + \underset{\sim}{j}75)\cdot(\underset{\sim}{i} - \underset{\sim}{j}2)/\sqrt{5}}{1/\sqrt{5}(\underset{\sim}{i} + \underset{\sim}{j}2)\cdot(\underset{\sim}{i} - \underset{\sim}{j}2)/\sqrt{5}}$$

$$= \frac{-\sqrt{5}(50 - 150)}{1 - 4} \qquad \underline{r_1 = -74.54} \qquad \text{(ans)}$$

$$r_2 = \frac{-\underset{\sim}{r}_3 \cdot (\underset{\sim}{r}_1{}^{u} \times \underset{\sim}{k})}{\underset{\sim}{r}_2{}^{u} \cdot (\underset{\sim}{r}_1{}^{u} \times \underset{\sim}{k})} = \frac{-(\underset{\sim}{i}50 + \underset{\sim}{j}75)\cdot(\underset{\sim}{i}2 - \underset{\sim}{j})/\sqrt{5}}{1/\sqrt{5}(\underset{\sim}{i}2 + \underset{\sim}{j})\cdot(\underset{\sim}{i}2 - \underset{\sim}{j})/\sqrt{5}}$$

$$= \frac{-\sqrt{5}(100 - 75)}{4 - 1} \qquad \underline{r_2 = -18.63} \qquad \text{(ans)}$$

check: $\underset{\sim}{r}_1 = r_1\underset{\sim}{r}_1{}^{u} = -\underset{\sim}{i}33.34 - \underset{\sim}{j}66.67$

$\underset{\sim}{r}_2 = r_2\underset{\sim}{r}_2{}^{u} = -\underset{\sim}{i}16.66 - \underset{\sim}{j}8.33$

$\underset{\sim}{r}_1 + \underset{\sim}{r}_2 + \underset{\sim}{r}_3 = 0$

2.34 See preceding problem.

$$r_1 = \frac{-(\underset{\sim}{i}5 - \underset{\sim}{j}8)\cdot(\underset{\sim}{i} - \underset{\sim}{j}2)/\sqrt{5}}{1/\sqrt{5}(\underset{\sim}{i} + \underset{\sim}{j}2)\cdot(\underset{\sim}{i} - \underset{\sim}{j}2)/\sqrt{5}}$$

$$= \frac{-\sqrt{5}(5 + 16)}{1-4} \qquad \underline{r_1 = 15.65} \text{ (ans)}$$

$$\underset{\sim}{r}_1 = r_1\underset{\sim}{r}_1^u = \underset{\sim}{i}7 + \underset{\sim}{j}14$$

$$r_2 = \frac{-(\underset{\sim}{i}5 - \underset{\sim}{j}8)\cdot(\underset{\sim}{i}2 - \underset{\sim}{j})/\sqrt{5}}{1/\sqrt{5}(\underset{\sim}{i}2 + \underset{\sim}{j})\cdot(\underset{\sim}{i}2 - \underset{\sim}{j})/\sqrt{5}}$$

$$= \frac{-\sqrt{5}(10 + 8)}{4 - 1} \qquad \underline{r_2 = -13.42} \text{ (ans)}$$

$$\underset{\sim}{r}_2 = r_2\underset{\sim}{r}_2^u = -\underset{\sim}{i}12 - \underset{\sim}{j}6$$

check: $\underset{\sim}{r}_1 + \underset{\sim}{r}_2 + \underset{\sim}{r}_3 = 0$

2.35 $\underset{\sim}{T} = \underset{\sim}{r} \times \underset{\sim}{F} = rF \sin\theta \perp \underset{\sim}{r}$ and $\underset{\sim}{F}$

$T = 180 \times 22 \sin 68^\circ = \underline{3672 \text{ Nmm CW}}$ (ans)

2.36 $\underset{\sim}{T} = \underset{\sim}{r} \times \underset{\sim}{F} = rF \sin\theta \perp r$ and F

$T = 180 \times 22 \sin 2 \text{ rad} = \underline{3601 \text{ Nmm CW}}$ (ans)

2.37 The crank vector is

$$\underset{\sim}{r}_1 = r_1(\underset{\sim}{i} \cos\theta_1 + \underset{\sim}{j} \sin\theta_1)$$
$$= r_1(\underset{\sim}{i}0.7660 + \underset{\sim}{j}0.6428$$
$$\underset{\sim}{r}_0^u = -\underset{\sim}{i}$$
$$\underset{\sim}{r}_1\cdot\underset{\sim}{r}_0^u = -0.7660\, r_1$$
$$\underset{\sim}{r}_0^u \times \underset{\sim}{k} = -\underset{\sim}{i} \times \underset{\sim}{k} = \underset{\sim}{j}$$

slider position is:

$$\underset{\sim}{r}_0 = [-\underset{\sim}{r}_1\cdot\underset{\sim}{r}_0^u \mp \sqrt{r_2^2 - [\underset{\sim}{r}_1\cdot(\underset{\sim}{r}_0^u \times \underset{\sim}{k})]^2}]\underset{\sim}{r}_0^u$$
$$= [0.7660 r_1 \mp \sqrt{(1.5r_1)^2 - [r_1(\underset{\sim}{i}0.7660 + \underset{\sim}{j}0.6428)\cdot\underset{\sim}{j}]^2}\,](-\underset{\sim}{i})$$
$$= -[0.7660 \mp 1.3553]\underset{\sim}{i}\, r_1$$

$r_0 = i0.5893r_1$ or $-i2.1213r_1$

(The second solution applies to Fig. 2.8 .)

The connecting rod vector is

$$\underset{\sim}{r}_2 = -[\underset{\sim}{r}_1\cdot\underset{\sim}{r}_0^u \times \underset{\sim}{k})](\underset{\sim}{r}_0^u \times \underset{\sim}{k}) \pm \sqrt{r_2^2 - [\underset{\sim}{r}_1\cdot(\underset{\sim}{r}_0^u \times \underset{\sim}{k})]^2}\, \underset{\sim}{r}_0^u$$
$$= -r_1[0.6428](\underset{\sim}{j}) \pm \sqrt{(1.5r_1)^2 - (0.6428r_1)^2}\,(-\underset{\sim}{i})$$
$$= r_1[-0.6428\underset{\sim}{j} \mp 1.3553\underset{\sim}{i}]$$

(The second solution applies to Fig. 2-8)

Note that $\underset{\sim}{r}_0 + \underset{\sim}{r}_1 + \underset{\sim}{r}_2 = 0$.

2.38 Refer to the above solution.

$$\mathbf{r}_1 = r_1(\mathbf{i}\cos 140^\circ + \mathbf{j}\sin 140^\circ)$$

$$= r_1(-\mathbf{i}0.7660 + \mathbf{j}\ 0.6428)$$

$$\mathbf{r}_0{}^u = -\mathbf{i}$$

$$\mathbf{r}_1 \cdot \mathbf{r}_0{}^u = +0.7660 r_1$$

$$\mathbf{r}_0{}^u \times \mathbf{k} = \mathbf{j}$$

$$\mathbf{r}_0 = [-0.7660 r_1 \mp \sqrt{(1.5 r_1)^2 - (0.6428 r_1)^2}](-\mathbf{i})$$

$$= \mathbf{i}\ r_1(0.7660 \pm 1.3553)$$

$$= \mathbf{i}\ r_1 2.1213 \text{ or } -\mathbf{i}\ r_1 0.5893^*$$

$$\mathbf{r}_2 = -[0.6428 r_1](\mathbf{j}) \pm \sqrt{(1.5 r_1)^2 - (0.6428 r_1)^2}(-\mathbf{i})$$

$$= r_1[-\mathbf{j}\ 0.6428 \mp \mathbf{i}\ 1.3553]^*$$

*The second solution applies if the linkage is assembled so that the vectors form a clockwise loop as in Fig. 2-11. In both cases, $\mathbf{r}_0 + \mathbf{r}_1 + \mathbf{r}_2 = 0$

2.39 See figure.

2.40

For $\mathbf{r}_0 = -\mathbf{i}120$ and $\mathbf{r}_1 = 60\angle\theta_1 = 30^\circ$, the triangle solution yields $\angle B = 23.79^\circ$ $\quad \mathbf{r}_d = 74.36\angle\theta_d = 156.21^\circ$

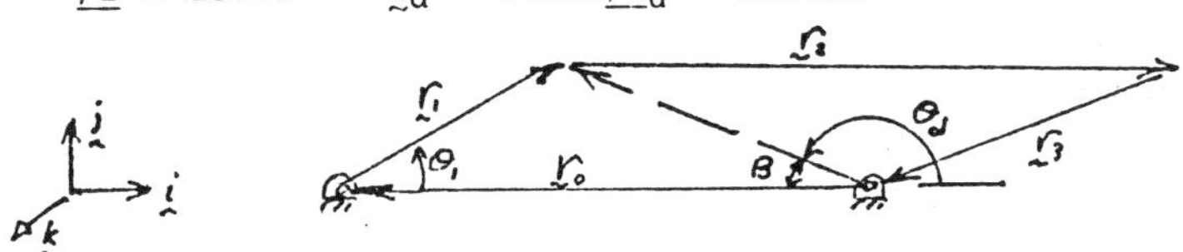

$$\mathbf{r}_d{}^u = \mathbf{i}\cos\theta_d + \mathbf{j}\sin\theta_d = -\mathbf{i}0.915 + \mathbf{j}0.4034$$

$$\text{Define } a = \frac{r_3{}^2 - r_2{}^2 + r_d{}^2}{2r_d} = \frac{80^2 - 140^2 + r_d{}^2}{2r_d}$$

$$= -51.58$$

$$\mathbf{r}_d{}^u \times \mathbf{k} = \mathbf{i}0.4034 + \mathbf{j}0.915$$

$$\mathbf{r}_2 = \mp\sqrt{r_3{}^2 - a^2}\,(\mathbf{r}_d{}^u \times \mathbf{k}) + \mathbf{r}_d{}^u(a - r_d)^*$$

$$= \mp(\mathbf{i}24.67 + \mathbf{j}55.96) + \mathbf{i}115.23 - \mathbf{j}50.80$$

$$\underline{\mathbf{r}_2 = \mathbf{i}139.90 + \mathbf{j}5.15} = 140\angle 2.1^\circ \quad \text{(ans)}$$

$$\mathbf{r}_3 = \pm\sqrt{r_3{}^2 - a^2}\,(\mathbf{r}_d{}^u \times \mathbf{k}) - a\mathbf{r}_d{}^u \quad *$$

$$= \pm(\mathbf{i}24.67 + \mathbf{j}55.96) - \mathbf{i}47.19 + \mathbf{j}20.81$$

$$\underline{\mathbf{r}_3 = -\mathbf{i}71.85 - \mathbf{j}35.15} = 80\angle -153.9 \quad \text{(ans)}$$

*The lower signs apply to the position shown. (Vector loop $\mathbf{r}_d\mathbf{r}_2\mathbf{r}_3$ is clockwise).

2.39 Distance from the head-dead-center position is plotted against crank angle.

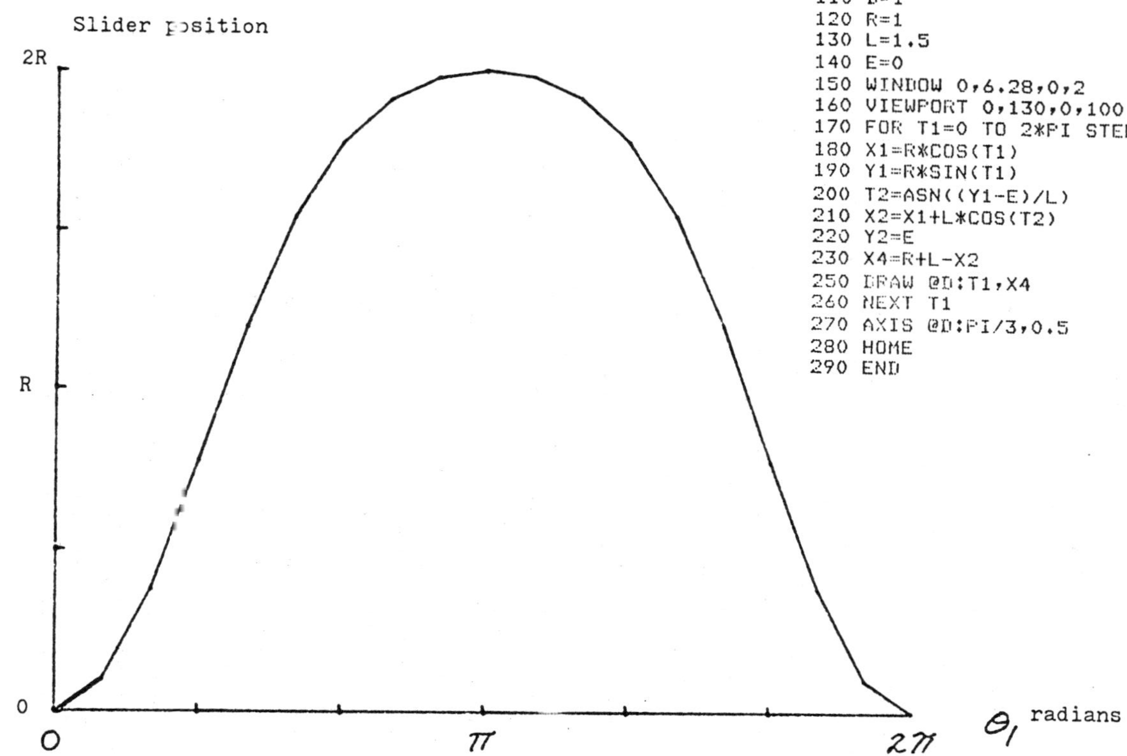


```
100 PAGE
105 REM SLIDER CRANK L/R=1.5
110 D=1
120 R=1
130 L=1.5
140 E=0
150 WINDOW 0,6.28,0,2
160 VIEWPORT 0,130,0,100
170 FOR T1=0 TO 2*PI STEP PI/9
180 X1=R*COS(T1)
190 Y1=R*SIN(T1)
200 T2=ASN((Y1-E)/L)
210 X2=X1+L*COS(T2)
220 Y2=E
230 X4=R+L-X2
250 DRAW @D:T1,X4
260 NEXT T1
270 AXIS @D:PI/3,0.5
280 HOME
290 END
```

2.41 Refer to the preceding problem.

For $\underset{\sim}{r}_0 = -\underset{\sim}{i}120$ and $\underset{\sim}{r}_1 = 60\angle\theta_1 = 60(\underset{\sim}{i}\cos 60^o + \underset{\sim}{j}\sin 60^o)$

$\underset{\sim}{r}_d = -\underset{\sim}{i}90 + \underset{\sim}{j}51.962 = 103.918 \angle 150^o$

$\underset{\sim}{r}_d^u = -\underset{\sim}{i}0.8660 + \underset{\sim}{j}0.5$

$\underset{\sim}{r}_d^u \times \underset{\sim}{k} = \underset{\sim}{i}0.5 + \underset{\sim}{j}0.8660$

$$a = \frac{r_3^2 - r_2^2 - r_d^2}{2r_d} = \frac{80^2 - 140^2 + r_d^2}{2r_d} = -11.553$$

$\underset{\sim}{r}_2 = \mp\sqrt{r_3^2 - a^2}(\underset{\sim}{r}_d^u \times \underset{\sim}{k}) + \underset{\sim}{r}_d^u(a - r_d)$ *

$= \mp 79.161(\underset{\sim}{i}0.5 + \underset{\sim}{j}0.866) + (-\underset{\sim}{i}0.866 + \underset{\sim}{j}0.5)(-115.471)$

$\underset{\sim}{r}_2 = \underset{\sim}{i}139.578 + \underset{\sim}{j}10.818 = 140 \angle 4.432^o$ (ans)

$r_3 = \pm\sqrt{r_3^2 - a^2}(\underset{\sim}{r}_d^u \times \underset{\sim}{k}) - a r_d^u$ *

$= \pm 79.161(\underset{\sim}{i}0.5 + \underset{\sim}{j}0.866) + 11.553(-\underset{\sim}{i}0.866 + \underset{\sim}{j}0.5)$

$r_3 = -\underset{\sim}{i}49.586 - \underset{\sim}{j}62.777 = 80 \angle -128.30^o$ (ans)

*The Lower signs apply if vector loop $\underset{\sim}{r}_d \underset{\sim}{r}_2 \underset{\sim}{r}_3$ is clockwise.

2.42

$\underset{\sim}{r}_0 = -\underset{\sim}{i}120 \qquad \underset{\sim}{r}_1 = 60\angle 30^o = \underset{\sim}{i}51.96 + \underset{\sim}{j}30$

$\underset{\sim}{r}_d = \underset{\sim}{r}_0 + \underset{\sim}{r}_1 = -\underset{\sim}{i}68.04 + \underset{\sim}{j}30 = 74.36\angle 156.21^o$

$\underset{\sim}{r}_3 \cdot \underset{\sim}{r}_3 = (\underset{\sim}{r}_d + \underset{\sim}{r}_2)\cdot(\underset{\sim}{r}_d + \underset{\sim}{r}_2) \qquad r_3^2 = r_d^2 + 2r_d r_2 \cos\phi_b + r_2^2$

$\cos\phi_b = (r_3^2 - r_2^2 - r_d^2)/(2r_d r_2)$

$= (80^2 - 140^2 - 74.36^2)/(2 \times 74.36 \times 140)$

$\phi_b = 154.10^o \qquad \theta_2 = \theta_d - \phi_b = 2.11^o$ (ans)

$\underset{\sim}{r}_3 = -(\underset{\sim}{r}_d + \underset{\sim}{r}_2) = -(-\underset{\sim}{i}68.04 + \underset{\sim}{j}30 + \underset{\sim}{i}139.91 + \underset{\sim}{j}5.16)$

$\underset{\sim}{r}_3 = -\underset{\sim}{i}71.87 - \underset{\sim}{j}35.15 = 80\angle 206.07^o$ (ans)

2.43

$\underset{\sim}{r}_0 = -\underset{\sim}{i}120 \qquad \underset{\sim}{r}_1 = 60\angle 60^o = \underset{\sim}{i}30 + \underset{\sim}{j}51.96$

$\underset{\sim}{r}_d = \underset{\sim}{r}_0 + \underset{\sim}{r}_1 = -\underset{\sim}{i}90 + \underset{\sim}{j}51.96 = 103.92\angle 150^o$

$\underset{\sim}{r}_3 \cdot \underset{\sim}{r}_3 = (\underset{\sim}{r}_d + \underset{\sim}{r}_2)\cdot(\underset{\sim}{r}_d + \underset{\sim}{r}_2) \qquad r_3^2 = r_d^2 + 2r_d r_2 \cos\phi_b + r_2^2$

$\cos\phi_b = (r_3^2 - r_2^2 - r_d^2)/(2r_d r_2)$

$= (80^2 - 140^2 - 103.92^2)/(2 \times 103.92 \times 140)$

$\phi_b = 145.57^o \qquad \theta_2 = \theta_d - \phi_b = 150 - 145.57 = 4.43^o$ (ans)

$\underset{\sim}{r}_3 = -(\underset{\sim}{r}_d + \underset{\sim}{r}_2) = -(-\underset{\sim}{i}90 + \underset{\sim}{j}51.96 + \underset{\sim}{i}139.58 + \underset{\sim}{j}10.82)$

$= -\underset{\sim}{i}49.58 - \underset{\sim}{j}62.78 = 80\angle 231.70^o$ (ans)

2.44 Refer to Prob. 2.40

The upper set of signs in the r_2 and r_3 equations anply when the vector loop $\tilde{r}_d\tilde{r}_2\tilde{r}_3$ is counterclockwise where

$$\tilde{r}_d = \tilde{r}_1 + \tilde{r}_2 = 74.36(-\tilde{i}0.915 + \tilde{j}0.4034)$$

$$\tilde{r}_2 = -(\tilde{i}24.67 + \tilde{j}55.96) + \tilde{i}115.23 - \tilde{j}50.80$$

$$\tilde{r}_2 = \tilde{i}90.56 - \tilde{j}106.67 = 140\underline{/-49.69^o} \text{ (ans)}$$

$$\tilde{r}_3 = +(\tilde{i}24.67 + j55.96) - i47.19 + j20.81$$

$$\tilde{r}_3 = -\tilde{i}22.52 + \tilde{j}76.77 = 80\underline{/106.35^o} \quad \text{(ans)}$$

2.45

As in solution 2.42, $\tilde{r}_d = \tilde{i}68.04 + \tilde{j}30 = 74.36\underline{/156.21^o}$ and ϕ_b = 154.10°. Since $\tilde{r}_d\tilde{r}_2\tilde{r}_3$ forms a counterclockwise loop,

$$\theta_2 = \theta_d + \phi_b = 156.21 + 154.10 = 310.31^o \quad \text{(ans)}$$

$$\tilde{r}_3 = -(\tilde{r}_d+\tilde{r}_2) = -(-\tilde{i}68.04+\tilde{j}30+\tilde{i}140\cos 310.31^o+\tilde{j}140\sin 310.31^o)$$

$$\tilde{r}_3 = -\tilde{i}22.53 + \tilde{j}76.76 = 80\underline{/106.36^o}$$

2.46 a) $\tilde{R}_1 = R_1(\cos\theta_1 + j\sin\theta_1)$

$\tilde{R}_1 = 200(0.5 + j0.866) = 100 + j173.2$

$\tilde{R}_2 = R_2(\cos\theta_2 + j\sin\theta_2)$

$\tilde{R}_2 = 75 - j129.9$ (ans)

b) $\tilde{R}_0 = \tilde{R}_2 - \tilde{R}_1$

$\tilde{R}_0 = -25 - j303.1$ (ans)

c) $\tilde{R}_0 = 304.1\ e^{-j1.653}$ (ans)

2.47 $\tilde{R}_1 = R_1 e^{j\theta_1}$

$$\frac{d\tilde{R}_1}{dt} = j\frac{d\theta_1}{dt} e^{j\theta_1} R_1$$

$$= \frac{d\theta_1}{dt} R_1 e^{j(\theta_1+\pi/2)}$$

$$= 120 \times 200\ e^{j5\pi/6}$$

$$= 24\ e^{j5\pi/6} \text{ m/s}$$

$$= 24 \text{ m/s } \underline{/150^o}$$

$$= -20.78 + j12 \text{ m/s} \quad \text{(ans)}$$

2.48

$\tilde{R}_0 = 320$

$\tilde{R}_1 = 170\underline{/80^o} = 29.52 + j167.41$

$\tilde{R}_2 = \tilde{R}_0 + \tilde{R}_1 = 349.52 + j167.41 = 387.5\ e^{j0.447}$ (ans)

2.49

$\tilde{R}_0 = 320$

$\tilde{R}_i = 170\underline{/110^o} = -58.14 + j159.75$

$\tilde{R}_2 = R_0 + R_1 = 261.85 + j159.75$

$= 306.74 \text{ mm } \underline{/31.4^o}$

$= \underline{306.74\ e^{j0.548}}$ (ans)

2.50 $\tilde{r}_1 = 0.7660r_1 + j0.6428r_1 = r_1\underline{/40^o}$

$\tilde{r}_0 + \tilde{r}_1 + \tilde{r}_2 = 0$

Imaginary parts: $\tilde{r}_{2y} = -r_{1y} = -0.6428r_1$

$r_2^2 = (1.5r_1)^2 = r_{2x}^2 + r_{2y}^2$

$r_{2x}^2 = (1.5r_1)^2 - (-0.6428r_1)^2$

$r_{2x} = \pm 1.3552r_1$ $\underline{r_2 = \pm 1.355r_1 - j0.6428r_1}$

Real parts

$\tilde{r}_0 = -r_1 - r_{2x} = (-0.7660 \mp 1.355)r_1$

$\underline{r_0 = -2.1213r_1 \text{ or } 0.5893r_1}$

The first solution applies to Fig. 2.[illegible].

2.51 $r_2 = 1.5r_1$

$\tilde{r}_1 = r_1\underline{/140^o} = r_1(-0.7660 + j0.6428)$

$\tilde{r}_0 + r_1 + r_2 = 0$

Imaginary parts:

$r_{2y} = -r_{1y} = -0.6428r_1$

$r_2^2 = r_{2x}^2 + r_{2y}^2$

$r_{2x}^2 = (1.5r_1)^2 - (-0.6428r_1)^2$ $r_{2x} = \pm 1.3552r_1$

$\underline{r_2 = (\pm 1.3552 - j0.6428)r_1}$

Real parts:

$r_0 = -r_{1x} - r_{2x} = (0.7660 \mp 1.3552)r_1$

$\underline{\tilde{r}_0 = -0.5892r_1 \quad \text{or} \quad 2.1212r_1}$

The first solution for r_2 and r_0 applies if the mechanism is initially assembled as in Fig. 2.[illegible].

2.52 Let displacement of link 3 be identified by $r_3 = S_2A$ which satisfies

$$r_3^2 + 2\cos\gamma(r_0 - r_1\cos\theta);\ r_3 + r_0^2 + r_1^2 - r_2^2 - 2r_0r_1\cos\theta = 0$$

for $a = 1$, $b = 2\cos\gamma(r_0 - r_1\cos\theta) = 2\cos 40^o(200-100\cos 60^o) = 229.81$

$c = r_0^2+r_1^2-r_2^2-2r_0r_1\cos\theta = 200^2+100^2-300^2-2\times200\times100\times\cos 60^o$

$= -60000 \quad r_3 = \sqrt{(-b \pm b^2 - 4ac)}/(2a)$

$r_3 = (-229.81 \pm\sqrt{229.81^2 + 4\times60000})/2 = 155.66$ (ans)

(only the positive root applies)

2.53 The following procedure may be used:

1. Store r_0, r_1, r_2, γ, $\theta_{initial}$
2. Print θ
3. Calculate roots of quadratic in r_3 (see solution 2.52)
4. Test values of r_3; print root in region of interest (positive root in this case)
5. Go to step 2, if $\theta < \theta_{max}$

The results are shown below. A programmable calculator was used. Values of θ(deg) and r_3(mm) are shown alternately. Note that $r_3(-\theta) = r_3(\theta)$ so only values of θ up to 180° need be shown.

```
          0.              105.
216.4283881       71.07208162
         15.              120.
212.1819563       46.5571684
         30.              135.
199.8152187       26.54815934
         45.              150.
180.4107106       11.8828836
         60.              165.
155.6548644       2.979904584
         75.              180.
127.6610401      -.0000000002
         90.
 98.7295236
```

2.54

Let link 1 lie in the yz-plane and link 3 in the xy-plane. An outline of a sphere of radius r_2 with center at S_2 is drawn in the yz-plane. The intersection of the sphere and the horizontal plane containing the path of S_2 is projected to the xy-plane, locating S_2. Displacement is given by $AS_2=r_3=156$mm. See figure.

2.54

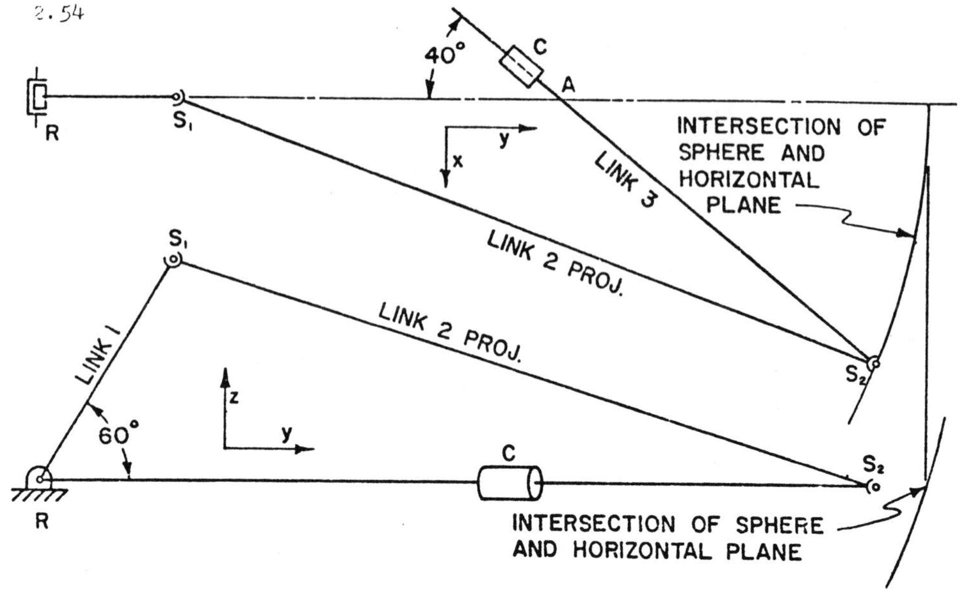

40°
C
A
R
S_1
y
x
LINK 3
INTERSECTION OF SPHERE AND HORIZONTAL PLANE
LINK 2 PROJ.
S_1
LINK 2 PROJ.
LINK 1
S_2
z
y
60°
C
S_2
R
INTERSECTION OF SPHERE AND HORIZONTAL PLANE

2.55

$A := \begin{bmatrix} 4 \\ 3 \\ 1 \end{bmatrix}$ $B := \begin{bmatrix} 2 \\ -1 \\ -5 \end{bmatrix}$ $C := \begin{bmatrix} 16 \\ -8 \\ -10 \end{bmatrix}$ $D := \begin{bmatrix} -9 \\ -5 \\ 7 \end{bmatrix}$

$E := \begin{bmatrix} 16 \\ -8 \\ -10 \end{bmatrix}$

$A + B + C + D = \begin{bmatrix} 13 \\ -11 \\ -7 \end{bmatrix}$ $A - B - C - D - E = \begin{bmatrix} -21 \\ 25 \\ 19 \end{bmatrix}$

$B \cdot C = 90$ $C \cdot B = 90$ $|B| = 5.477$ $|C| = 20.494$

$(|B|)^2 = 30$ $B \cdot B = 30$ $B \times B = \begin{bmatrix} 0 \\ 0 \\ 0 \end{bmatrix}$

$C \times D = \begin{bmatrix} -106 \\ -22 \\ -152 \end{bmatrix}$ $D \times C = \begin{bmatrix} 106 \\ 22 \\ 152 \end{bmatrix}$ $C \cdot (C \times D) = 0$

2.56

$A := \begin{bmatrix} 4 \\ 3 \\ 1 \end{bmatrix}$ $\quad B := \begin{bmatrix} 2 \\ -1 \\ -5 \end{bmatrix}$ $\quad C := \begin{bmatrix} 16 \\ -8 \\ -10 \end{bmatrix}$ $\quad D := \begin{bmatrix} -9 \\ -5 \\ 7 \end{bmatrix}$

$E := \begin{bmatrix} 16 \\ -8 \\ -10 \end{bmatrix}$

$A + B + C + E = \begin{bmatrix} 38 \\ -14 \\ -24 \end{bmatrix}$ $\quad A + B - C - D - E = \begin{bmatrix} -17 \\ 23 \\ 9 \end{bmatrix}$

$D \cdot E = -174$ $\quad E \cdot D = -174$ $\quad |D| = 12.45$ $\quad |E| = 20.494$

$(|D|)^2 = 155$ $\quad D \cdot D = 155$ $\quad D \times D = \begin{bmatrix} 0 \\ 0 \\ 0 \end{bmatrix}$

$E \times D = \begin{bmatrix} -106 \\ -22 \\ -152 \end{bmatrix}$ $\quad D \times E = \begin{bmatrix} 106 \\ 22 \\ 152 \end{bmatrix}$ $\quad E \cdot (E \times D) = 0$

2.57

$$A := \begin{bmatrix} 4 \\ 3 \\ 1 \end{bmatrix} \qquad B := \begin{bmatrix} 2 \\ -1 \\ -5 \end{bmatrix} \qquad C := \begin{bmatrix} 16 \\ -8 \\ -10 \end{bmatrix} \qquad D := \begin{bmatrix} -9 \\ -5 \\ 7 \end{bmatrix}$$

$$E := \begin{bmatrix} 16 \\ -8 \\ -10 \end{bmatrix}$$

$$D + B + C + E = \begin{bmatrix} 25 \\ -22 \\ -18 \end{bmatrix} \qquad B + B - C - D - E - A = \begin{bmatrix} -23 \\ 16 \\ 2 \end{bmatrix}$$

$$B \cdot D = -48 \qquad D \cdot B = -48 \qquad |D| = 12.45 \qquad |B| = 5.477$$

$$(|E|)^2 = 420 \qquad E \cdot E = 420 \qquad E \times E = \begin{bmatrix} 0 \\ 0 \\ 0 \end{bmatrix}$$

$$B \times D = \begin{bmatrix} -32 \\ 31 \\ -19 \end{bmatrix} \qquad E \cdot (E \times B) = 0$$

2.58 Define a vector

$\mathbf{c} = \mathbf{r}_0 + \mathbf{r}_1$

where $\mathbf{c}$ is known for any given value of θ. Then,

$\mathbf{c} + \mathbf{r}_2 + \mathbf{r}_3 = 0$

and, considering the plane of rotation of link 3, there are five unknowns: r_{2x}, r_{2y}, r_{2z}, r_{3x} and r_{3z}

Equating the vector components in each coordinate direction and noting the link lengths, we have the following five equations:

$c_x + r_{2x} + r_{3x} = 0$

$c_y + r_{2y} = 0$

$r_{2z} + r_{3z} = 0$

$r_{2x}^2 + r_{2y}^2 + r_{2z}^2 = r_2^2$

$r_{3x}^2 + r_{3z}^2 = r_3^2$

The five equations can be reduced to a single equation for r_{2x} in terms of known values. The solution and plots of vector components of $\mathbf{r}_2$ and $\mathbf{r}_3$, which were obtained using mathematics software, are shown in figure S2.58.

2.59 See solution 2.58 above, and equations and plots in figure S2.59.

52,58

RSSR linkage. Link 1 rotates in xy plane; link 3 rotates in xz plane

$r_0 := 32 \quad r_1 := 10 \quad r_2 := 28 \quad r_3 := 20 \qquad \theta := 0, \frac{\pi}{6} .. 2\cdot\pi$

$c_x(\theta) := -r_0 + r_1 \cdot \cos(\theta) \qquad c_y(\theta) := r_1 \cdot \sin(\theta)$

$$r_{2x}(\theta) := \frac{-r_2^{\,2} + c_y(\theta)^2 + r_3^{\,2} - c_x(\theta)^2}{2 \cdot c_x(\theta)} \qquad r_{3x}(\theta) := -c_x(\theta) - r_{2x}(\theta)$$

$$r_{2y}(\theta) := -c_y(\theta) \qquad r_{2z}(\theta) := \left[r_3^{\,2} - r_{3x}(\theta)^2 \right]^{.5} \qquad r_{3z}(\theta) := -r_{2z}(\theta)$$

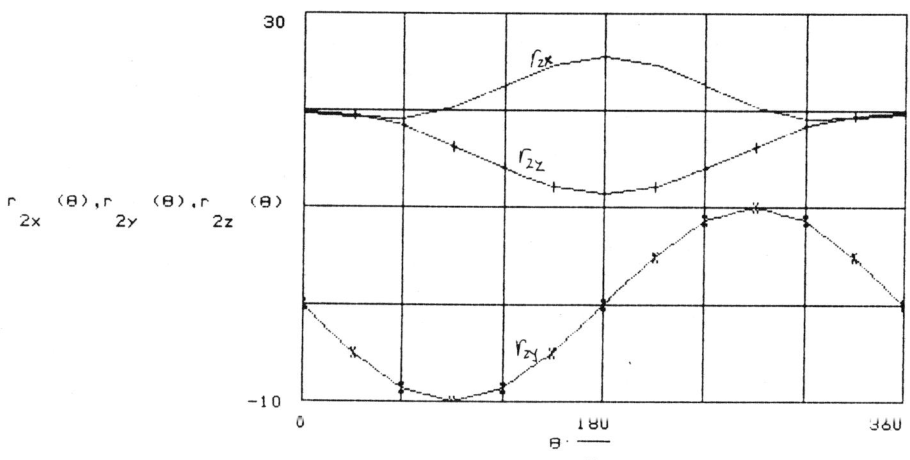

S 2.58 contd

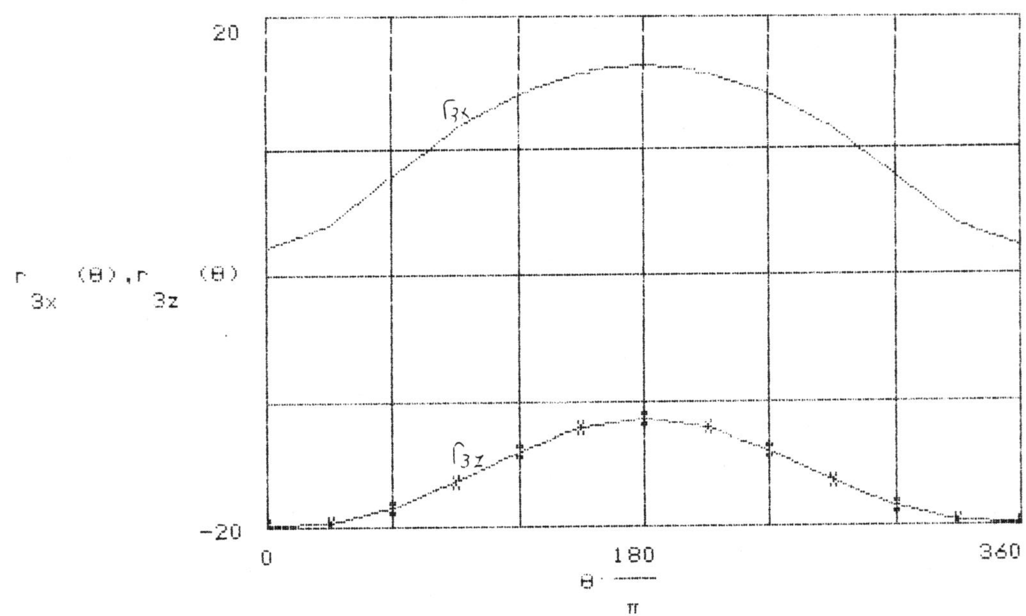

52.59

RSSR linkage. Link 1 rotates in xy plane; link 3 rotates in xz plane

$r_0 := 62 \quad r_1 := 20 \quad r_2 := 55 \quad r_3 := 45 \qquad \theta := 0, \frac{\pi}{6} .. 2\cdot\pi$

$c_x(\theta) := -r_0 + r_1\cdot\cos(\theta) \qquad c_y(\theta) := r_1\cdot\sin(\theta)$

$$r_{2x}(\theta) := \frac{-r_2^{\,2} + c_y(\theta)^2 + r_3^{\,2} - c_x(\theta)^2}{2\cdot c_x(\theta)} \qquad r_{3x}(\theta) := -c_x(\theta) - r_{2x}(\theta)$$

$$r_{2y}(\theta) := -c_y(\theta) \qquad r_{2z}(\theta) := \left[r_3^{\,2} - r_{3x}(\theta)^2\right]^{.5} \qquad r_{3z}(\theta) := -r_{2z}(\theta)$$

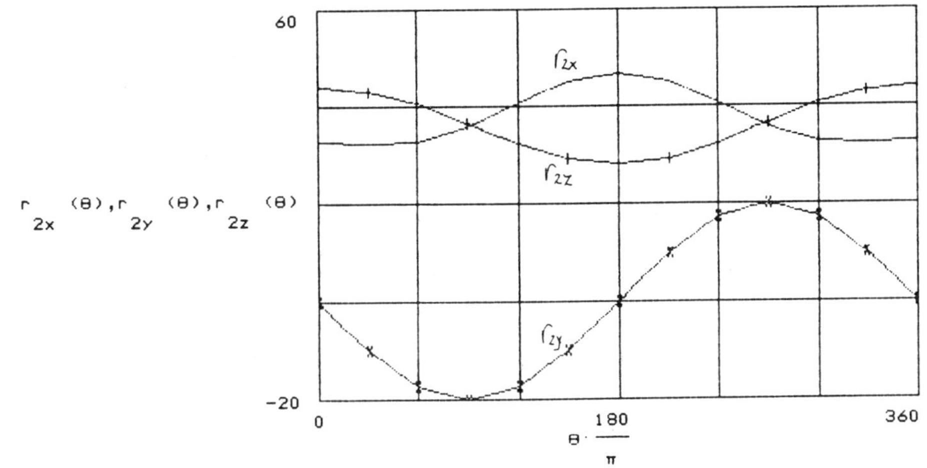

52,59 contd

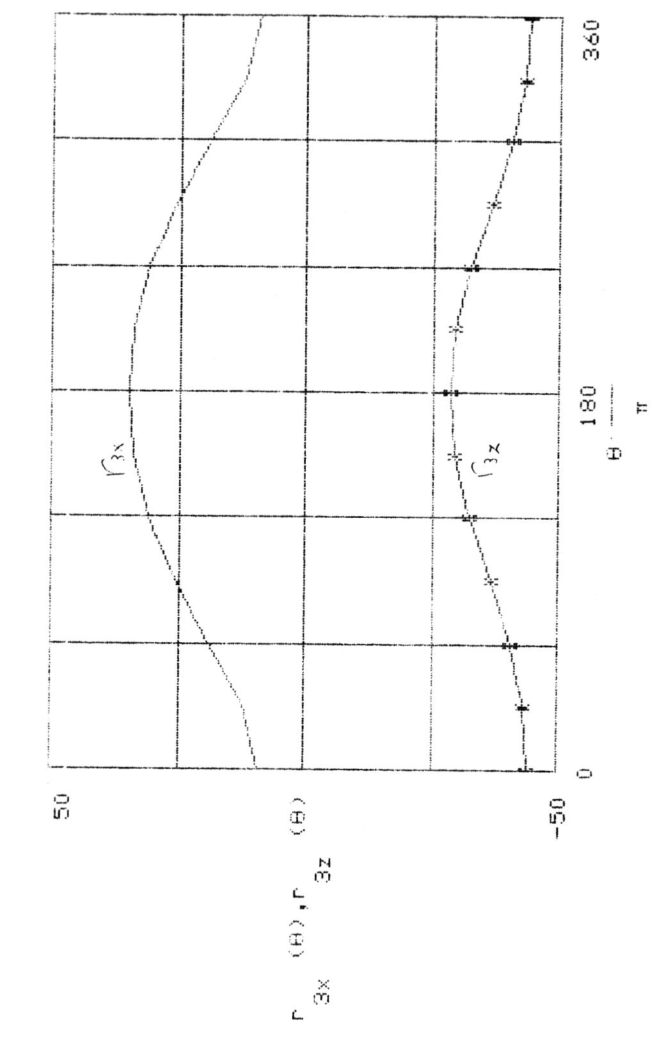

2.60 through 2.66

Refer to text section 2.7. These problems are solved by the Newton-Raphson method, using a first approximation based on a sketch of the linkage.

$$X = \begin{bmatrix} \theta_2 \\ \theta_3 \end{bmatrix}$$

where the angles are in radians. When the value of X_{new} does not change significantly from the previous iteration, this value is taken as the solution.

2.60

Position analysis of a four-bar linkage using the Newton-Raphson method.

$r_0 := 120 \quad r_1 := 60 \quad r_2 := 140 \quad r_3 := 80 \quad \theta_0 := -\pi \quad \theta_1 := \frac{\pi}{3}$

$\theta_2 := -1 \quad \theta_3 := 1$ *(First approx)*

$r_{dx} := r_0 \cdot \cos\left[\theta_0\right] + r_1 \cdot \cos\left[\theta_1\right] \quad r_{dy} := r_0 \cdot \sin\left[\theta_0\right] + r_1 \cdot \sin\left[\theta_1\right]$

$F_1 := r_2 \cdot \cos\left[\theta_2\right] + r_3 \cdot \cos\left[\theta_3\right] + r_{dx}$

$F_2 := r_2 \cdot \sin\left[\theta_2\right] + r_3 \cdot \sin\left[\theta_3\right] + r_{dy}$

$F := \begin{bmatrix} F_1 \\ F_2 \end{bmatrix} \quad F = \begin{bmatrix} 28.867 \\ 1.473 \end{bmatrix} \quad X := \begin{bmatrix} \theta_2 \\ \theta_3 \end{bmatrix}$

$G := \begin{bmatrix} -r_2 \cdot \sin\left[\theta_2\right] & -r_3 \cdot \sin\left[\theta_3\right] \\ r_2 \cdot \cos\left[\theta_2\right] & r_3 \cdot \cos\left[\theta_3\right] \end{bmatrix} \quad X_{new} := X - G^{-1} \cdot F \quad X_{new} = \begin{bmatrix} -1.132 \\ 1.197 \end{bmatrix}$

$\theta_2 := -1.132 \quad \theta_3 := 1.197$ *(2nd approx)*

$r_{dx} := r_0 \cdot \cos\left[\theta_0\right] + r_1 \cdot \cos\left[\theta_1\right] \quad r_{dy} := r_0 \cdot \sin\left[\theta_0\right] + r_1 \cdot \sin\left[\theta_1\right]$

$F_1 := r_2 \cdot \cos\left[\theta_2\right] + r_3 \cdot \cos\left[\theta_3\right] + r_{dx}$

$F_2 := r_2 \cdot \sin\left[\theta_2\right] + r_3 \cdot \sin\left[\theta_3\right] + r_{dy}$

$F := \begin{bmatrix} F_1 \\ F_2 \end{bmatrix} \quad F = \begin{bmatrix} -1.309 \\ -0.3 \end{bmatrix} \quad X := \begin{bmatrix} \theta_2 \\ \theta_3 \end{bmatrix}$

$G := \begin{bmatrix} -r_2 \cdot \sin\left[\theta_2\right] & -r_3 \cdot \sin\left[\theta_3\right] \\ r_2 \cdot \cos\left[\theta_2\right] & r_3 \cdot \cos\left[\theta_3\right] \end{bmatrix} \quad X_{new} := X - G^{-1} \cdot F \quad X_{new} = \begin{bmatrix} -1.125 \\ 1.192 \end{bmatrix}$

2.60 cont'd

$\theta_2 := -1.125 \qquad \theta_3 := 1.192$ *(3rd approx)*

$r_{dx} := r_0 \cdot \cos\left(\theta_0\right) + r_1 \cdot \cos\left(\theta_1\right) \qquad r_{dy} := r_0 \cdot \sin\left(\theta_0\right) + r_1 \cdot \sin\left(\theta_1\right)$

$F_1 := r_2 \cdot \cos\left(\theta_2\right) + r_3 \cdot \cos\left(\theta_3\right) + r_{dx}$

$F_2 := r_2 \cdot \sin\left(\theta_2\right) - r_3 \cdot \sin\left(\theta_3\right) + r_{dy}$

$F := \begin{bmatrix} F_1 \\ F_2 \end{bmatrix} \qquad F = \begin{bmatrix} -0.051 \\ -0.027 \end{bmatrix} \qquad X := \begin{bmatrix} \theta_2 \\ \theta_3 \end{bmatrix}$

$G := \begin{bmatrix} -r_2 \cdot \sin\left(\theta_2\right) & -r_3 \cdot \sin\left(\theta_3\right) \\ r_2 \cdot \cos\left(\theta_2\right) & r_3 \cdot \cos\left(\theta_3\right) \end{bmatrix} \qquad X_{new} := X - G^{-1} \cdot F \qquad X_{new} = \begin{bmatrix} -1.125 \\ 1.192 \end{bmatrix}$ *(ans)*

2.61

Position analysis of a four-bar linkage using the Newton-Raphson method.

$r_0 := 3 \quad r_1 := 1 \quad r_2 := 3.6 \quad r_3 := 2.1 \quad \theta_0 := -\pi \quad \theta_1 := \frac{\pi}{4}$

$\theta_2 := 0 \quad \theta_3 := 4$ *(1st approx)*

$r_{dx} := r_0 \cdot \cos[\theta_0] + r_1 \cdot \cos[\theta_1] \qquad r_{dy} := r_0 \cdot \sin[\theta_0] + r_1 \cdot \sin[\theta_1]$

$F_1 := r_2 \cdot \cos[\theta_2] + r_3 \cdot \cos[\theta_3] + r_{dx}$

$F_2 := r_2 \cdot \sin[\theta_2] + r_3 \cdot \sin[\theta_3] + r_{dy}$

$F := \begin{bmatrix} F_1 \\ F_2 \end{bmatrix} \qquad F = \begin{bmatrix} -0.066 \\ -0.882 \end{bmatrix} \qquad X := \begin{bmatrix} \theta_2 \\ \theta_3 \end{bmatrix}$

$G := \begin{bmatrix} -r_2 \cdot \sin[\theta_2] & -r_3 \cdot \sin[\theta_3] \\ r_2 \cdot \cos[\theta_2] & r_3 \cdot \cos[\theta_3] \end{bmatrix}$

$G = \begin{bmatrix} 0 & 1.589 \\ 3.6 & -1.373 \end{bmatrix} \qquad |G| = -5.721 \qquad X_{new} := X - G^{-1} \cdot F$

$G^{-1} = \begin{bmatrix} 0.24 & 0.278 \\ 0.629 & 0 \end{bmatrix} \qquad X_{new} = \begin{bmatrix} 0.261 \\ 4.041 \end{bmatrix}$

$\theta_2 := .261 \quad \theta_3 := 4.041$ *(2nd approx)*

$r_{dx} := r_0 \cdot \cos[\theta_0] + r_1 \cdot \cos[\theta_1] \qquad r_{dy} := r_0 \cdot \sin[\theta_0] + r_1 \cdot \sin[\theta_1]$

$F_1 := r_2 \cdot \cos[\theta_2] + r_3 \cdot \cos[\theta_3] + r_{dx}$

$F_2 := r_2 \cdot \sin[\theta_2] + r_3 \cdot \sin[\theta_3] + r_{dy}$

$F := \begin{bmatrix} F_1 \\ F_2 \end{bmatrix} \qquad F = \begin{bmatrix} -0.121 \\ -0.008 \end{bmatrix} \qquad X := \begin{bmatrix} \theta_2 \\ \theta_3 \end{bmatrix}$

2.61 cont'd

$$G := \begin{bmatrix} -r_2 \cdot \sin(\theta_2) & -r_3 \cdot \sin(\theta_3) \\ r_2 \cdot \cos(\theta_2) & r_3 \cdot \cos(\theta_3) \end{bmatrix}$$

$$G = \begin{bmatrix} -0.929 & 1.644 \\ 3.478 & -1.306 \end{bmatrix} \qquad |G| = -4.505 \qquad X_{new} := X - G^{-1} \cdot F$$

$$G^{-1} = \begin{bmatrix} 0.29 & 0.365 \\ 0.772 & 0.206 \end{bmatrix} \qquad X_{new} = \begin{bmatrix} 0.299 \\ 4.136 \end{bmatrix}$$

$\theta_2 := .299 \qquad \theta_3 := 4.136$ *(3rd approx)*

$$r_{dx} := r_0 \cdot \cos(\theta_0) + r_1 \cdot \cos(\theta_1) \qquad r_{dy} := r_0 \cdot \sin(\theta_0) + r_1 \cdot \sin(\theta_1)$$

$$F_1 := r_2 \cdot \cos(\theta_2) + r_3 \cdot \cos(\theta_3) + r_{dx}$$

$$F_2 := r_2 \cdot \sin(\theta_2) + r_3 \cdot \sin(\theta_3) + r_{dy}$$

$$F := \begin{bmatrix} F_1 \\ F_2 \end{bmatrix} \qquad F = \begin{bmatrix} 0.003 \\ 0.007 \end{bmatrix} \qquad X := \begin{bmatrix} \theta_2 \\ \theta_3 \end{bmatrix}$$

$$G := \begin{bmatrix} -r_2 \cdot \sin(\theta_2) & -r_3 \cdot \sin(\theta_3) \\ r_2 \cdot \cos(\theta_2) & r_3 \cdot \cos(\theta_3) \end{bmatrix}$$

$$G = \begin{bmatrix} -1.06 & 1.761 \\ 3.44 & -1.144 \end{bmatrix} \qquad |G| = -4.844 \qquad X_{new} := X - G^{-1} \cdot F$$

$$G^{-1} = \begin{bmatrix} 0.236 & 0.364 \\ 0.71 & 0.219 \end{bmatrix} \qquad X_{new} = \begin{bmatrix} 0.296 \\ 4.132 \end{bmatrix}$$

2.61 concl.

$\theta_2 := .296 \qquad \theta_3 := 4.132$ (4th approx)

$r_{dx} := r_0 \cdot \cos[\theta_0] + r_1 \cdot \cos[\theta_1] \qquad r_{dy} := r_0 \cdot \sin[\theta_0] + r_1 \cdot \sin[\theta_1]$

$F_1 := r_2 \cdot \cos[\theta_2] + r_3 \cdot \cos[\theta_3] + r_{dx}$

$F_2 := r_2 \cdot \sin[\theta_2] + r_3 \cdot \sin[\theta_3] + r_{dy}$

$F := \begin{bmatrix} F_1 \\ F_2 \end{bmatrix} \qquad F = \begin{bmatrix} -9.873 \cdot 10^{-4} \\ 0.001 \end{bmatrix} \qquad X := \begin{bmatrix} \theta_2 \\ \theta_3 \end{bmatrix}$

$G := \begin{bmatrix} -r_2 \cdot \sin[\theta_2] & -r_3 \cdot \sin[\theta_3] \\ r_2 \cdot \cos[\theta_2] & r_3 \cdot \cos[\theta_3] \end{bmatrix}$

$G = \begin{bmatrix} -1.05 & 1.756 \\ 3.443 & -1.152 \end{bmatrix} \qquad |G| = -4.838 \qquad X_{new} := X - G^{-1} \cdot F$

$G^{-1} = \begin{bmatrix} 0.238 & 0.363 \\ 0.712 & 0.217 \end{bmatrix} \qquad X_{new} = \begin{bmatrix} 0.296 \\ 4.132 \end{bmatrix}$ (ans)

2,62

Multiloop linkage - displacement analysis

$r_1 := 1 \quad r_2 := 5.2 \quad r_F := 5 \quad r_{DE} := 4 \quad r_{CD} := 1.5 \quad \theta_1 := \frac{\pi}{6}$

$r_{CE} := r_{DE} - r_{CD}$

Estimates: $\theta_2 := 1 \quad r_D := 4$

$\theta_3 := 3 \quad r_E := 6$

Given

$$r_1 \cdot \cos\left(\theta_1\right) + r_2 \cdot \cos\left(\theta_2\right) + r_{CE} \cdot \cos\left(\theta_3\right) - r_E \approx 0$$

$$r_1 \cdot \sin\left(\theta_1\right) + r_2 \cdot \sin\left(\theta_2\right) + r_{CE} \cdot \sin\left(\theta_3\right) \approx 0$$

$$r_{DE} \cdot \cos\left(\theta_3\right) - r_E + r_F \approx 0$$

$$r_D + r_{DE} \cdot \sin\left(\theta_3\right) \approx 0$$

$$X := \mathrm{Find}\left(\theta_2, \theta_3, r_D, r_E\right) \qquad X = \begin{bmatrix} 0.313 \\ 5.286 \\ 3.36 \\ 7.17 \end{bmatrix}$$

2.63

Multiloop linkage - displacement analysis

$r_1 := 1 \quad r_2 := 5.2 \quad r_F := 5 \quad r_{DE} := 4 \quad r_{CD} := 1.5 \quad \theta_1 := \frac{\pi}{2}$

$r_{CE} := r_{DE} - r_{CD}$

Estimates: $\theta_2 := 1 \quad r_D := 4$

$\theta_3 := 5 \quad r_E := 6$

Given

$$r_1 \cdot \cos(\theta_1) + r_2 \cdot \cos(\theta_2) + r_{CE} \cdot \cos(\theta_3) - r_E \approx 0$$

$$r_1 \cdot \sin(\theta_1) + r_2 \cdot \sin(\theta_2) + r_{CE} \cdot \sin(\theta_3) \approx 0$$

$$r_{DE} \cdot \cos(\theta_3) - r_E + r_F \approx 0$$

$$r_D + r_{DE} \cdot \sin(\theta_3) \approx 0$$

$X := \mathrm{Find}(\theta_2, \theta_3, r_D, r_E)$

$$X = \begin{bmatrix} 0.293 \\ 4.698 \\ 4 \\ 4.944 \end{bmatrix}$$

2.64

Multiloop linkage - displacement analysis

$r_1 := 1 \quad r_2 := 5.2 \quad r_F := 5 \quad r_{DE} := 4 \quad r_{CD} := 1.5 \quad \theta_1 := 2 \cdot \frac{\pi}{3}$

$r_{CE} := r_{DE} - r_{CD}$

Estimates: $\theta_2 := 1 \quad r_D := 4$

$\theta_3 := 5 \quad r_E := 6$

Given

$$r_1 \cdot \cos(\theta_1) + r_2 \cdot \cos(\theta_2) + r_{CE} \cdot \cos(\theta_3) - r_E \approx 0$$

$$r_1 \cdot \sin(\theta_1) + r_2 \cdot \sin(\theta_2) + r_{CE} \cdot \sin(\theta_3) \approx 0$$

$$r_{DE} \cdot \cos(\theta_3) - r_E + r_F \approx 0$$

$$r_D + r_{DE} \cdot \sin(\theta_3) \approx 0$$

$$X := \mathrm{Find}(\theta_2, \theta_3, r_D, r_E) \qquad X = \begin{bmatrix} 0.289 \\ 4.362 \\ 3.756 \\ 3.625 \end{bmatrix}$$

2.65

Multiloop linkage - displacement analysis

$r_1 := 20 \quad r_2 := 115 \quad r_F := 100 \quad r_{DE} := 85 \quad r_{CD} := 40 \quad \theta_1 := \frac{\pi}{4}$

$r_{CE} := r_{DE} - r_{CD}$

Estimates: $\theta_2 := 1 \quad r_D := 80$

$\theta_3 := 4 \quad r_E := 100$

Given

$$r_1 \cdot \cos[\theta_1] + r_2 \cdot \cos[\theta_2] + r_{CE} \cdot \cos[\theta_3] - r_E \approx 0$$

$$r_1 \cdot \sin[\theta_1] + r_2 \cdot \sin[\theta_2] + r_{CE} \cdot \sin[\theta_3] \approx 0$$

$$r_{DE} \cdot \cos[\theta_3] - r_E + r_F \approx 0$$

$$r_D + r_{DE} \cdot \sin[\theta_3] \approx 0$$

$$X := \mathrm{Find}[\theta_2, \theta_3, r_D, r_E]$$

$$X = \begin{bmatrix} 0.16 \\ 5.476 \\ 61.386 \\ 158.794 \end{bmatrix}$$

2.66

Multiloop linkage - displacement analysis

$r_1 := 20 \quad r_2 := 125 \quad r_F := 100 \quad r_{DE} := 85 \quad r_{CD} := 35 \quad \theta_1 := 2\cdot\frac{\pi}{3}$

$r_{CE} := r_{DE} - r_{CD}$

Estimates: $\theta_2 := 1 \quad r_D := 80$

$\theta_3 := 4 \quad r_E := 100$

Given

$$r_1\cdot\cos\left[\theta_1\right] + r_2\cdot\cos\left[\theta_2\right] + r_{CE}\cdot\cos\left[\theta_3\right] - r_E \approx 0$$

$$r_1\cdot\sin\left[\theta_1\right] + r_2\cdot\sin\left[\theta_2\right] + r_{CE}\cdot\sin\left[\theta_3\right] \approx 0$$

$$r_{DE}\cdot\cos\left[\theta_3\right] - r_E + r_F \approx 0$$

$$r_D + r_{DE}\cdot\sin\left[\theta_3\right] \approx 0$$

$$X := \mathrm{Find}\left[\theta_2, \theta_3, r_D, r_E\right]$$

$$X = \begin{bmatrix} 0.288 \\ 4.72 \\ 84.998 \\ 100.63 \end{bmatrix}$$

CHAPTER THREE

3.1 The velocity of point p:

$$\underset{\sim}{v} = \underset{\sim}{\omega} \times \underset{\sim}{r} = \begin{vmatrix} \underset{\sim}{i} & \underset{\sim}{j} & \underset{\sim}{k} \\ 10 & 15 & 20 \\ -100 & 50 & 60 \end{vmatrix}$$

$$= \underset{\sim}{i}(15x60-20x50) + \underset{\sim}{j}(-100x20-10x60)$$

$$+ \underset{\sim}{k}(10x50+15x100)$$

$$= -\underset{\sim}{i}100 - \underset{\sim}{j}2600 + \underset{\sim}{k}2000$$

3.2 The velocity of point p:

$$\underset{\sim}{v} = \underset{\sim}{\omega} \times \underset{\sim}{r} = \begin{vmatrix} \underset{\sim}{i} & \underset{\sim}{j} & \underset{\sim}{k} \\ 200 & 0 & -150 \\ -100 & 50 & 60 \end{vmatrix}$$

$$= \underset{\sim}{i}(0+150x50) + \underset{\sim}{j}(150x100-200x60)$$

$$+ \underset{\sim}{k}(200x50-0)$$

$$= \underset{\sim}{i}7500 + \underset{\sim}{j}3000 + \underset{\sim}{k}10000$$

3.3 $\omega = \frac{2\pi}{60} 800 = 83.8$ rad/sec.

$v = R\omega = (15 \text{ in})(83.8 \text{ rad/sec}) = 1260$ in/sec.

3.4 a) (100 ft/min)/(2π/12 ft/rev) = 191 rev/min.

Range: 191 to 764 rev/min.

b) 127 to 509 rev/min.

3.5 $n \text{ rev/min} = \frac{(S \text{ ft/min})(12 \text{ in/ft})}{\pi D \text{ in/rev}}$ $\qquad n = \frac{12\,S}{\pi D}$

Valid for miller, also, at low feed rates)

3.6 The velocity of point P in the fixed system is given by:

$$\dot{R} = \dot{R}_o + \dot{r}_r + \omega \times r \qquad \text{where}$$

$$\dot{R}_o = I80 - J90$$

$$\dot{r}_r = i50 + j45$$

$$\omega \times r = \begin{vmatrix} i & j & k \\ 0 & 0 & 100 \\ 3 & 4 & 0 \end{vmatrix}$$

$$= -i400 + j300$$

Correcting for rotation,

$$i = I\cos 30^\circ + J\sin 30^\circ$$

$$j = -I\sin 30^\circ + J\cos 30^\circ$$

Adding terms, the velocity of P is:

$$\dot{R} = -I395.6 + J33.8$$

3.7 The velocity of P is found as in the above problem except

$$\omega \times r = \begin{vmatrix} i & j & k \\ 0 & 0 & -95 \\ 3 & 4 & 0 \end{vmatrix} = i(95 \times 4) + j(-95 \times 3)$$

$$= i380 - j285$$

$$= (I\cos 30^\circ + J\sin 30^\circ)380 - (-I\sin 30^\circ + J\cos 30^\circ)285$$

$$= I471.59 - J56.82$$

3.7 (cont'd.)

The velocity of P is given by

$$\dot{R} = \dot{R}_o + r_r + \omega \times r$$

$$= I\,572.4 - J\,82.8$$

3.8 $V = \omega r$ $\qquad \omega = 800\pi/30$

$V_B = i500\omega \qquad V_C = i250\omega$

$V_{CB} = V_C - V_B = -i250\omega = -20,940$ mm/s

for rolling in the x(i) direction.

3.9 $V = \omega r$ $\qquad \omega = 800\pi/30$

$V_B = i16\omega$ (for rolling in the i direction)

$V_C = i8\omega$

$V_{CB} = V_C - V_B = -i8\omega = -i670.2$ in/s

3.10 to 3.14 See figures and computer output.

3.15 Let $r_0=3$, $r_1=1$, $r_2=3.5$ and $r_3=2$

$L_{max} + L_{min} < L_a + L_b \qquad 3.5 + 1 < 3 + 2$

The linkage is a crank rocker. Position analysis used the equations of section 2.3.3 (see example prob) although the cross product method could have been used.

Using $\omega_1 \times r_1 + \omega_2 \times r_2 + \omega_3 \times r_3 = 0$

$\omega_3/\omega_1 = -(r_{2y}r_{1x} - r_{1y}r_{2x})/(r_{2y}r_{3x} - r_{3y}r_{2x})$

Note that the proportions correspond to example problem and the results agree for $\theta_1 = 45^o$. See plot.

3.10

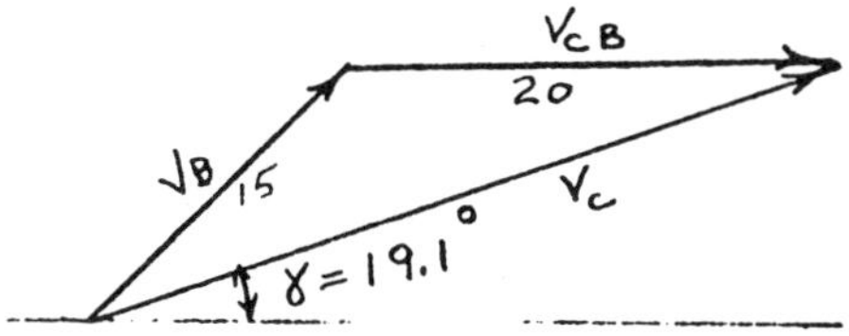

$V_{c_x} = 15 \cos 45° + 20 = 30.6$

$V_{cy} = 15 \sin 45° = 10.6$

$V_c = 32.4 \quad \gamma = 19.1°$

3.11

$V_{B_x} = 30 + 10 \cos 315° = 37.07$

$V_{By} = 10 \sin 315 = 7.07$

$V_B = 37.7 \quad \gamma = 10.8°$

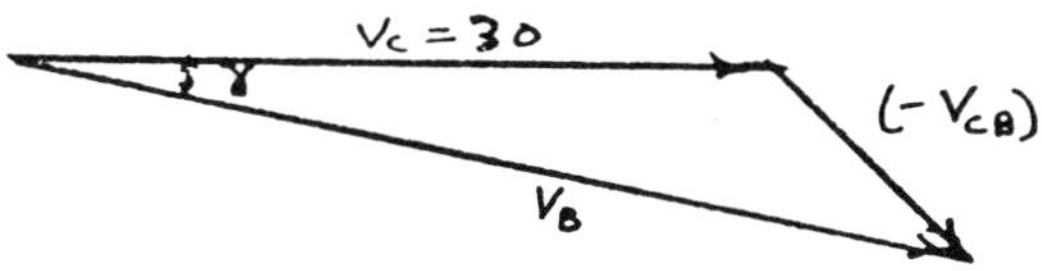

3.12

$V_{cB} = 20$

$V_c = 28.28$

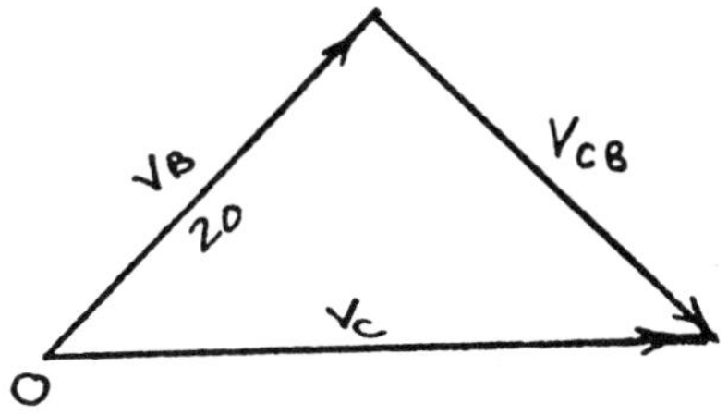

3.13

```
LO =                30

L1 =                10

L2 =                35

L3 =                20

THETA 1=            45

OMEGA 1=            30

ALPHA 1=            0

THE POINT IS ON LINK                3

THE RADIUS VECTOR IS                0

THE ANGLE FROM LINK IS              0

THE PHASE IS

CRANK ROCKER


POSITION ANALYSIS BY THE
DOT PRODUCT METHOD



THETA 2=            16.3527593 DEGREES
THETA 3=            -122.192105 DEGREES

VELOCITY ANALYSIS BY
CROSS PRODUCT METHOD



OMEGA 2= -2.8701557 RADIANS/SECOND


OMEGA 3= 10.8623432 RADIANS/SECOND


THE VELOCITY OF POINT C IS

-183.848762 (I)
115.740366 (J)

OR IN POLAR FORM

217.246263 ∠147.807896°
```

3.14

```
LO =            30

L1 =            10

L2 =            35

L3 =            20

THETA 1=        60

OMEGA 1=        30

ALPHA 1=        0

THE POINT IS ON LINK            3

THE RADIUS VECTOR IS            0

THE ANGLE FROM LINK IS          0

THE PHASE IS    -

CRANK ROCKER


POSITION ANALYSIS BY THE
DOT PRODUCT METHOD



THETA 2=        15.464503 DEGREES
THETA 3=        -115.889368 DEGREES

VELOCITY ANALYSIS BY
CROSS PRODUCT METHOD



OMEGA 2= -.818433807 RADIANS/SECOND


OMEGA 3= 14.01510[illegible] RADIANS/SECOND


THE VELOCITY OF POINT C IS

-252.169632 (I)
122.391892 (J)

OR IN POLAR FORM

280.302156 ∠154.110132°
```

3.15

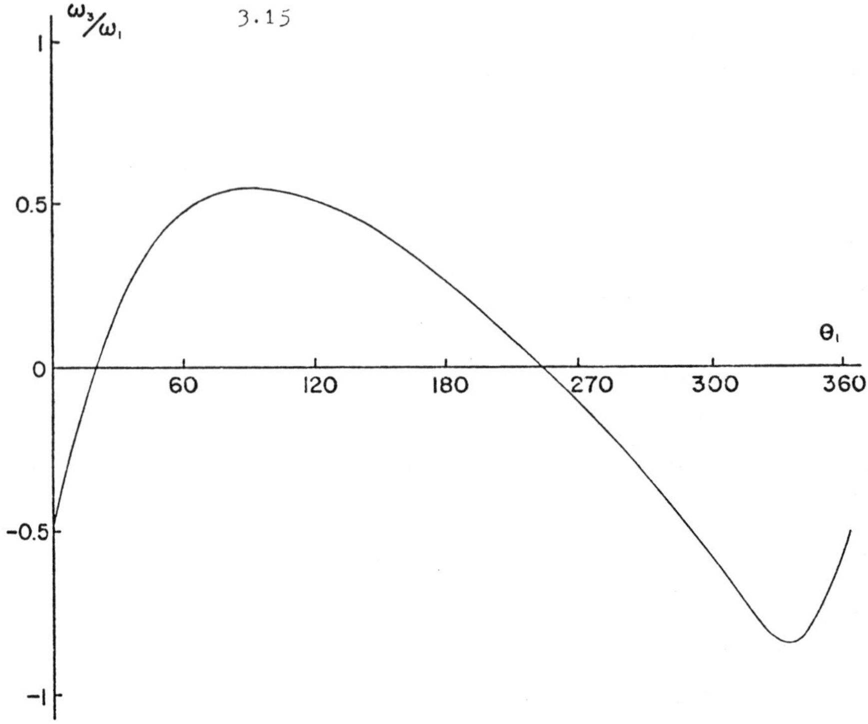

3.16

```
LO =              3
L1 =              1
L2 =              3.5
L3 =              2
THETA 1=          30
OMEGA 1=          1
ALPHA 1=          0
THE POINT IS ON LINK            2
THE RADIUS VECTOR IS            1.5
THE ANGLE FROM LINK IS          20
THE PHASE IS     -
CRANK ROCKER #1

POSITION ANALYSIS BY THE
DOT PRODUCT METHOD

THETA 2=          18.510494 DEGREES
THETA 3=          -126.333008 DEGREES

VELOCITY ANALYSIS BY
CROSS PRODUCT METHOD

OMEGA 2= -.199182268 RADIANS/SECOND

OMEGA 3= .172963175 RADIANS/SECOND

THE VELOCITY OF POINT D IS

-.313966362 (I)
.632236972 (J)
```

OR IN POLAR FORM $v_D/(\omega_1 r_1) =$

.705902589 ∠116.408812

Link lengths are divided by r_1; angular velocities by ω_1 and velocities by $\omega_1 r_1$.

3.16 See computer output.

3.17

$\underset{\sim}{R}_o + \underset{\sim}{R}_1 = \underset{\sim}{R}_2$ (a)

Noting that R_o is fixed and that R_1 is constant, differentiation yields

$$0 + \underset{\sim}{\omega}_1 \times \underset{\sim}{R}_1 = \underset{\sim}{\omega}_2 \times \underset{\sim}{R}_2 + \dot{\underset{\sim}{R}}_{2r} \quad \text{where } \underset{\sim}{\omega}_1 = \omega_1 \underset{\sim}{k}$$

$$\underset{\sim}{\omega}_2 = \underset{\sim}{\omega}_2 k \text{ and } \dot{R}_{2r} = \dot{R}_2 \underset{\sim}{R}_2^u \text{ (sliding velocity)}$$

$$\text{Thus:} \quad \omega_1 \begin{vmatrix} \underset{\sim}{i} & \underset{\sim}{j} & \underset{\sim}{k} \\ 0 & 0 & 1 \\ R_{1x} & R_{1y} & 0 \end{vmatrix} = \omega_2 \begin{vmatrix} \underset{\sim}{i} & \underset{\sim}{j} & \underset{\sim}{k} \\ 0 & 0 & 1 \\ R_{2x} & R_{2y} & 0 \end{vmatrix} + \dot{R}_2 \frac{\underset{\sim}{i}R_{2x}+\underset{\sim}{j}R_{2y}}{R_2}$$

from which

$$\omega_1(-\underset{\sim}{i}R_{1y}+\underset{\sim}{j}R_{1x})=\omega_2(-\underset{\sim}{i}R_{2y}+\underset{\sim}{j}R_{2x})+\dot{R}_2(\underset{\sim}{i}R_{2x}+\underset{\sim}{j}R_{2y})/R_2$$

Noting that $R_{1x}=R_1\cos\theta_1$ etc. and equating the $\underset{\sim}{i}$ terms, we obtain:

$$-\omega_1 R_1 \sin\theta_1 = -\omega_2 R_2 \sin\theta_2 + \dot{R}_2 \cos\theta_2 \quad \text{(b)}$$

From the $\underset{\sim}{j}$ terms:

$$\omega_1 R_1 \cos\theta_1 = \omega_2 R_2 \cos\theta_2 + \dot{R}_2 \sin\theta_2 \quad \text{(c)}$$

Equations (b) and (c) are identical to equations 3.28 and 3.29 where $\dot{R}_2=V_{B1B2}$

From equation (a), $\underset{\sim}{R}_2=\underset{\sim}{R}_o+\underset{\sim}{R}_1=40\underset{\sim}{i}+(19.319\underset{\sim}{j}+5.176\underset{\sim}{i})$

$\underset{\sim}{R}_2 = 49.134\text{mm} \angle \theta_2 = 23.15^o$

Simultaneous solution of equation (b) and (c) for $\omega_1=20$ yields $\omega_2=5.03$ rad/s (ccw) and $\dot{R}_2=-314.6$ mm/s

3.18 The procedure may follow that of solution 3.17. As an alternative, the vector elimination method may be used. The displacement equation: $\underset{\sim}{R}_o+\underset{\sim}{R}_1=\underset{\sim}{R}_2$ is differentiated, noting the directions.

$$\underset{\sim}{\omega}_1 \times \underset{\sim}{R}_1 = \underset{\sim}{\omega}_2 \times \underset{\sim}{R}_2 + \dot{\underset{\sim}{R}}_{2r}$$

$$\omega_1 R_1(\underset{\sim}{k} \times \underset{\sim}{R}_1^u) = \omega_2 R_2(\underset{\sim}{k} \times \underset{\sim}{R}_2^u) + \dot{R}_2 \underset{\sim}{R}_2^u$$

The dot product of each term with $\underset{\sim}{k} \times \underset{\sim}{R}_2^u$ is formed

$$\omega_1 R_1(\underset{\sim}{k} \times \underset{\sim}{R}_1^u)\cdot(\underset{\sim}{k} \times \underset{\sim}{R}_2^u) = \omega_2 R_2(\underset{\sim}{k} \times \underset{\sim}{R}_2^u)\cdot(\underset{\sim}{k} \times \underset{\sim}{R}_2^u)+\dot{R}_2\underset{\sim}{R}_2^u\cdot(\underset{\sim}{k} \times \underset{\sim}{R}_2^u)$$

Noting $\underset{\sim}{R}_2^u\cdot(\underset{\sim}{k} \times \underset{\sim}{R}_2^u) = 0$ and

$$(\underset{\sim}{A} \times \underset{\sim}{B})\cdot(\underset{\sim}{C} \times \underset{\sim}{D}) = (\underset{\sim}{A}\cdot\underset{\sim}{C})(\underset{\sim}{B}\cdot\underset{\sim}{D})-(\underset{\sim}{A}\cdot\underset{\sim}{D})(\underset{\sim}{B}\cdot\underset{\sim}{C}) \text{ from which}$$

$$(\underset{\sim}{k} \times \underset{\sim}{R}_1^u)\cdot(\underset{\sim}{k} \times \underset{\sim}{R}_2^u)=(\underset{\sim}{k}\cdot\underset{\sim}{k})(\underset{\sim}{R}_1^u\cdot\underset{\sim}{R}_2^u)-(\underset{\sim}{k}\cdot\underset{\sim}{R}_2^u)(\underset{\sim}{R}_1^u\cdot\underset{\sim}{k})$$

$$= \underset{\sim}{R}_1^u\cdot\underset{\sim}{R}_2^u = \cos\theta_1\cos\theta_2 + \sin\theta_1\sin\theta_2$$

$$= \cos(\theta_1 - \theta_2)$$

$$\text{Thus } \frac{\omega_2}{\omega_1} = \frac{R_1}{R_2}\cos(\theta_1 - \theta_2)$$

3.18 (cont'd.)

The program which follows computes R_2(R2) and angular velocities ω(W) in terms of angles θ(T radians or H degrees).

```
120 PRINT @D:                            "ANGULAR VELOCITY RATIO"
160 W1=1
170 R1=1
180 R0=3
185 PRINT @D:" R0/R1
190 PRINT @D:R0
195 PRINT @D:"R2/R1                       T2                    W2
200 FOR T1=0 TO 2*PI STEP PI/12
210 R2=(R0↑2+R1↑2+2*R0*R1*COS(T1))↑0.5
220 T2=ASN(R1*SIN(T1)/R2)
230 W2=W1*R1*COS(T1-T2)/R2
240 H1=180*T1/PI
245 H2=180*T2/PI
247 PRINT @D:H1,H2,W2
250 NEXT T1
```

3.19 a) $O_2B = R_2 = \sqrt{R_o^2+R_1^2+2R_oR_1\cos\theta_1}$

$O_2B = \sqrt{50^2+30^2+2x50x30\cos\theta_1} = 79.36$ mm

b) $\tan\theta_2 = R_1\sin\theta_1/(R_o+R_1\cos\theta_1)$

$= 30\sin\theta_1/(50+30\cos\theta_1) \quad \theta_2 = 5.615^o$

c) $R_o + R_1 e^{j\theta_1} = R_2e^{j\theta_2}$

$j\omega_1R_1e^{j\theta_1} = j\omega_2R_2e^{j\theta_2} + V_{B_1B_2}\,e^{j\theta_2}$

Substituting $e^{j\theta}=\cos\theta+j\sin\theta$ and separating the real and imaginary parts,

$\omega_1R_1\sin\theta_1 = -\omega_2R_2\sin\theta_2 + V_{B_1B_2}\cos\theta_2$

$\omega_1R_1\cos\theta_1 = \omega_2R_2\cos\theta_2 + V_{B_1B_2}\sin\theta_2$

Solving simultaneously,

$\omega_2 = \omega_1R_1\cos(\theta_1-\theta_2)/R_2$

$= 10x30\cos(\theta_1-\theta_2)R_2$

$= 3.730$ rad/s

3.20 Refer to solution 3.19 where $\theta_1=30^o$

a) $O_2B = R_2 = 77.45$ mm

b) $\theta_2 = 11.167^o$

c) $\omega_2 = 3.666$ rad/s

3.21 Refer to solution 3.19 where $\theta_1 = 45^o$

a) $O_2B = R_2 = 74.31$ mm

b) $\theta_2 = 16.587^o$

c) $\omega_2 = 3.551$ rad/s

3.22 For $\theta_1 = 45^o$, diagonal O_3B is given by

$$\underset{\sim}{r}_d = \underset{\sim}{r}_o + \underset{\sim}{r}_1 = -30 + 10\cos 45^o + j10\cos 45^o$$
$$= -22.929 + j7.071 = 23.994\underline{/\theta_d} = 162.86$$

$\theta_2 = 16.35^o$ and $\theta_3 = 237.81^o$ if r_d, r_2, r_3 form a clockwise loop.

$\underset{\sim}{r}_o + \underset{\sim}{r}_1 + \underset{\sim}{r}_2 + \underset{\sim}{r}_3 = 0$ is written as

$r_o + r_1 e^{j\theta_1} + r_2 e^{j\theta_2} + r_3 e^{j\theta_3} = 0$ and differentiated:

$$j\omega_1 r_1 e^{j\theta_1} + j\omega_2 r_2 e^{j\theta_2} + j\omega_3 r_3 e^{j\theta_3} = 0$$

Dividing by $je^{j\theta_3}$:

$$\omega_1 r_1 e^{j(\theta_1-\theta_3)} + \omega_2 r_2 e^{j(\theta_2-\theta_3)} + \omega_3 r_3 = 0$$

Substituting $e^{j\theta} = \cos\theta + j\sin\theta$ and equating the imaginary components:

$\omega_1 r_1 \sin(\theta_1-\theta_3) + \omega_2 r_2 \sin(\theta_2-\theta_3) = 0$ from which

$$\omega_2 = -\omega_1 r_1 \sin(\theta_1-\theta_3)/[r_2 \sin(\theta_2-\theta_3)]$$
$$= -30\text{x}10\sin(45-237.81)/[35\sin(16.35-237.81)]$$
$$= -2.870 \text{ rad/s (i.e. } 2.870 \text{ rad/s cw)}$$

Equating the real components:

$\omega_1 r_1 \cos(\theta_1-\theta_3) + \omega_2 r_2 \cos(\theta_2-\theta_3) + \omega_3 r_3 = 0$ from which

$$\omega_3 = -[\omega_1 r_1 \cos(\theta_1-\theta_3) + \omega_2 r_2 \cos(\theta_2-\theta_3)]/r_3$$
$$= -[30\text{x}10\cos(45-237.81) - 2.870\text{x}35\cos(16.35-237.81)]/20$$
$$= 10.862 \text{ rad/s ccw}$$

$$V_c = \frac{d}{dt}(-\underset{\sim}{r}_3) = -\frac{d}{dt} r_3 e^{j\theta_3} = -j\omega_3 r_3 e^{j\theta_3} = -j\omega_3 r_3(\cos\theta_3 + j\sin\theta_3)$$
$$= -j10.862\text{x}20(\cos 237.81^o + j\sin 237.81^o)$$

$$V_c = -183.85 + j115.73 = 217.2 \text{ mm/s } \underline{/147.8^o}$$

3.23 See program listing and output.

3.23 Listings of programs to print and plot angular velocity ratio are given. The columns of data refer, respectively, to θ_1° , θ_2° and ω_2/ω_1 .

```
LIST
100 PAGE
110 D=32
120 PRINT @D:                      ANGULAR VELOCITY RATIO"
160 W1=1
170 R1=1
180 R0=3
185 PRINT @D:" R0/R1
190 PRINT @D:R0
195 PRINT @D:" T1                   T2               W2
200 FOR T1=0 TO 2*PI STEP PI/12
210 R2=(R0↑2+R1↑2+2*R0*R1*COS(T1))↑0.5
220 T2=ASN(R1*SIN(T1)/R2)
230 W2=W1*R1*COS(T1-T2)/R2
240 H1=180*T1/PI
245 H2=180*T2/PI
247 PRINT @D:H1,H2,W2
250 NEXT T1
```

```
                    ANGULAR VELOCITY RATIO
R0/R1
3
T1                   T2                 W2
0                    0                   0.25
15                   3.73386716217       0.246764199757
30                   7.36925978757       0.236775475217
45                   10.7990804918       0.21915320425
60                   13.897886248        0.192307692308
75                   16.5100151745       0.153767014426
90                   18.4349488229       0.1
105                  19.4112241516       0.0264639038663
120                  19.1066053509      -0.0714285714286
135                  17.1392722378      -0.194762960347
150                  13.186785432       -0.332665886176
165                  7.25144491523      -0.451374071914
180                  1.918465387E-12    -0.5
195                 -7.25144491522      -0.451374071914
210                 -13.186785432       -0.332665886176
225                 -17.1392722378      -0.194762960347
240                 -19.1066053509      -0.0714285714286
255                 -19.4112241516      0.0264639038662
270                 -18.4349488229      0.1
285                 -16.5100151745      0.153767014426
300                 -13.897886248       0.192307692308
315                 -10.7990804919      0.21915320425
330                 -7.36925978757      0.236775475217
345                 -3.73386716217      0.246764199757
360                 -3.836930773E-12    0.25
```

3.23 continued

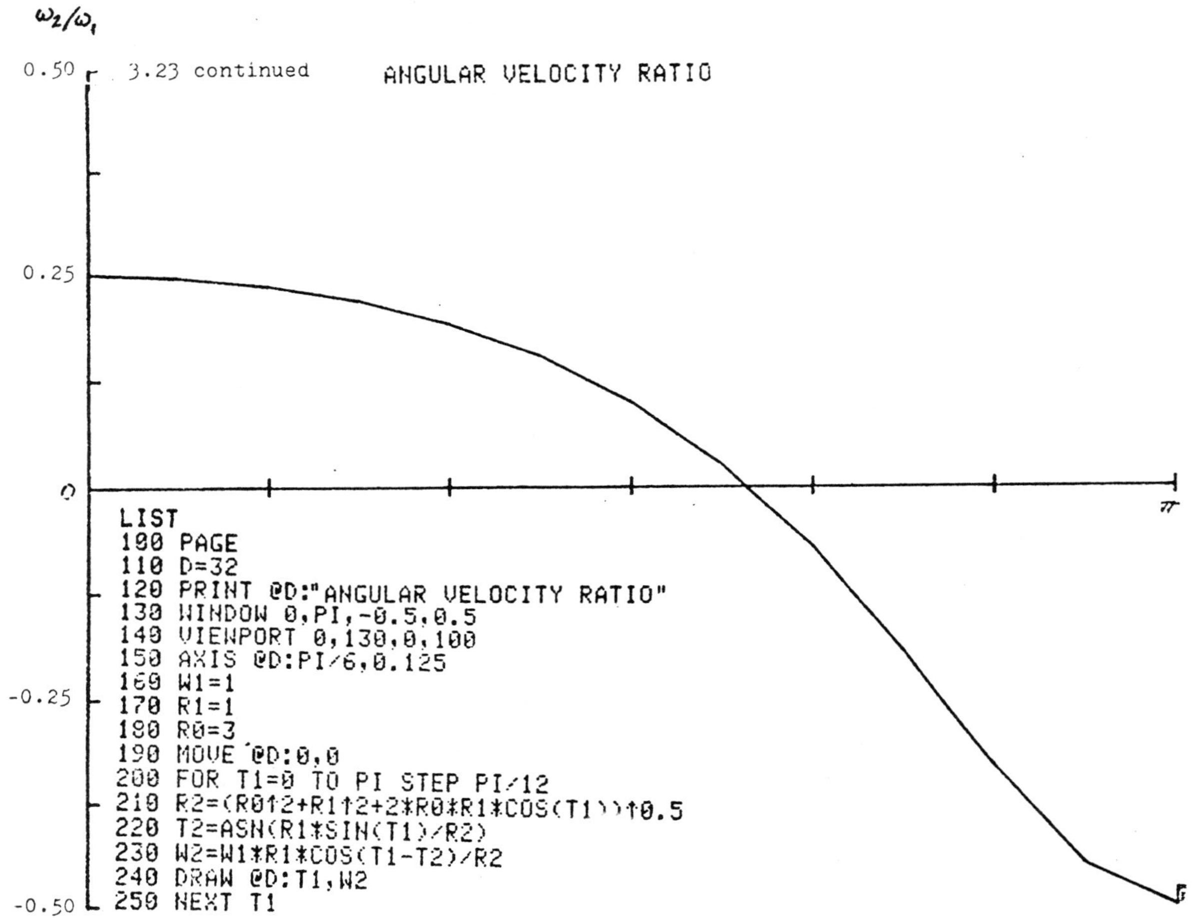


```
LIST
100 PAGE
110 D=32
120 PRINT @D:"ANGULAR VELOCITY RATIO"
130 WINDOW 0,PI,-0.5,0.5
140 VIEWPORT 0,130,0,100
150 AXIS @D:PI/6,0.125
160 W1=1
170 R1=1
180 R0=3
190 MOVE @D:0,0
200 FOR T1=0 TO PI STEP PI/12
210 R2=(R0↑2+R1↑2+2*R0*R1*COS(T1))↑0.5
220 T2=ASN(R1*SIN(T1)/R2)
230 W2=W1*R1*COS(T1-T2)/R2
240 DRAW @D:T1,W2
250 NEXT T1
```

3.24 $\underset{\sim}{r}_o + \underset{\sim}{r}_1 + \underset{\sim}{r}_2 + \underset{\sim}{r}_3 = 0$ (a) where

$R_1S_1 = \underset{\sim}{r}_1 = -\underset{\sim}{i}25 + \underset{\sim}{j}0 + \underset{\sim}{k}35$

$S_1S_2 = \underset{\sim}{r}_2 = \underset{\sim}{i}5 + \underset{\sim}{j}80 + \underset{\sim}{k}5$

$S_2R_2 = \underset{\sim}{r}_3 = \underset{\sim}{i}0 + \underset{\sim}{j}15 - \underset{\sim}{k}40$

$R_2R_1 = \underset{\sim}{r}_o = \underset{\sim}{i}20 - \underset{\sim}{j}95 + \underset{\sim}{k}0$

Differentiating equation (a),

$$\dot{\underset{\sim}{r}}_1 + \dot{\underset{\sim}{r}}_2 + \dot{\underset{\sim}{r}}_3 = \underset{\sim}{\omega}_1 \times \underset{\sim}{r}_1 + \underset{\sim}{\omega}_2 \times \underset{\sim}{r}_2 + \underset{\sim}{\omega}_3 \times \underset{\sim}{r}_3 = 0 \qquad \text{(b)}$$

or $\underset{\sim}{V}_{s1} + \underset{\sim}{V}_{s2S1} = \underset{\sim}{V}_{s2}$ (c)

$$\underset{\sim}{\omega}_1 = -\underset{\sim}{j}25 \qquad \underset{\sim}{V}_{s1} = \underset{\sim}{\omega}_1 \times \underset{\sim}{r}_1 = \begin{vmatrix} \underset{\sim}{i} & \underset{\sim}{j} & \underset{\sim}{k} \\ 0 & -25 & 0 \\ -25 & 0 & 35 \end{vmatrix} = -\underset{\sim}{i}875 + \underset{\sim}{j}0 - \underset{\sim}{k}625 \text{mm/s} \qquad \text{(d)}$$

$$\underset{\sim}{V}_{s2} = \underset{\sim}{\omega}_2 \times \underset{\sim}{r}_2 = \begin{vmatrix} \underset{\sim}{i} & \underset{\sim}{j} & \underset{\sim}{k} \\ \omega_{2x} & \omega_{2y} & \omega_{2z} \\ 5 & 80 & 5 \end{vmatrix} = \underset{\sim}{i}(5\omega_{2y} - 80\omega_{2z}) + \underset{\sim}{j}(5\omega_{2z} - 5\omega_{2x}) + \underset{\sim}{k}(80\omega_{2x} - 5\omega_{2y}) \qquad \text{(e)}$$

$$\underset{\sim}{\omega}_3 = \omega_3\underset{\sim}{i} \qquad V_{s2} = \underset{\sim}{\omega}_3 \times (-\underset{\sim}{r}_3) = \begin{vmatrix} \underset{\sim}{i} & \underset{\sim}{j} & \underset{\sim}{k} \\ \omega_3 & 0 & 0 \\ 0 & -15 & 40 \end{vmatrix} = \underset{\sim}{i}0 - \underset{\sim}{j}40\omega_3 - \underset{\sim}{k}15\omega_3 \qquad \text{(f)}$$

Substituting Eqs. (d), (e) and (f) in (b) or (c) and equating the $\underset{\sim}{i}$ components:

$$-875 + 5\omega_{2y} - 80\omega_{2z} = 0 \qquad \text{(g)}$$

From the $\underset{\sim}{j}$ components: $0 + 5\omega_{2z} - 5\omega_{2x} = -40\omega_3$ (h)

From the $\underset{\sim}{k}$ components: $-625 + 80\omega_{2x} - 5\omega_{2y} = -15\omega_3$ (i)

In addition, $\underset{\sim}{\omega}_2 \cdot \underset{\sim}{r}_2 = 0 = 5\omega_{2x} + 80\omega_{2y} + 5\omega_{2z}$ (j)

Equations (g) through (j) may be solved simultaneously:

$$\begin{bmatrix} 0 & 5 & -80 & 0 \\ -5 & 0 & 5 & 40 \\ 80 & -5 & 0 & 15 \\ 5 & 80 & 5 & 0 \end{bmatrix} \begin{bmatrix} \omega_{2x} \\ \omega_{2y} \\ \omega_{2z} \\ \omega_3 \end{bmatrix} = \begin{bmatrix} 875 \\ 0 \\ 625 \\ 0 \end{bmatrix} \qquad \text{(k)}$$

A programmable calculator was used to obtain the three components of ω_2 and ω_3:

$$\underset{\sim}{\omega}_2 = \underset{\sim}{i}7.397 + \underset{\sim}{j}0.220 - \underset{\sim}{k}10.924 \text{ rad/s} \qquad \text{(l)}$$

$$\underset{\sim}{\omega}_3 = \underset{\sim}{i}2.290 \text{ rad/s} \qquad \text{(m)}$$

3.24 (Cont'd.)

$$\underset{\sim}{V}_{s2} = \underset{\sim}{\omega}_3 \times (-\underset{\sim}{r}_3) = \begin{vmatrix} \underset{\sim}{i} & \underset{\sim}{j} & \underset{\sim}{k} \\ 2.290 & 0 & 0 \\ 0 & -15 & 40 \end{vmatrix} = \underset{\sim}{i}0 - \underset{\sim}{j}91.6 - \underset{\sim}{k}34.35 \text{ mm/s} \quad (n)$$

$$\underset{\sim}{V}_{s2S1} = \underset{\sim}{\omega}_2 \times \underset{\sim}{r}_2 = \begin{vmatrix} \underset{\sim}{i} & \underset{\sim}{j} & \underset{\sim}{k} \\ 7.397 & 0.220 & -10.924 \\ 5 & 80 & 5 \end{vmatrix} = \underset{\sim}{i}875 - \underset{\sim}{j}91.6 + \underset{\sim}{k}590.7 \text{ mm/s} \quad (o)$$

Note that eqs. (e), (n) and (o) satisfy eq. (c).

3.25 See solution 3.24

$\underset{\sim}{\omega}_1 \times \underset{\sim}{r}_1 + \underset{\sim}{\omega}_2 \times \underset{\sim}{r}_2 + \underset{\sim}{\omega}_3 \times \underset{\sim}{r}_3 = 0$ (a) where $\omega_3 = -\underset{\sim}{i}10$

$$\underset{\sim}{\omega}_1 \times \underset{\sim}{r}_1 = \begin{vmatrix} \underset{\sim}{i} & \underset{\sim}{j} & \underset{\sim}{k} \\ 0 & \omega_1 & 0 \\ -25 & 0 & 35 \end{vmatrix} = \underset{\sim}{i}35\omega_1 + \underset{\sim}{k}25\omega_1 \quad (b)$$

$$\underset{\sim}{\omega}_2 \times \underset{\sim}{r}_2 \quad \begin{vmatrix} \underset{\sim}{i} & \underset{\sim}{j} & \underset{\sim}{k} \\ \omega_{2x} & \omega_{2y} & \omega_{2z} \\ 5 & 80 & 5 \end{vmatrix} = \underset{\sim}{i}(5\omega_{2y}-80\omega_{2z})+\underset{\sim}{j}(5\omega_{2z}-5\omega_{2x}) + \underset{\sim}{k}(80\omega_{2x} - 5\omega_{2y}) \quad (c)$$

$$\underset{\sim}{\omega}_3 \times \underset{\sim}{r}_3 \quad \begin{vmatrix} \underset{\sim}{i} & \underset{\sim}{j} & \underset{\sim}{k} \\ -10 & 0 & 0 \\ 0 & 15 & -40 \end{vmatrix} = -\underset{\sim}{j}400 - \underset{\sim}{k}150 \quad (d)$$

Substituting (b), (c) and (d) in (a), the $\underset{\sim}{i}$ components yield:

$35\omega_1 + 5\omega_{2y} - 80\omega_{2z} = 0$ (e)

The $\underset{\sim}{j}$ components: $5\omega_{2z} - 5\omega_{2x} - 400 = 0$ (f)

The $\underset{\sim}{k}$ components: $25\omega_1 + 80\omega_{2x} - 5\omega_{2y} - 150 = 0$ (g)

Also, $\omega_2 \cdot r_2 = 0 = 5\omega_{2x} + 80\omega_{2y} + 5\omega_{2z}$ (h)

(i.e., the rotation of link 2 about its own axis is taken as zero.)

Equations (e) through (h) may be solved simultaneously:

$$\begin{bmatrix} 0 & 5 & -80 & 35 \\ -5 & 0 & 5 & 0 \\ 80 & -5 & 0 & 25 \\ 5 & 80 & 5 & 0 \end{bmatrix} \begin{bmatrix} \omega_{2x} \\ \omega_{2y} \\ \omega_{2z} \\ \omega_1 \end{bmatrix} = \begin{bmatrix} 0 \\ 400 \\ 150 \\ 0 \end{bmatrix}$$

from which

$\underset{\sim}{\omega}_2 = -\underset{\sim}{i}32.300 - \underset{\sim}{j}0.963 + \underset{\sim}{k}47.700$ rad/s

$\underset{\sim}{\omega}_1 = \underset{\sim}{j}109.167$ rad/s

3.25 (Cont'd.)

$$\underset{\sim}{V}_{s1} = \underset{\sim}{\omega}_1 \times \underset{\sim}{r}_1 = \begin{vmatrix} \underset{\sim}{i} & \underset{\sim}{j} & \underset{\sim}{k} \\ 0 & 109.167 & 0 \\ -25 & 0 & 35 \end{vmatrix} = \underset{\sim}{i}3820.8 + \underset{\sim}{k}2729.2$$

$$\underset{\sim}{V}_{s2} = \underset{\sim}{\omega}_3 \times (-\underset{\sim}{r}_3) = \begin{vmatrix} \underset{\sim}{i} & \underset{\sim}{j} & \underset{\sim}{k} \\ -10 & 0 & 0 \\ 0 & -15 & 40 \end{vmatrix} = \underset{\sim}{j}400 + \underset{\sim}{k}150$$

3.26 a) Several different methods of solution may be used. The linkage may be described by the vector equation $\underset{\sim}{r}_o+\underset{\sim}{r}_1+\underset{\sim}{r}_2=0$. Using complex number notation, we write:

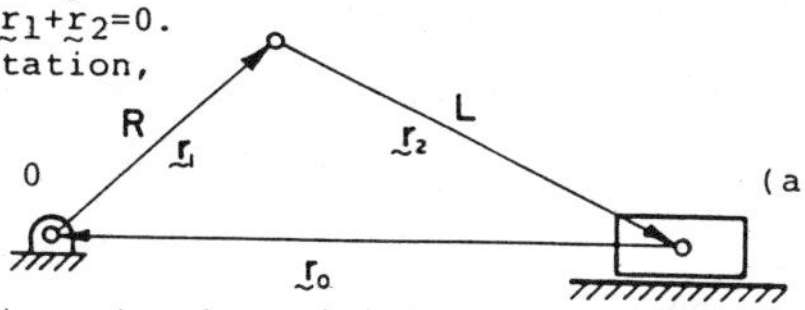

$$-r_o + r_1e^{j\theta_1} + r_2e^{j\theta_2} = 0 \qquad \text{(a)}$$

From Eulers identity:

$$-r_o + r_1(\cos\theta_1 + j\sin\theta_1) + r_2(\cos\theta_2 + j\sin\theta_2) = 0 \qquad \text{(b)}$$

Equating the imaginary parts of eq. (b),

$$r_1\sin\theta_1 + r_2\sin\theta_2 = 0 \qquad \text{(c)}$$

from which

$$\theta_2 = \arcsin(-\sin\theta_1/(r_2/r_1)) \qquad \text{(d)}$$

Differentiating eq. (a),

$$-\dot{r}_o + jr_1\omega_1e^{j\theta_1} + jr_2\omega_2e^{j\theta_2} = 0 \qquad \text{(e)}$$

or

$$-\dot{r}_o + jr_1\omega_1(\cos\theta_1 + j\sin\theta_1) + jr_2\omega_2(\cos\theta_2 + j\sin\theta_2) = 0 \qquad \text{(f)}$$

Equating the imaginary parts:

$$r_1\omega_1\cos\theta_1 + r_2\omega_2\cos\theta_2 = 0 \qquad \text{(g)}$$

from which

$$\omega_2 = -r_1\omega_1\cos\theta_1/(r_2\cos\theta_2) \qquad \text{(h)}$$

Equating the real parts:

$$-\dot{r}_o - r_1\omega_1\sin\theta_1 - r_2\omega_2\sin\theta_2 = 0 \qquad \text{(i)}$$

or

$$\dot{r}_o = -r_1\omega_1\sin\theta_1 - r_2\omega_2\sin\theta_2 \qquad \text{(j)}$$

Differentiating eq. (e)

$$-\ddot{r}_o + jr_1e^{j\theta_1}(j\omega_1^2) + jr_2e^{j\theta_2}(j\omega_2^2 + \alpha_2) = 0 \qquad \text{(k)}$$

or

$$-\ddot{r}_o - \omega_1^2r_1(\cos\theta_1 + j\sin\theta_1) + r_2(-\omega_2^2 + j\alpha_2)(\cos\theta_2 + j\sin\theta_2) = 0 \qquad \text{(l)}$$

3.26a (Cont'd.)

Equating the imaginary parts:

$$-\omega_1{}^2 r_1 \sin\theta_1 - \omega_2{}^2 r_2 \sin\theta_2 + \omega_2 r_2 \cos\theta_2 = 0 \qquad \text{(m)}$$

from which

$$\alpha_2 = (\omega_1{}^2 r_1 \sin\theta_1 + \omega_2{}^2 r_2 \sin\theta_2)/(r_2 \cos\theta_2) \qquad \text{(n)}$$

Equating the real parts:

$$\ddot{r}_o = -\omega_1{}^2 r_1 \cos\theta_1 - \omega_2{}^2 r_2 \cos\theta_2 - \alpha_2 r_2 \sin\theta_2 \qquad \text{(o)}$$

To find maximum piston velocity, set

$\ddot{r}_o = 0$ with $r_1 = 1$, $r_2 = L/R = 2$ and $\omega_1 = 1$

Equations (d), (h) and (n) were substituted in (o). Solving $\ddot{r}_o(\theta_1) = 0$ for $0 < \theta_1 < \pi$ leads to $\theta_1 = 1.122$ rad (64.3^o) corresponding to $|\dot{r}_o|_{max}$

b) Substituting the result in eq. (j),

$\dot{r}_o = -1.120 r_1\omega_1$ ($|v|_{max} = 1.120\ R\omega$)

The results were obtained using a programmable calculator subroutine for zeros of a function, utilizing the bisection method. Results agree with Fig. 3.15B.

3.27 a) Using the program based on the equations in solution 3.26 with $r_2/r_1 = 1.5$, $\ddot{r}(\theta_1) = 0$ leads to the root $\theta_1 = 1.069$ rad (61.24^o)

b) Substituting in eq. (j),

$\dot{r}_o = -1.223 r_1\omega_1$ (i.e. $|v_{max}| = 1.223\ R\omega$). The results agree with Fig. 3.15B, interpolating between the L/R=1.4 and 1.6 curves.

3.28 $\omega_2 = \dfrac{bc}{BC} = \dfrac{12.4 \text{ in/sec}}{3 \text{ in}} = 4.1$ rad/sec cw.

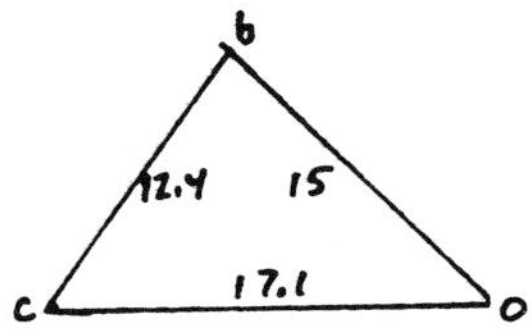

3.29 See sketch $\omega_2 = 3.1$ rad/sec CC.

3.30 " " $\omega_2 = 5$ rad/sec CW.

3.31 " " $v_D = 155$ in/sec, $\omega_2 = 60$ rad/sec CW.

3.32 " " $v_D = 116$ in/sec, $\omega_2 = 4.7$ rad/sec CW,
$\omega_3 = 39$ rad/sec CW.

3.33 a) " " $v_{CB} = 20$ in/sec, $\omega_2 = 4.4$ rad/sec CW.
b) " " $v_{CB} = 20$ in/sec, $\omega_2 = 4.4$ rad/sec CC.

3.34 " " Sliding velocity is 31.4 in/sec.

3.35 " " " " " 32 in/sec.

3.36 a) $v = R^{\omega} \sin \theta \left\{ 1 + \frac{R}{L} \frac{\cos \theta}{\left[1 - (\frac{R}{L})^2 \sin^2\theta\right]^{1/2}} \right\}$ = 161.3 in/sec to left.

or 157.7 by approximate equation.

b) $v = 0$ for $\theta = 0$ or 180° by either equation.

3.37 and b) $v_{CB} = \omega R$ as shown.

3.38 $v_E = 259$ in/sec as shown.

3.39 Follower velocity = 7.1 and sliding velocity = 12.6 in/sec as shown.

3.40 Follower velocity = 5 and sliding velocity = 28.5 in/sec as shown.

3.41 See sketch. $\omega_2 = 3.45$ rad/sec CC, $\omega_3 = 12.9$ rad/sec CW,
$v_D = 17.5$, $v_{\Pi} = 30$, $v_{\Gamma} = 12.9$ in/sec.

3.42 Sliding velocity = 45 in/sec as shown. $\omega_2 = 4.3$ rad/sec CC.

3.29

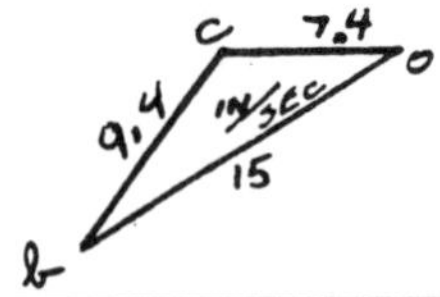

$$\omega_2 = \frac{bc}{BC} = \frac{9.4\,\text{in/sec}}{3\ \text{in}} = 3.1\ \frac{\text{rad}}{\text{sec}}$$

3.30

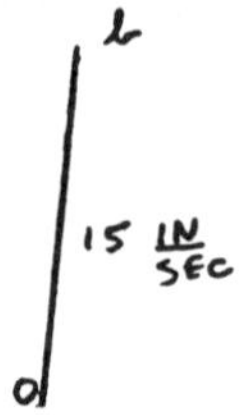

$$\omega_2 = \frac{bc}{BC} = \frac{15\,\text{in/sec}}{3\ \text{in}} = 5\ \text{rad/sec}$$

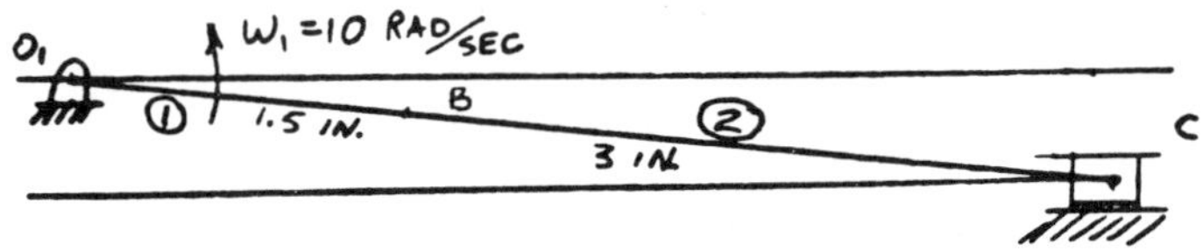

3.31

BC = 3 IN.
BD = 2 IN.

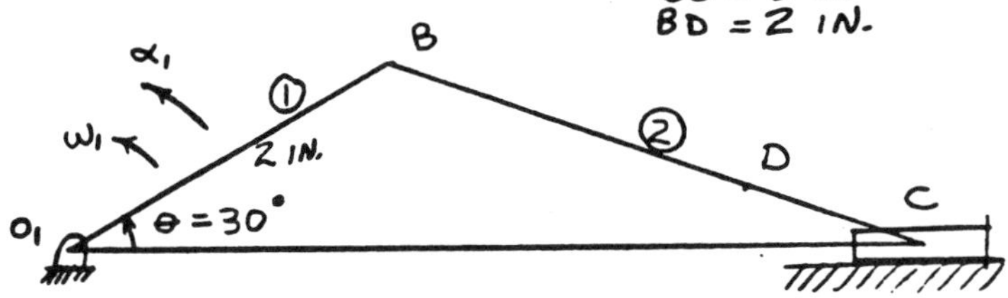

$$\omega_2 = \frac{bc}{BC} = \frac{180}{3} = 60\ \text{rad/sec}$$

$$\frac{bd}{bc} = \frac{BD}{BC} \qquad bd = 120$$

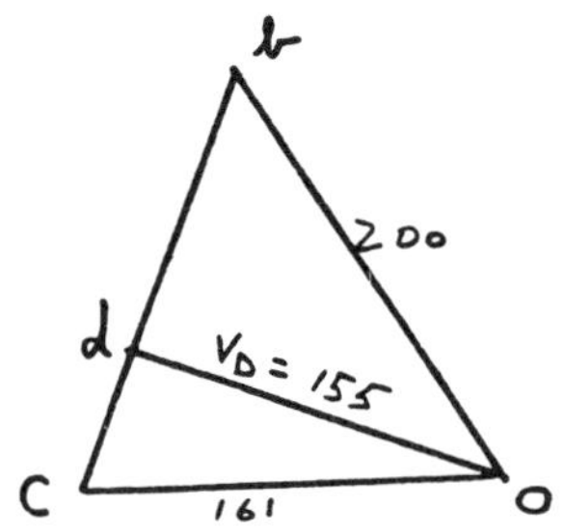

3.32

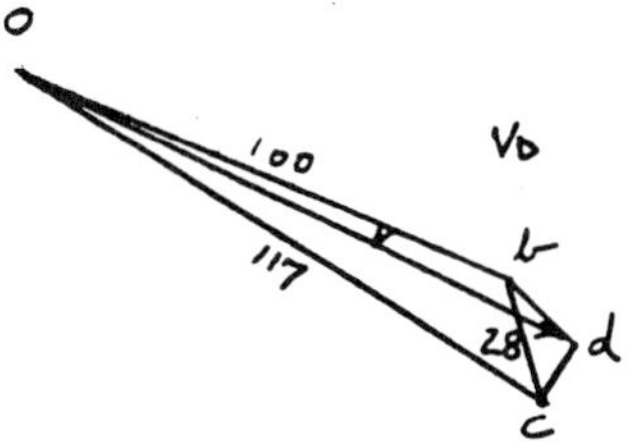

$$\omega_2 = \frac{bc}{BC} = \frac{28}{6} = 4.7 \text{ rad/sec}$$

$$\omega_3 = \frac{oc}{O_3C} = \frac{117}{3} = 39 \text{ rad/sec}$$

$$V_D = 115 \text{ in/sec}$$

3.33

a)

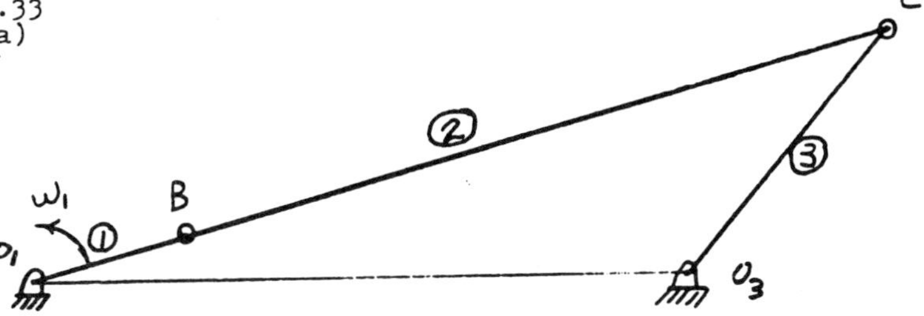

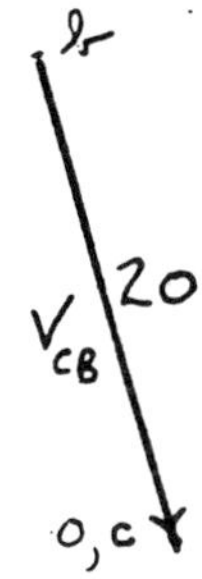

$$\omega_2 = \frac{bc}{BC} = \frac{20 \text{in/sec}}{4.5} = 4.4 \text{ rad/sec}$$

b)

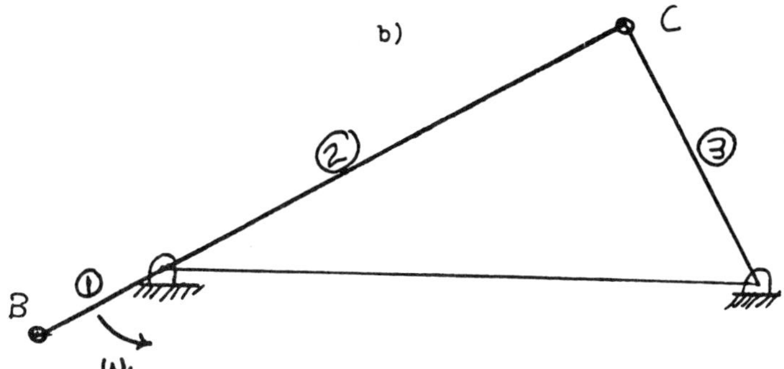

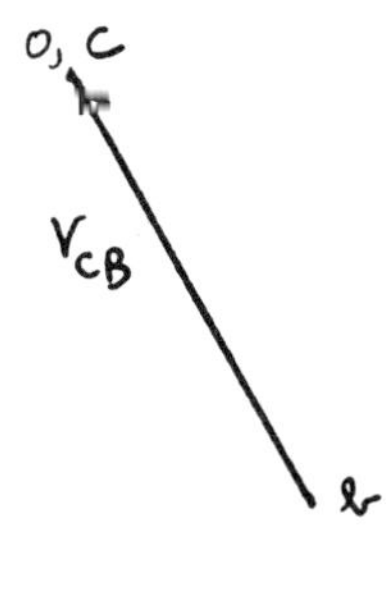

$$\omega_2 = \frac{20 \text{in/sec}}{4.5 \text{ in}} = 4.4 \text{ rad/sec}$$

3.34

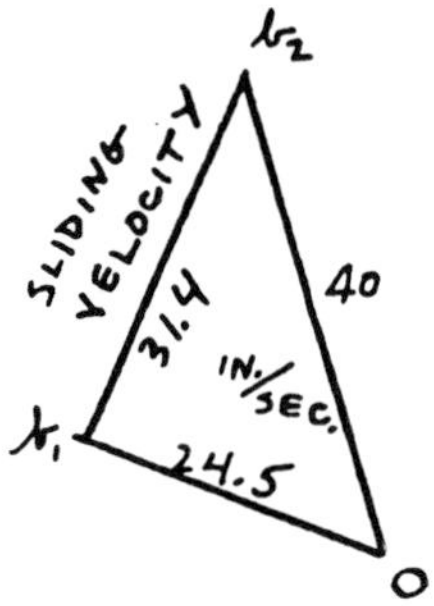

$O_1B_1 = [(2\sin 105^\circ)^2 + (4 - 2\cos 105^\circ)^2]^{1/2} = 4.9$

$\omega_1 = ob_1 / O_1B_1 = 24.5/4.9 = 5$ rad/s CC

3.35

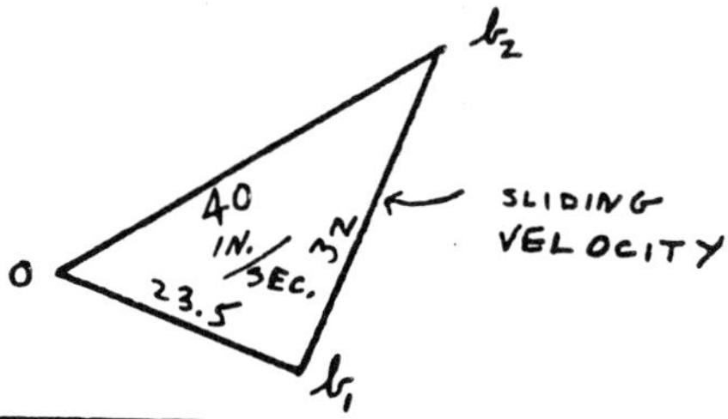

3.37

a)

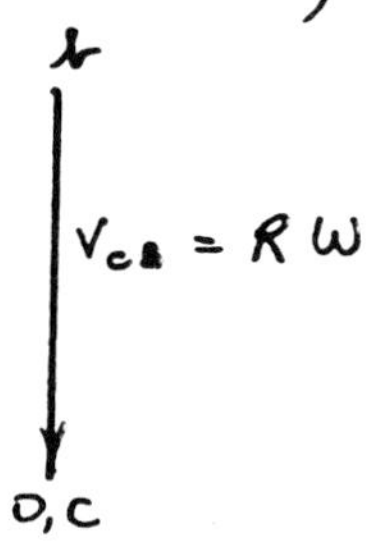

b)

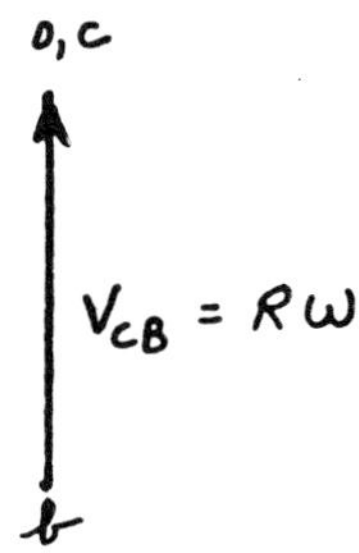

3.38

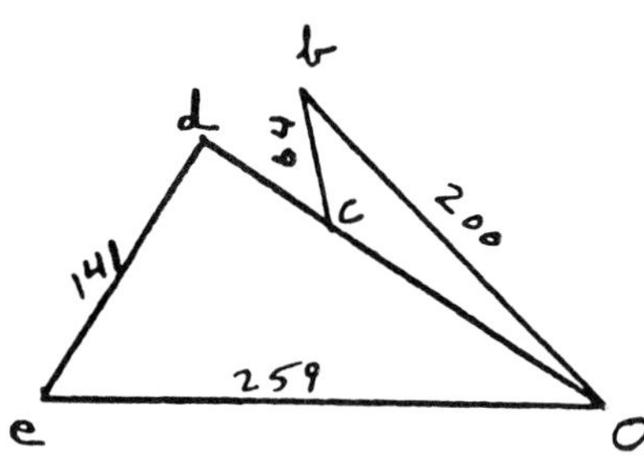

$\frac{od}{oc} = \frac{O_3D}{O_3C}$

$od = 150\frac{6}{4} = 225$ in/sec

3.39

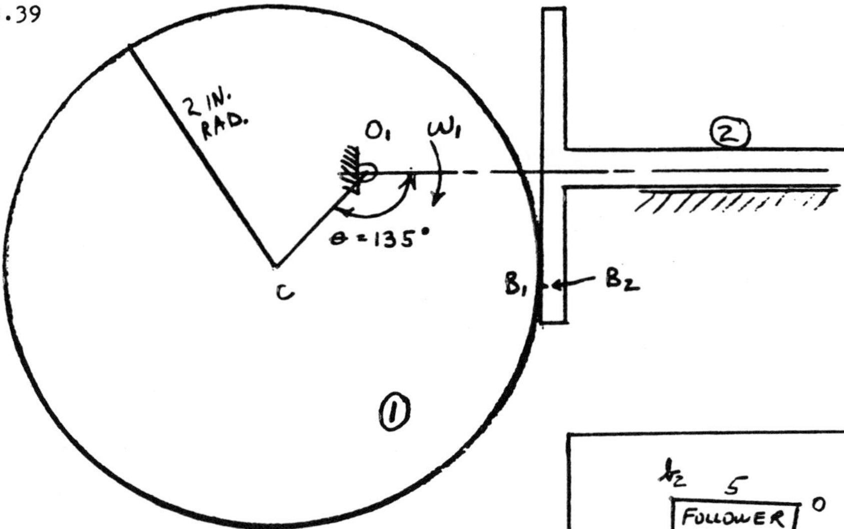

2 IN. RAD.
O1
ω1
2
θ = 135°
C
B1
B2
1

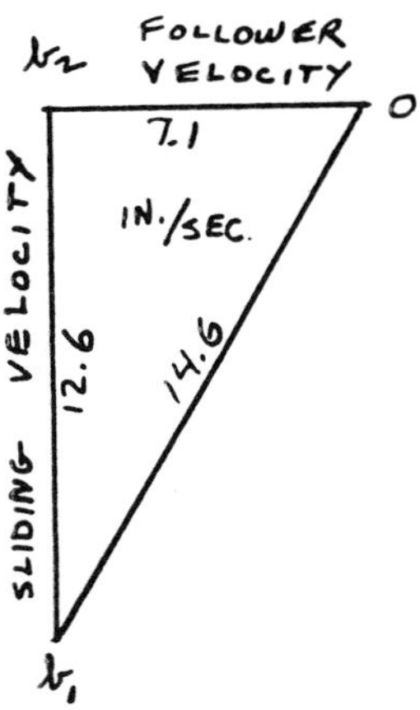

b2
FOLLOWER VELOCITY
o
7.1
IN./SEC.
SLIDING VELOCITY
12.6
14.6
b1

3.40

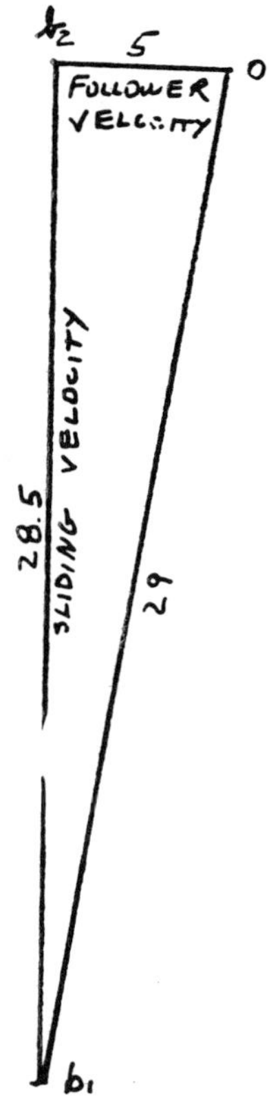

b2
5
o
FOLLOWER VELOCITY
SLIDING VELOCITY
28.5
29
b1

3.41

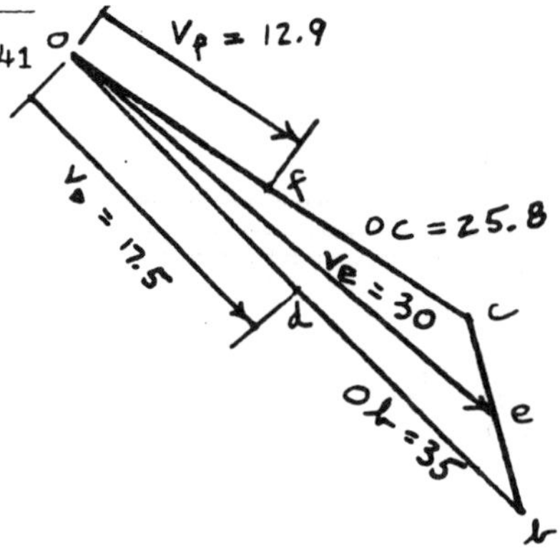

Based on O_3C = 2 in

$\omega_2 = \frac{bc}{BC} = \frac{11.2}{3.25} = 3.45$ rad/sec

$\omega_3 = \frac{oc}{O_3C} = \frac{25.8}{2} = 12.9$ rad/sec

3.42

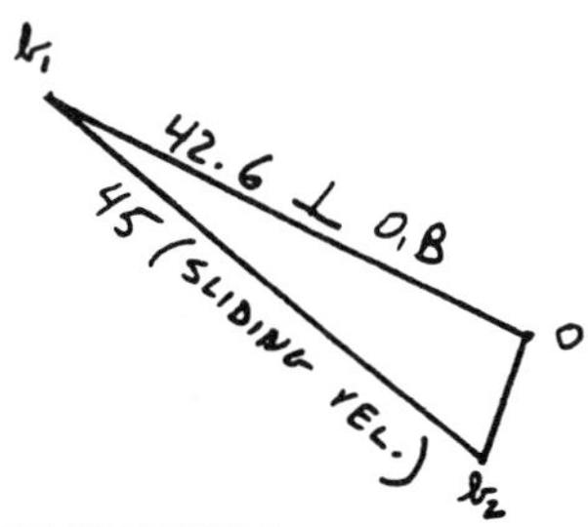

$ob_1 = \omega_1 O_1B = 42.6$ in/sec

$\perp O_1B$

$\omega_2 = \frac{9.5}{2.2} = 4.3$

3.44

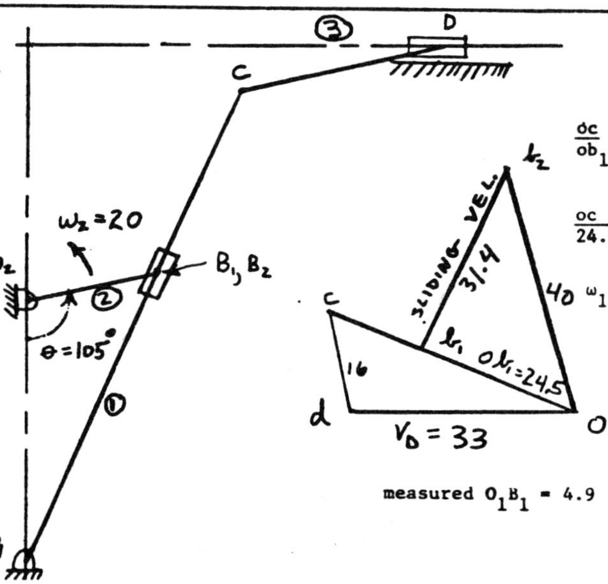

$\frac{oc}{ob_1} = \frac{O_1C}{O_1B_1}$

$\frac{oc}{24.5} = \frac{8}{4.9}$ oc = 40

$\omega_1 = \frac{ob_1}{O_1B_1} = \frac{24.5}{4.9} = 5$

measured O_1B_1 = 4.9

3.45

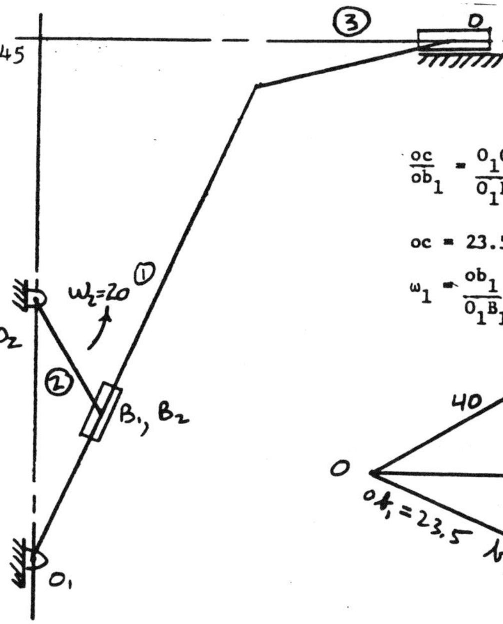

$$\frac{oc}{ob_1} = \frac{O_1C}{O_1B_1}$$

$$oc = 23.5\,\frac{8}{2.5} = 75.2$$

$$\omega_1 = \frac{ob_1}{O_1B_1} = \frac{23.5}{2.5} = 9.4$$

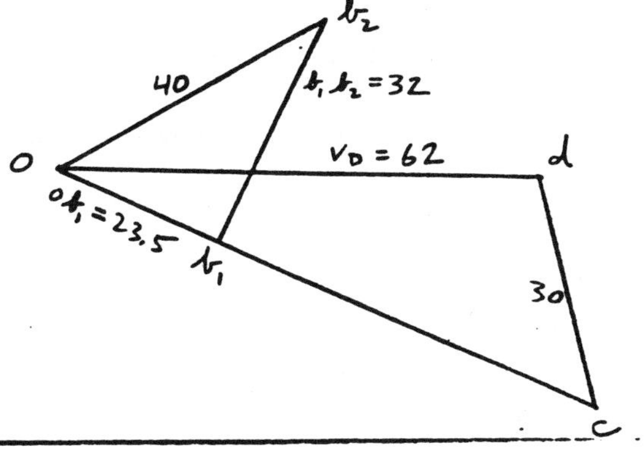

3.47

$O_1B_1 = 2.2$ in.

$ob_1 = \omega_1 O_1 B_1 = 66$ in/sec

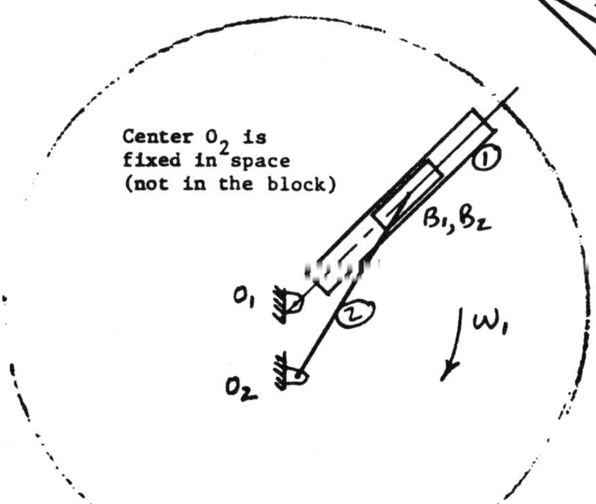

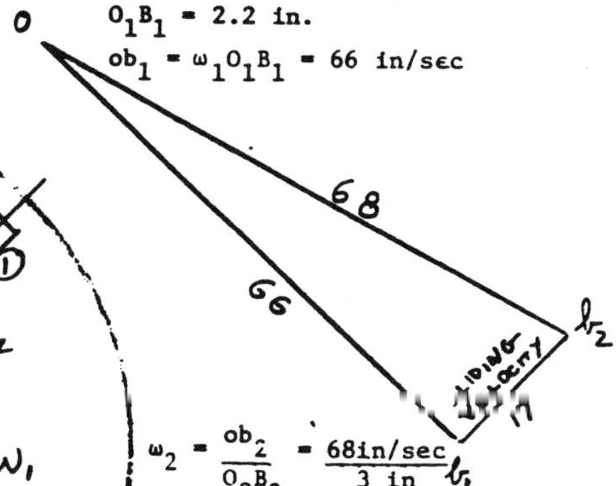

$$\omega_2 = \frac{ob_2}{O_2B_2} = \frac{68\text{in/sec}}{3\text{ in}}$$

$$= 22.7 \text{ rad/sec}$$

3.43 ω_1 = (.1047)(300) = 31.4 rad/sec ω_2 = -104.7 rad/sec

Sliding velocity = $|\omega_2 - \omega_1|$ (dist. to pitch point) = 163 in/sec.

(oriented 20° from line of centers.)

3.44 v_D = 33 in/sec as shown.

3.45 v_D = 62 in/sec as shown.

3.46 v = R ωsin θ

a) v = 1(10) sin 135° = 7.07 in/sec to left.

b) v = 1(10) sin 30° = 5 in/sec to left.

3.47 Sliding velocity = 17 in/sec as shown; ω_2 = 22.7 rad/sec CW.

3.48 ω_2 = 21 rad/sec CC; v_D = 240 in/sec as shown.

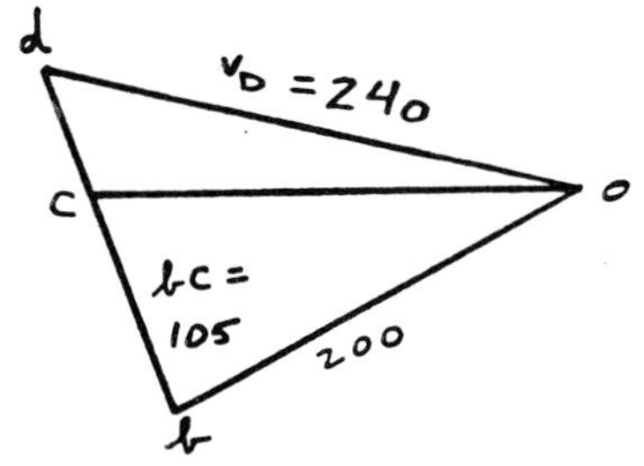

$$\frac{bd}{bc} = \frac{BD}{BC} = \frac{8}{5}$$

$$\omega_2 = \frac{bc}{BC} = \frac{105}{5} = 21 \frac{rad}{sec}$$

3.49 $v_C = v_{13} = \omega_1\, 0_1 13$ = 210 in/sec to left.

$$V_c = V_{13} = \omega_1 0113$$
$$= (100\ \frac{\text{rad}}{\text{sec}})\ (2.1\ \text{in}) = 210 \frac{\text{in}}{\text{sec}}$$

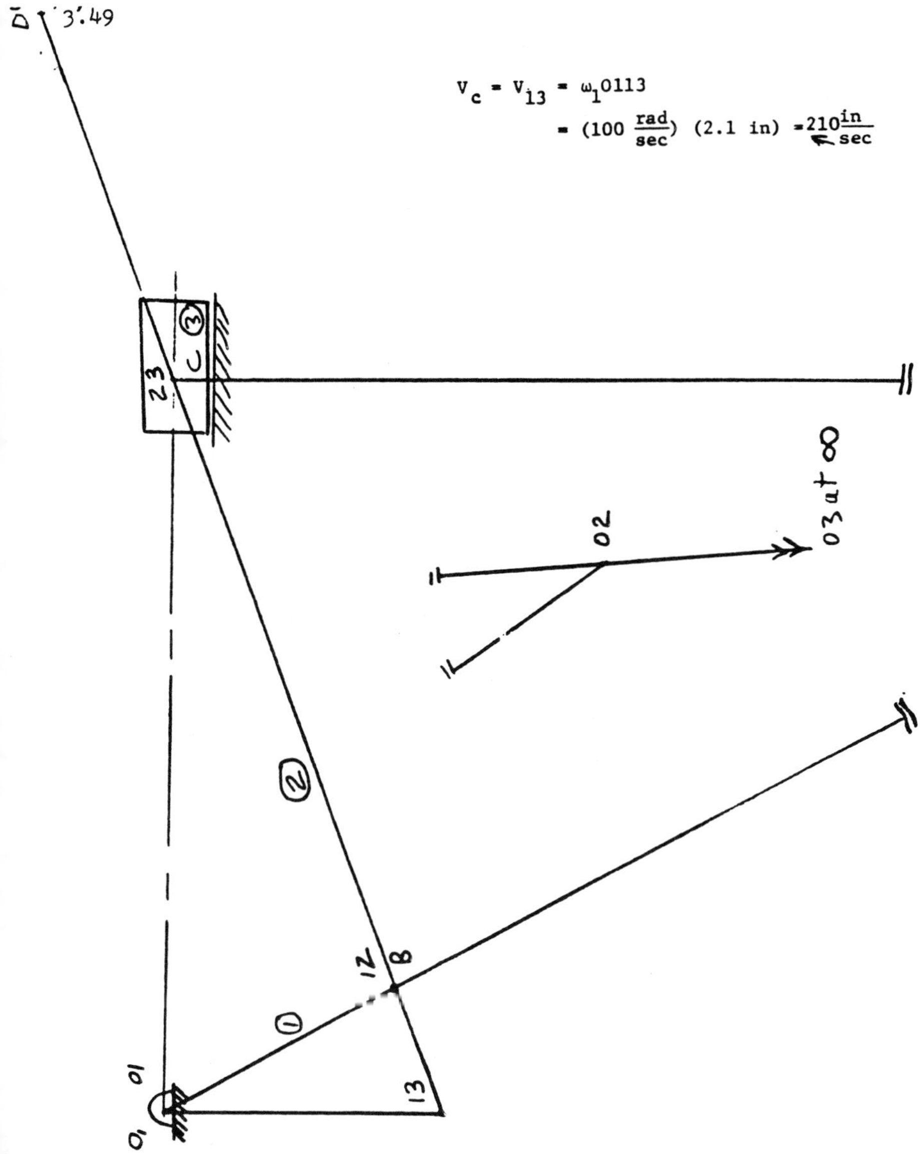

3.50

a) $\omega_2 = \frac{V_B}{0212}$

b) $\frac{\omega_2}{\omega_1} = -\frac{0112}{02121}$

c) $V_C = \omega_2 02C$

d) $V_D = \omega_2 O_2 D$

e) $\omega_2 = -\frac{0112}{0212}\,\omega 1$

$= -\frac{2}{9.5}(100)$

$= 21$ rad/sec ↺

$V_C = 21\ \frac{rad}{sec}\ (10\ in) = 210\frac{in}{sec}$

$V_D = 21\ \frac{rad}{sec}\ (11.4\ in) = 240\ \frac{in}{sec}$ $\perp O_2 D$

3.51

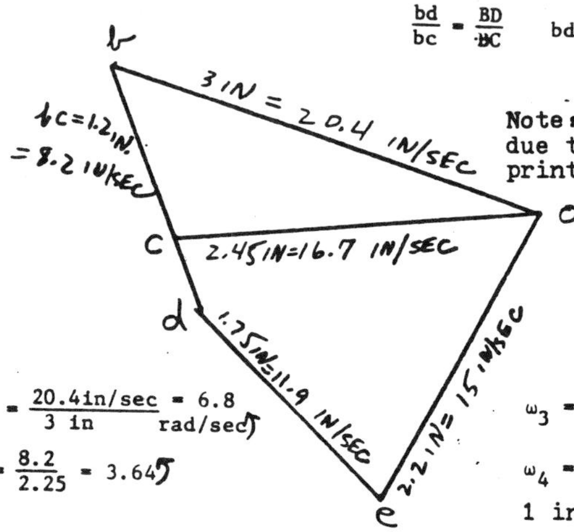

$\frac{bd}{bc} = \frac{BD}{BC}$ $\quad bd = 1.2\frac{3.25}{2.25}$

length bd = 1.73

Note: Scale may change due to reduction in printing.

$\omega_1 = \frac{ob}{O_1B} = \frac{20.4\ in/sec}{3\ in} = 6.8$ rad/sec ↺

$\omega_2 = \frac{bc}{BC} = \frac{8.2}{2.25} = 3.64$ ↺

$\omega_3 = \frac{oc}{O_3C} = \frac{16.7}{2} = 8.35$ ↺

$\omega_4 = \frac{de}{DE} = \frac{11.9}{2.75} = 4.33$ ↺

1 in = 6.8 in/s

3.52

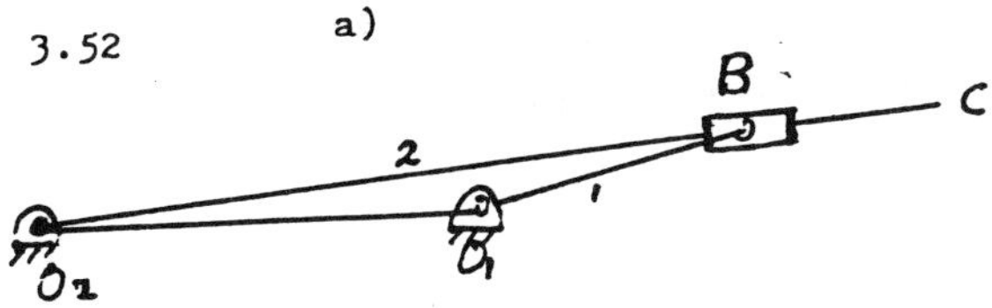

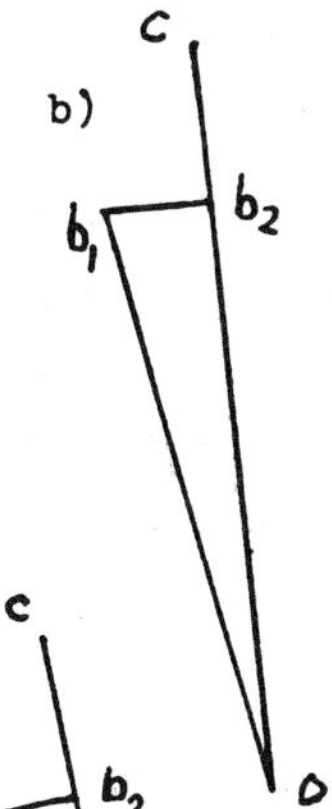

3.53

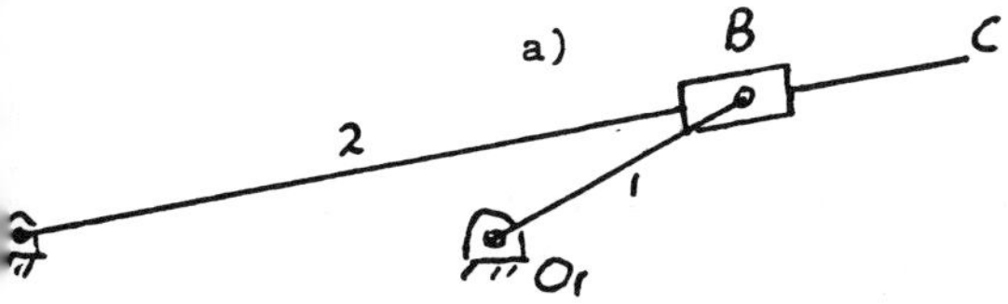

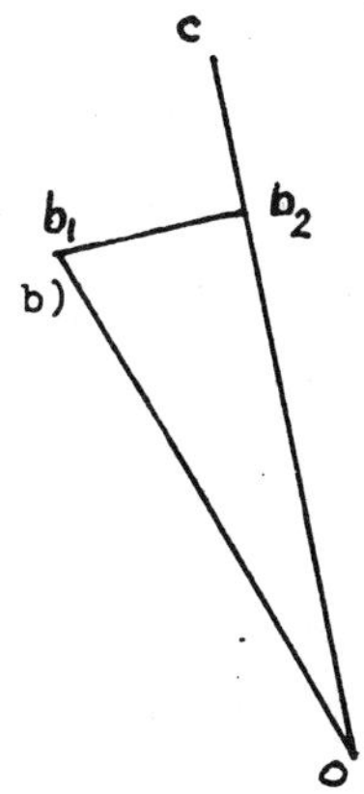

3.54

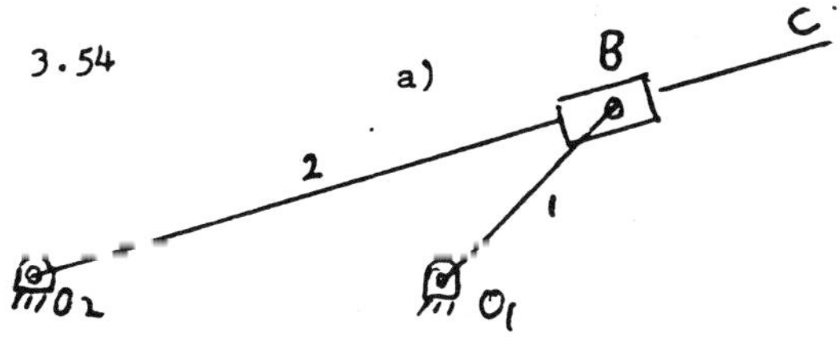

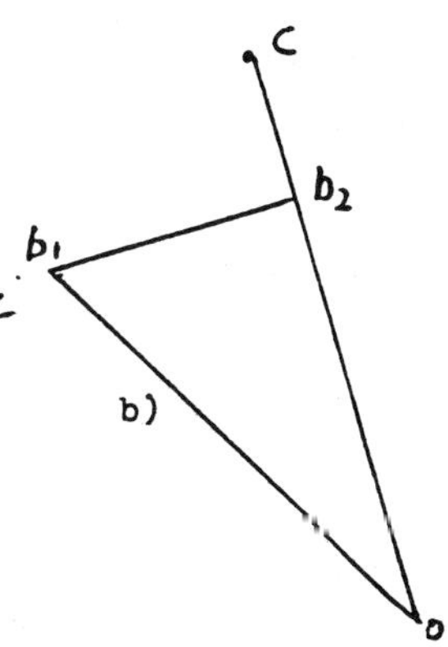

3.52 to 3.54 See figures.

$ob_2 = ob_1 + b_1b_2$ where

$V_{B2} = ob_2 \perp O_2B_2$ $V_{B1} = ob_1 = \omega_1 O_1B_1 \perp O_1B_1$ $V_{B2B1} = b_1b_2 \| \text{link } 2$

$\omega_2 = ob_2/O_2B_2$ $oc/ob_2 = O_2C/O_2B_2$

3.52 $ob_1 = 10x30 = 300$ mm/s

$V_{B2} = 296$ mm/s $V_C = 372$ mm/s

$\omega_2 = 3.73$ rad/s cc

3.53 $ob_1 = 300$ mm/s $V_{B2} = 284$ mm/s

$V_C = 367$ mm/s $\omega_2 = 3.67$ rad/s cc

3.54 $ob_1 = 300$ mm/s $V_{B2} = 264$ mm/s

$V_C = 355$ mm/s $\omega_2 = 3.55$ rad/s cc

3.55 $\omega_2 = 0$ when $ob_2 = 0$

Then, ob_1 is parallel to b_1b_2 from which link 1 $\perp$ link 2.

Then, $\sin(180^o - \theta_1) = O_1B/O_1O_2 = 30/50$

$\theta_1 = \pm 143.13^o$

3.56 a) See figure. $ob = V_B = \omega_1 O_1B = 500x0.12 = 60$ m/s

b) $\omega_2 = bc/BC = 58.67$ m/s/0.2 m $= 293.3$ rad/s cw.

c) See figure. $oc = V_C = 24.64$ m/s $bc = V_{CB} = 58.67$ m/s

$bd/bc = BD/BC$ $bd = 58.67x40/200 = 11.73$ m/s

$V_D = od = 48$ m/s

d) See figure. The velocity of B $V_B = ob = \omega_1 O_1B$. Using the law of sines, we find

$$\frac{bc}{\sin(90-\theta)} = \frac{ob}{\sin(90-\phi)} \quad \text{from which}$$

$$\frac{bc}{\cos\theta} = \frac{ob}{\cos\phi} \qquad V_{CB} = bc = ob\,\frac{\cos\theta}{\cos\phi} = \omega_1 O_1B\,\frac{\cos\theta}{\cos\phi}$$

where $BC\sin\phi = O_1B\sin\theta$ $\sin\phi = \frac{O_1B}{BC}\sin\theta$

$$\phi = \sin^{-1}\left(\frac{O_1B}{BC}\sin\theta\right)$$

$$\frac{oc}{\sin(\theta+\phi)} = \frac{ob}{\sin(90-\phi)} \qquad V_C = oc = \frac{\sin(\theta+\phi)}{\cos\phi}\,\omega_1 O_1B$$

3.56

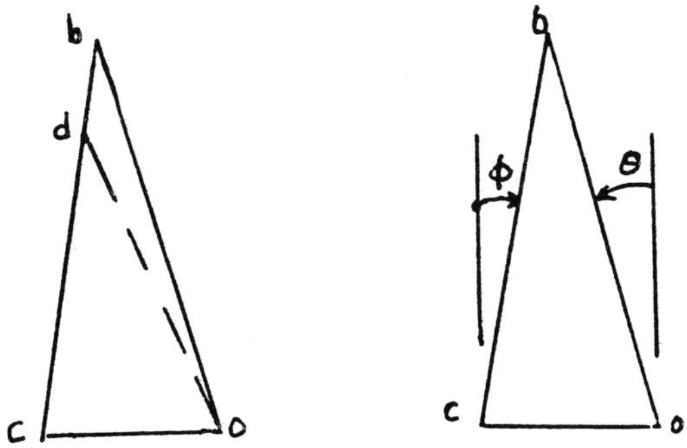

.57

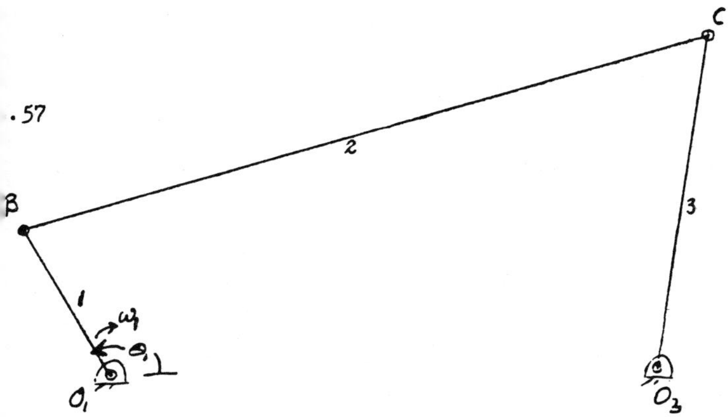

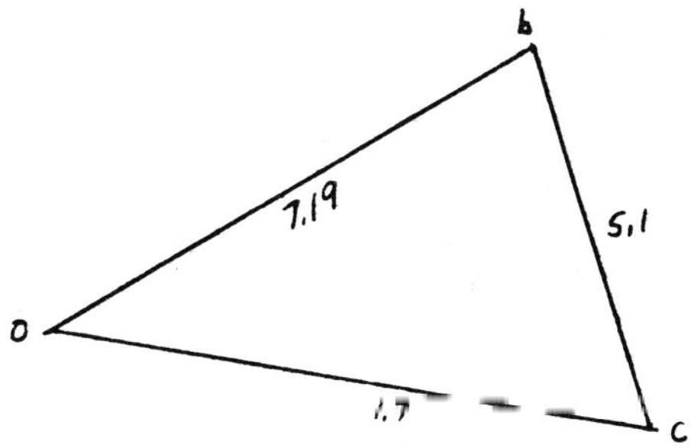

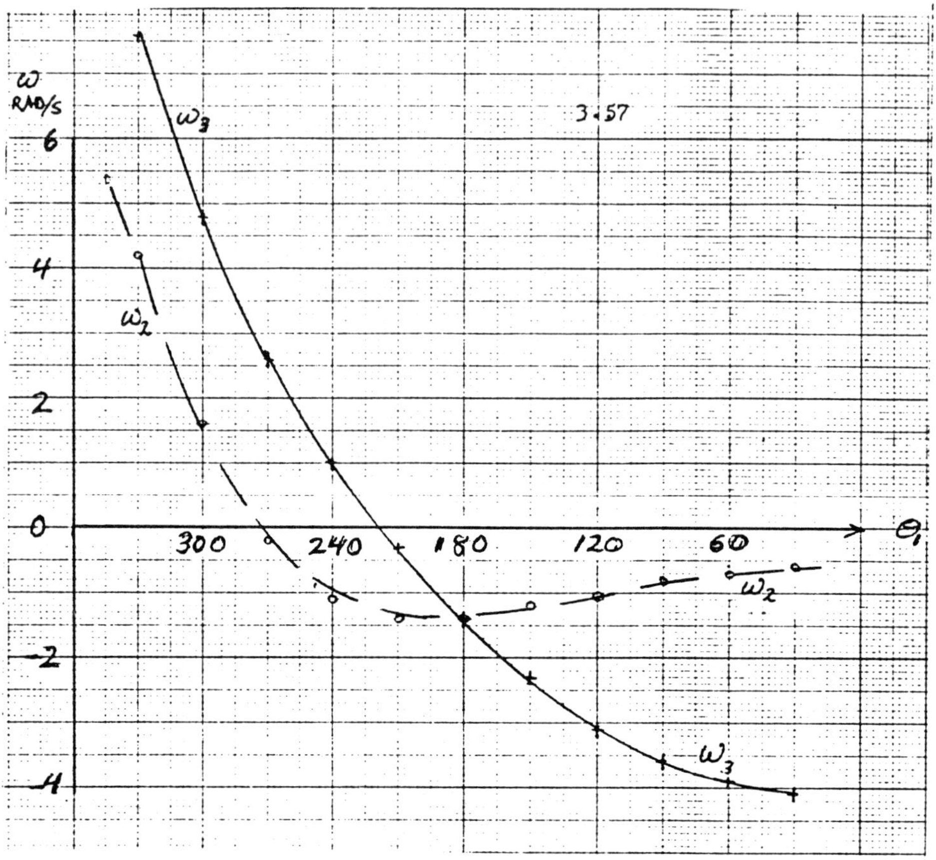
ω
RAD/S
6
4
2
0
-2
-4
ω_3
3.57
ω_2
300
240
180
120
60
θ_1
ω_2
ω_3

3.57 See figure. Since $L_{max} + L_{min} = L_a + L_b$, this is a change point or crossover position linkage. The solution follows the procedure of section 3.4. Results are plotted for $\theta_1 = 330^o$, 300°, etc., for clockwise rotation of link 1. It is assumed that the assembly configuration corresponds to the solution figure. $\theta_1 = 0$ is the crossover position. The velocity polygon checks results at $\theta_1=120^o$. $V_{CB} = \omega_2 BC$ and $V_C = \omega_3 O_3 C$. See plotted results.

3.58 a) See figure. $\underset{\sim}{ob_1} = \underset{\sim}{ob_2} + \underset{\sim}{b_2 b_1}$

$ob_2 = \omega_2 O_2 B_2 = 500 \text{x} 0.150 = 75$ m/s $\perp O_2 B_2$

b) $oc/ob_1 = O_1C/O_1B_1$ $\quad oc = 21 \text{x} 400/221 = 38$ m/s

c) $\omega_1 = ob_1/O_1B_1 = 21/0.221 = 95$ rad/s cw

d) See figure.

e) See figure.

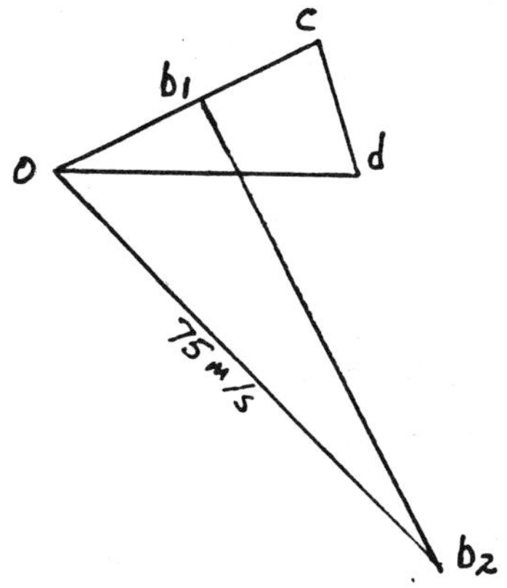

3.59 a) At $\theta_2 = 120^\circ$ $O_2B \perp$ link 1 (limiting position)

$$v_{B2} = v_D = \omega_1 = 0$$

b) $\Delta\theta_2 = 135 - 120 = 15^\circ = 0.2618$ rad

$\Delta t = \Delta\theta_2/\omega_2 = 0.2618/500 = .5236\text{x}10^{-3}$s

$\alpha_{1\,avg} = \Delta\omega_1/\Delta t = (95-0)/.5236\text{x}10^{-3} = 181{,}000$ rad/s^2 cw

$a_{D_{avg}} = \Delta V_D/\Delta t = (40-0)/.5236\text{x}10^{-3} = 76000$ m/s^2

3.60 At the limiting positions,

$\omega_2 = 0 = \omega_1 R_1 \cos(\theta_1 - \theta_2)/R_2$

$\cos(\theta_1-\theta_2) = 0 \qquad \theta_1 - \theta_2 = 90^\circ$ or 270°

$O_2B \perp BO_1 \qquad \dfrac{\sin\theta_2}{30} = \dfrac{\sin 90^\circ}{50}$

$\theta_2 = 36.87^\circ \qquad BO_1O_2 + \theta_2 = 90^\circ \qquad \theta_1 = 180^\circ - BO_1O_2$

$\theta_1 = 180^\circ - 90^\circ + \theta_2 \qquad \theta_1 = 126.87^\circ$ and -126.87°

3.61

$$\underset{\sim}{r}_0 + \underset{\sim}{r}_1 + \underset{\sim}{r}_2 + \underset{\sim}{r}_3 = 0 \qquad (a)$$

from which

$$-r_o + r_1 e^{j\theta_1} + r_2 e^{j\theta_2} + r_3 e^{j\theta_3} = 0 \qquad (b)$$

Substituting $e^{j\theta} = \cos\theta + j\sin\theta$, we obtain

$$-r_o + r_1(\cos\theta_1 + j\sin\theta_1) + r_2(\cos\theta_2 + j\sin\theta_2) + r_3(\cos\theta_3 + j\sin\theta_3) = 0 \qquad (c)$$

Equating the imaginary parts of c,

$$r_1\sin\theta_1 + r_2\sin\theta_2 + r_3\sin\theta_3 = 0 \qquad (d)$$

Equating the real parts of c,

$$-r_o + r_1\cos\theta_1 + r_2\cos\theta_2 + r_3\cos\theta_3 = 0 \qquad (e)$$

Position:

Equations d and e may be solved simultaneously to find position angles θ_2 and θ_3 if θ_1 is given. There are two assembly configurations.

3.61 a) (Cont'd.)

Angular velocity:

Equation b is differentiated with respect to time to obtain

$$j\omega_1 r_1 e^{j\theta_1} + j\omega_2 r_2 e^{j\theta_2} + j\omega_3 r_3 e^{j\theta_3} = 0 \qquad \text{(f)}$$

Multiplying equation f by $e^{-j\theta_3}/j$:

$$\omega_1 r_1 e^{j(\theta_1-\theta_3)} + \omega_2 r_2 e^{j(\theta_2-\theta_3)} + \omega_3 r_3 = 0 \qquad \text{(g)}$$

Substituting $e^{j\theta} = \cos\theta + j \sin\theta$ in eqn. g,

$$\omega_1 r_1[\cos(\theta_1-\theta_3) - j\sin(\theta_1-\theta_3)] + \omega_2 r_2[\cos(\theta_2-\theta_3) - j\sin(\theta_2-\theta_3)]$$
$$+ \omega_3 r_3 = 0 \qquad \text{(h)}$$

Equating the imaginary parts of h:

$\omega_1 r_1 \sin(\theta_1-\theta_3) + \omega_2 r_2 \sin(\theta_2-\theta_3) = 0$ from which

$$\omega_2 = -[\omega_1 r_1 \sin(\theta_1-\theta_3)]/[r_2 \sin(\theta_2-\theta_3)] \qquad \text{(i)}$$

Equating the real parts of h:

$$\omega_3 = -[\omega_1 r_1 \cos(\theta_1-\theta_3) + \omega_2 r_2 \cos(\theta_2 - \theta_3)]/r_3 \qquad \text{(j)}$$

b) There are several possible methods. From equation d,

$$\sin\theta_3 = \frac{1}{r_3}(-r_1 \sin\theta_1 - r_2 \sin\theta_2) \qquad \text{(k)}$$

Equation k may be used to eliminate θ_3 in equation e: The result is

$$f(\theta_2) = 0 = r_o - r_1\cos\theta_1 - r_2\cos\theta_2$$
$$- r_3\cos[\arcsin(-\frac{r_1}{r_3} \sin\theta_1 - \frac{r_2}{r_3} \sin\theta_2)] \qquad \text{(l)}$$

Since the arcsin is double valued, both the computed result and 180° minus the computed arcsin should be considered. The solutions may be obtained using a calculator programmed to solve $f(\theta_2) = 0$, yielding $\theta_2 = 288.3°$ and $11.8°$. For the second value, using equation k,

$$\theta_3 = \arcsin(-\frac{2}{3} \sin 60° - \frac{4.5}{3} \sin 11.8°) = -62.1 \text{ and } 242.1°$$

The second value applies to the configuration in the sketch.

From equation i,

$$\omega_2 = -[-50x2x\sin(60-242.1)]/[4.5x\sin(11.8-242.1)]$$
$$= 1.06 \text{ rad/s cc}$$

From equation j,

$$\omega_3 = -[-50x2x\cos(60-242.1)+1.06x4.5x\cos(11.8-242.1)]/3$$
$$= -32.3 \text{ (32.3 rad/s cw)}$$

3.61 (c)

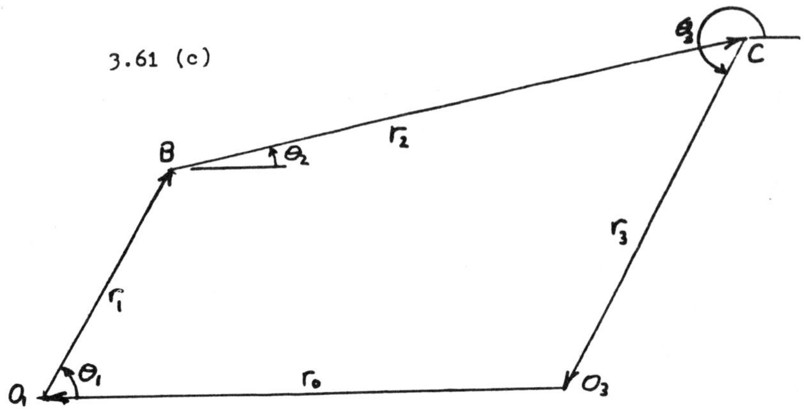
B
θ_2
r_2
C
θ_3
r_3
r_1
θ_1
O_1
r_0
O_3

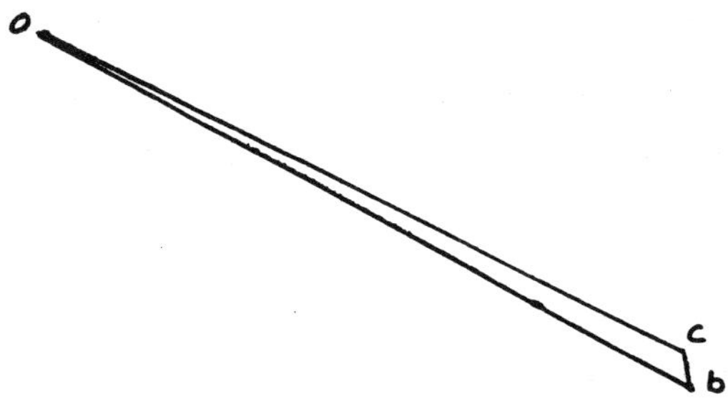
o
c
b

3.62

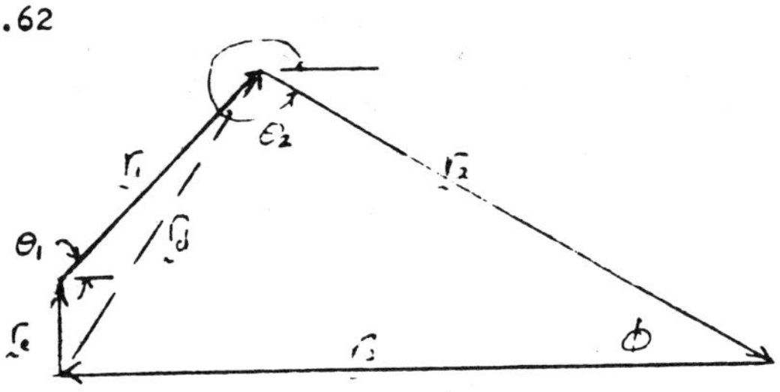
θ_2
r_1
r_2
θ_1
ϕ

3.61 c) (Cont'd.)

See figure.

$ob = \omega_1 r_1 = 50 \times 2 = 100 \perp O_1B$

$bc \perp BC \qquad oc \perp O_3C$

$bc = 5 \qquad oc = 97$

$\omega_2 = bc/BC = 5/4.5 = 1.1$ rad/s cc

$\omega_3 = oc/O_3C = 97/3 = 32.3$ rad/s cw

3.62 Position Analysis:

The linkage of Figure P3.2 may be represented by vectors as in the solution figure.

$\tilde{r}_1 = \tilde{i}r_{1x} + \tilde{j}r_{1y} = \tilde{i}(1.5\cos\theta_1) + \tilde{j}(1.5\sin\theta_1) = \tilde{i}1.0607 + j1.0607$

and

$\tilde{r}_d = \tilde{r}_e + \tilde{r}_1 = \tilde{i}(1.5\cos\theta_1) + \tilde{j}(0.5 + 1.5\sin\theta_1) = 1.887\angle 55.8^\circ$

Then $\quad r_d + r_2 + r_o = 0 \quad$ from which

$(-\tilde{r}_2)^2 = (\tilde{r}_o + \tilde{r}_d)\cdot(\tilde{r}_o + \tilde{r}_d)$

Substituting and solving the quadratic,

$\tilde{r}_o = -\tilde{i}3.6233$. Solving triangle $\tilde{r}_o\tilde{r}_d\tilde{r}_2$,

$\phi = 31.341^\circ$ or $\theta_2 = 328.659^\circ$ and

$\tilde{r}_2 = \tilde{i}2.5623 - \tilde{j}1.5604$

Velocities:

From the position equation: $\tilde{r}_o + \tilde{r}_e + \tilde{r}_1 + \tilde{r}_2 = 0$ (a)

we have $\dot{\tilde{r}}_o + \dot{\tilde{r}}_e + \dot{\tilde{r}}_1 + \dot{\tilde{r}}_2$ from which

$\dot{\tilde{r}}_o + \tilde{\omega}_1 \times \tilde{r}_1 + \tilde{\omega}_2 \times \tilde{r}_2 = 0$ where (b)

$$\tilde{\omega}_1 \times \tilde{r}_1 = \begin{vmatrix} \tilde{i} & \tilde{j} & \tilde{k} \\ 0 & 0 & \omega_1 \\ r_{1x} & r_{1y} & 0 \end{vmatrix} = \tilde{i}\omega_1 r_{1y} + \tilde{j}\omega_1 r_{1x} \qquad (c)$$

$$\tilde{\omega}_2 \times \tilde{r}_2 = \begin{vmatrix} \tilde{i} & \tilde{j} & \tilde{k} \\ 0 & 0 & \omega_2 \\ r_{2x} & r_{2y} & 0 \end{vmatrix} = -\tilde{i}\omega_2 r_{2y} + \tilde{j}\omega_2 r_{2x} \qquad (d)$$

Substituting:

$\tilde{i}\dot{r}_o - \tilde{i}\omega_1 r_{1y} + \tilde{j}\omega_1 r_{1x} - \tilde{i}\omega_2 r_{2y} + \tilde{j}\omega_2 r_{2x} = 0$ (e)

3.62 (Cont'd.)

From the $\underset{\sim}{j}$ components:

$\omega_1 r_{1x} + \omega_2 r_{2x} = 0 \quad \omega_2 = -\omega_1 r_{1x}/r_{2x}$ (f)

$\omega_2 = -10x1.0607/2.5623 = -4.139$ (4.139 rad/s cw)

From the the $\underset{\sim}{i}$ components:

$\dot{r}_o = \omega_1 r_{1y} + \omega_2 r_{2y}$ (g)

$v_c = \dot{r}_o = 10x1.0607 + (-4.139)(-1.5604) = 17.15$ in/s

positive $\dot{r}_o$ to the left (r_o decreasing the $-\underset{\sim}{i}$ direction).

3.63 The analysis is similar to the above problem, except

$\underset{\sim}{r}_1 = \underset{\sim}{i}(1.5\cos 120^o) + \underset{\sim}{j}(1.5\sin 120^o) = -\underset{\sim}{i}0.75 + \underset{\sim}{j}1.2990$

$\underset{\sim}{r}_d = \underset{\sim}{r}_e + \underset{\sim}{r}_1 = -\underset{\sim}{i}0.75 + \underset{\sim}{j}1.7990 = 1.9491\underline{/112.631^o}$

$\underset{\sim}{r}_d + \underset{\sim}{r}_2 + r_o = 0$ leads to

$r_o = -\underset{\sim}{i}1.6507$ and $\phi = 36.8470^o$ or $\theta_2 = 323.153^o$

From which $r_2 = \underset{\sim}{i}2.4007 - \underset{\sim}{j}1.7990$

From 3.62 eqn (f),

$\omega_2 = -\omega_1 r_{1x}/r_{2x} = -10x(-0.75)/2.4007 = 3.124$ rad/s cc

From eqn (g)

$\dot{r}_o = \omega_1 r_{1y} + \omega_2 r_{2y} = 10x1.299 + 3.124x(-1.799) = 7.37$ in/s

to the left.

3.64 Referring to solution figure 3.62, we may write

$\underset{\sim}{r}_o + \underset{\sim}{r}_e + \underset{\sim}{r}_1 + \underset{\sim}{r}_2 = 0,$ or in complex form,

$-r_o + jr_e + r_1 e^{j\theta_1} + r_2 e^{j\theta_2} = 0$

Differentiating with respect to time:

$-\dot{r}_o + j\omega_1 r_1 e^{j\theta_1} + j\omega_2 r_2 e^{j\theta_2} = 0$

Substituting $e^{j\theta} = \cos\theta + j\sin\theta_1$,

$-\dot{r}_o + j\omega_1 r_1(\cos\theta_1 + j\sin\theta_1) + j\omega_2 r_2(\cos\theta_2 + j\sin\theta_2) = 0$ (a)

Equating the imaginary parts of (a)

$\omega_1 r_1 \cos\theta_1 + \omega_2 r_2 \cos\theta_2 = 0$ from which

$\omega_2 = -\omega_1 r_1 \cos\theta_1/(r_2\cos\theta_2)$ (b)

3.64 (Cont'd.)

For $\theta_1 = 45^o$, $\theta_2 = 328.66^o$ and

$\omega_2 = -10x1.5xcos45^o/(3xcos328.66^o)$

$\omega = -4.14$ (i.e. 4.14 rad/s cw)

From the real parts of (a)

$-\dot{r}_o - \omega_1 r_1 sin\theta_1 - \omega_2 r_2 sin\theta_2 = 0$

$V_C = \dot{r}_o = -\omega_1 r_1 sin\theta_1 - \omega_2 r_2 sin\theta_2$ (c)

$= -10x1.5xsin45^o-(-4.14)x3xsin328.66^o$

$= -17.1$ (i.e. 17.1 in/s to the left).

3.65 The solution is similar to the above problem except $\theta_1 = 120^o$ and $\theta_2 = 323.15^o$.

From (b),

$\omega_2 = -10x1.5xcos120^o/(3xcos323.15^o)=3.124$rad/s cc

From (c)

$V_C = \dot{r}_o = -10x1.5xsin120^o-3.124x3xsin323.15^o$

$= -7.37$ (7.37 in/s to the left).

3.66 The problem was solved using a spreadsheet with formulas similar to those given in secton 3.5. Link positions, transmission angle, and angular velocities are given in spreadsheet table S3.66 Angular velocities are plotted in figure S3.66 .

3.67 The problem was solved using a spreadsheet with formulas similar to those given in secton 3.5. Link positions, transmission angle, and angular velocities are given in spreadsheet table S3.67. Angular velocities are plotted in figure S3.67

3.68 The problem was solved using a spreadsheet with formulas similar to those given in secton 3.5. Link positions, transmission angle, angular velocities, and the velocity of point D are given in spreadsheet table S3.68 Angular velocities are plotted in figure S3.68A. The x- and y-components of v_D and the resultant are plotted in figure S3.68B

3.69 The problem was solved using a spreadsheet with formulas similar to those given in secton 3.5. Link positions, transmission angle, angular velocities, and the velocity of point D are given in spreadsheet table S3.69. Angular velocities are plotted in figure S3.69A.

3.69 concl.

The x- and y-components of v_D and the resultant are plotted in figure S3.69B

53,66

Four-bar linkage.

Position analysis based on the vector cross product method.

If Rd-R2-R3 loop clockwise, config = 1; -1 if ccw.

R0	deg	R1	R2	R3	config
312.48	180	100	200	300	1

Theta1 deg	Rdx	Rdy	Rd	Rdux	Rduy	a
0	-212.48	0	212.48	-1	0	223.8981
5	-212.861	8.715574	213.0389	-.999163	.0409107	223.8689
10	-213.999	17.36482	214.7026	-.996724	.0808785	223.7914
15	-215.887	25.8819	217.4333	-.99289	.1190338	223.6944
20	-218.511	34.20201	221.1712	-.987971	.1546404	223.6202
25	-221.849	42.26183	225.8387	-.982335	.1871328	223.6178
30	-225.877	50	231.3453	-.976365	.2161272	223.7362
35	-230.565	57.35764	237.5921	-.970423	.2414122	224.0184
40	-235.876	64.27876	244.4771	-.964817	.2629235	224.4976
45	-241.769	70.71068	251.8976	-.959792	.280712	225.1955
50	-248.201	76.60444	259.7539	-.955525	.2949116	226.1219
55	-255.122	81.9152	267.9506	-.952125	.3057101	227.2761
60	-262.48	86.60254	276.3978	-.949646	.3133257	228.6482
65	-270.218	90.63078	285.0119	-.948094	.3179894	230.2216
70	-278.278	93.96926	293.7156	-.94744	.3199328	231.9742
75	-286.598	96.59258	302.4378	-.947627	.31938	233.8805
80	-295.115	98.48078	311.1132	-.948578	.3165432	235.9132
85	-303.764	99.61947	319.6824	-.950207	.3116201	238.0438
90	-312.48	100	328.0911	-.952418	.3047934	240.2439
95	-321.196	99.61947	336.2895	-.955116	.2962313	242.4855
100	-329.845	98.48078	344.2326	-.958203	.2860879	244.7416
105	-338.362	96.59258	351.8791	-.961586	.274505	246.9867
110	-346.682	93.96926	359.1916	-.965173	.2616132	249.1966
115	-354.742	90.63078	366.1362	-.968879	.247533	251.3487
120	-362.48	86.60254	372.6818	-.972626	.2323766	253.4223
125	-369.838	81.9152	378.8007	-.976338	.2162488	255.3981
130	-376.759	76.60444	384.4677	-.979949	.1992481	257.2588
135	-383.191	70.71068	389.6602	-.983397	.1814675	258.9886
140	-389.084	64.27876	394.3583	-.986627	.1629958	260.5733
145	-394.395	57.35764	398.5442	-.98959	.1439179	262.0004
150	-399.083	50	402.2025	-.992243	.1243155	263.259

53.66 contd

	H	I	J	K	L	M	N
1							
2							
3							
4	Omega1	Alpha1					
5	1	0					
6							
7	r2x	r2y	theta2	r3x	r3y	theta3	theta3
8			deg			deg	deg
9	-11.4181	199.6738	93.27283	223.8981	-199.674	-41.7268	318.2732
10	-2.65082	199.9824	90.75943	215.5114	-208.698	-44.0798	315.9202
11	7.099918	199.8739	87.9656	206.8993	-217.239	-46.3965	313.6035
12	17.57848	199.226	84.95763	198.3089	-225.108	-48.6215	311.3785
13	28.50626	197.9581	81.80565	190.0045	-232.16	-50.7024	309.2976
14	39.60593	196.0392	78.57824	182.2433	-238.301	-52.5927	307.4073
15	50.62334	193.4872	75.33798	175.2541	-243.487	-54.2549	305.7451
16	61.34337	190.3602	72.13852	169.2214	-247.718	-55.6622	304.3378
17	71.59817	186.745	69.02314	164.2774	-251.024	-56.7981	303.2019
18	81.26827	182.7443	66.02479	160.5011	-253.455	-57.6559	302.3441
19	90.27861	178.465	63.1669	157.9226	-255.069	-58.2369	301.7631
20	98.59144	174.0107	60.46486	156.5309	-255.926	-58.549	301.451
21	106.1981	169.4755	57.92758	156.2819	-256.078	-58.6047	301.3953
22	113.1113	164.9419	55.55907	157.1069	-255.573	-58.4199	301.5801
23	119.3576	160.4798	53.3598	158.9204	-254.449	-58.0125	301.9875
24	124.9727	156.1468	51.32785	161.6254	-252.739	-57.4013	302.5987
25	129.9961	151.9902	49.45986	165.1191	-250.471	-56.6058	303.3942
26	134.4689	148.0477	47.75173	169.2955	-247.667	-55.645	304.355
27	138.4308	144.3499	46.19913	174.0492	-244.35	-54.5379	305.4621
28	141.9193	140.9216	44.7979	179.2763	-240.541	-53.3027	306.6973
29	144.9683	137.7832	43.54435	184.8766	-236.264	-51.9568	308.0432
30	147.6076	134.9518	42.43544	190.7543	-231.544	-50.5171	309.4829
31	149.8631	132.4426	41.46887	196.8189	-226.412	-48.9997	311.0003
32	151.7561	130.2692	[illegible]	[illegible]	[illegible]	[illegible]	[illegible]
33	153.3036	128.4445	39.95776	209.1764	-215.047	-45.7928	314.2072
34	154.5182	126.9808	39.41286	215.3194	-208.896	-44.1325	315.8675
35	155.4083	125.8899	39.00952	221.3505	-202.494	-42.4527	317.5473
36	155.9781	125.1832	38.74947	227.2126	-195.894	-40.7666	319.2334
37	156.2278	124.8714	38.63501	232.8566	-189.15	-39.0871	320.9129
38	156.1539	124.9638	38.6689	238.2413	-182.321	-37.4261	322.5739
39	155.7493	125.4678	38.85405	243.3332	-175.468	-35.7954	324.2046
40	155.0035	126.388	39.1934	248.1073	-168.65	-34.2057	325.7943

S 3.66 contd.

	O	P	Q	R	S
1					
2					
3					
4					
5					
6					
7	Trans	r1x	r1y	Omega2	Omega3
8	angle				
9	45.00038	100	0	-.470633	-.470633
10	45.16074	99.61947	8.715574	-.532813	-.468801
11	45.63795	98.48078	17.36482	-.582491	-.455995
12	46.42083	96.59258	25.8819	-.61834	-.432271
13	47.49193	93.96926	34.20201	-.64015	-.398522
14	48.8291	90.63078	42.26183	-.648718	-.356324
15	50.40713	86.60254	50	-.645592	-.30767
16	52.19932	81.9152	57.35764	-.632751	-.254697
17	54.17878	76.60444	64.27876	-.612304	-.199447
18	56.31936	70.71068	70.71068	-.586261	-.143714
19	58.59623	64.27876	76.60444	-.556387	-.088961
20	60.98617	57.35764	81.9152	-.524137	-.036301
21	63.46771	50	86.60254	-.490645	.0134732
22	66.02101	42.26183	90.63078	-.456748	.0598415
23	68.62773	34.20201	93.96926	-.423029	.1025024
24	71.27084	25.8819	96.59258	-.389864	.1413174
25	73.93438	17.36482	98.48078	-.357468	.1762645
26	76.60325	8.715574	99.61947	-.32593	.2074002
27	79.26294	0	100	-.295251	.2348293
28	81.89941	-8.71557	99.61947	-.265364	.2586834
29	84.49883	-17.3648	98.48078	-.236156	.2791049
30	87.04744	-25.8819	96.59258	-.207486	.2962368
31	89.53143	-34.202	93.96926	-.179192	.3102159
32	91.93682	-42.2618	90.63078	-.151105	.32117
33	94.24939	-50	86.60254	-.123052	.3292169
34	96.45464	-57.3576	81.9152	-.094871	.3344653
35	98.53779	-64.2788	76.60444	-.066408	.3370183
36	100.4839	-70.7107	70.71068	-.037536	.3369772
37	102.2779	-76.6044	64.27876	-.008153	.3344469
38	103.905	-81.9152	57.35764	.0218046	.3295413
39	105.3506	-86.6025	50	.052356	.3223897

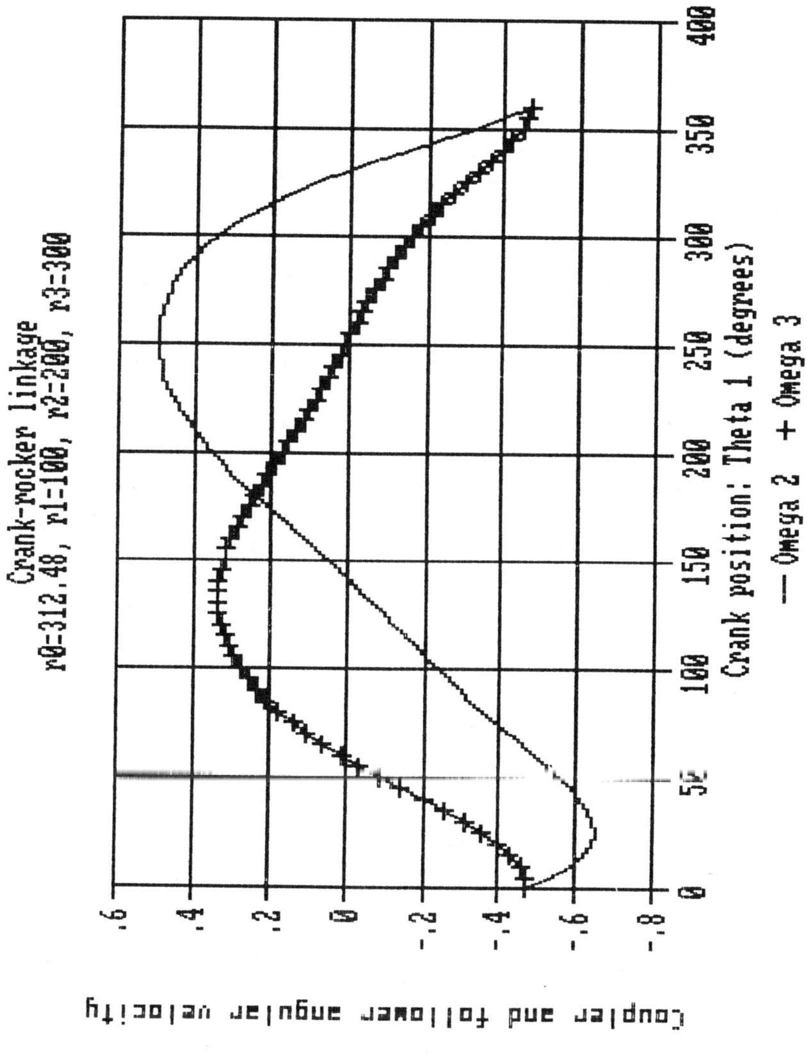

Crank-rocker linkage
r0=312.48, r1=100, r2=200, r3=300
.6
.4
.2
0
-.2
-.4
-.6
-.8
Coupler and follower angular velocity
0
50
100
150
200
250
300
350
400
Crank position: Theta 1 (degrees)
— Omega 2 + Omega 3

S 3.67

	A	B	C	D	E	F	G
1	Four-bar linkage.						
2	Position analysis based on the vector cross product method.						
3	If Rd-R2-R3 loop clockwise, config = 1; -1 if ccw.						
4	R0	deg	R1	R2	R3	config	
5	312.48	180	100	200	300	-1	
6							
7	Theta1	Rdx	Rdy	Rd	Rdux	Rduy	a
8	deg						
9	0	-212.48	0	212.48	-1	0	223.8981
10	5	-212.861	8.715574	213.0389	-.999163	.0409107	223.8689
11	10	-213.999	17.36482	214.7026	-.996724	.0808785	223.7914
12	15	-215.887	25.8819	217.4333	-.99289	.1190338	223.6944
13	20	-218.511	34.20201	221.1712	-.987971	.1546404	223.6202
14	25	-221.849	42.26183	225.8387	-.982335	.1871328	223.6178
15	30	-225.877	50	231.3453	-.976365	.2161272	223.7362
16	35	-230.565	57.35764	237.5921	-.970423	.2414122	224.0184
17	40	-235.876	64.27876	244.4771	-.964817	.2629235	224.4976
18	45	-241.769	70.71068	251.8976	-.959792	.280712	225.1955
19	50	-248.201	76.60444	259.7539	-.955525	.2949116	226.1219
20	55	-255.122	81.9152	267.9506	-.952125	.3057101	227.2761
21	60	-262.48	86.60254	276.3978	-.949646	.3133257	228.6482
22	65	-270.218	90.63078	285.0119	-.948094	.3179894	230.2216
23	70	-278.278	93.96926	293.7156	-.94744	.3199328	231.9742
24	75	-286.598	96.59258	302.4378	-.947627	.31938	233.8805
25	80	-295.115	98.48078	311.1132	-.948578	.3165432	235.9132
26	85	-303.764	99.61947	319.6824	-.950207	.3116201	238.0438
27	90	-312.48	100	328.0911	-.952418	.3047934	240.2439
28	95	-321.196	99.61947	336.2895	-.955116	.2962313	242.4855
29	100	-329.845	98.48078	344.2326	-.958203	.2860879	244.7416
30	105	-338.362	96.59258	351.8791	-.961586	.274505	246.9867
31	110	-346.682	93.96926	359.1916	-.965173	.2616132	249.1966
32	115	-354.742	90.63078	366.1362	-.968879	.247533	251.3487
33	120	-362.48	86.60254	372.6818	-.972626	.2323766	253.4223
34	125	-369.838	81.9152	378.8007	-.976338	.2162488	255.3981
35	130	-376.759	76.60444	384.4677	-.979949	.1992481	257.2588
36	135	-383.191	70.71068	389.6602	-.983397	.1814675	258.9886
37	140	-389.084	64.27876	394.3583	-.986627	.1629958	260.57[illegible]
38	145	-394.395	57.[illegible]	398.5442	-.98959	.1439179	262

S 3.67 cont'd

	H	I	J	K	L	M	N
1							
2							
3							
4	Omega1	Alpha1					
5	1	0					
6							
7	r2x	r2y	theta2	r3x	r3y	theta3	theta3
8			deg			deg	deg
9	-11.4181	-199.674	-93.2728	223.8981	199.6738	41.72679	41.72679
10	-18.9911	-199.096	-95.4488	231.8516	190.3807	39.3905	39.3905
11	-25.218	-198.404	-97.2437	239.2173	181.0389	37.11833	37.11833
12	-30.0117	-197.735	-98.6303	245.8991	171.8535	34.94886	34.94886
13	-33.3453	-197.201	-99.5975	251.856	162.9986	32.91053	32.91053
14	-35.2426	-196.87	-100.149	257.0918	154.6086	31.02165	31.02165
15	-35.7649	-196.776	-100.301	261.6424	146.7762	29.29156	29.29156
16	-34.9989	-196.914	-100.078	265.5637	139.5562	27.72236	27.72236
17	-33.0451	-197.251	-99.5103	268.9207	132.9724	26.31087	26.31087
18	-30.0113	-197.735	-98.6302	271.7806	127.0248	25.05044	25.05044
19	-26.0062	-198.302	-97.4714	274.2074	121.6975	23.9324	23.9324
20	-21.137	-198.88	-96.0666	276.2594	116.9647	22.94718	22.94718
21	-15.5078	-199.398	-94.4471	277.9878	112.7953	22.08517	22.08517
22	-9.21844	-199.787	-92.6418	279.4366	109.1567	21.33717	21.33717
23	-2.36493	-199.986	-90.6775	280.6429	106.0168	20.69475	20.69475
24	4.960691	-199.938	-88.5787	281.6374	103.3459	20.15044	20.15044
25	12.67005	-199.598	-86.3679	282.4451	101.1175	19.69775	19.69775
26	20.67825	-198.928	-84.0655	283.0862	99.30868	19.33124	19.33124
27	28.90372	-197.9	-81.6906	283.5763	97.90042	19.04646	19.04646
28	37.26822	-196.497	-79.2607	283.9273	96.87755	18.83992	18.83992
29	45.69691	-194.71	[illegible]	[illegible]	[illegible]	[illegible]	[illegible]
30	54.11848	-192.539	-74.3005	284.2434	95.94622	18.65208	18.65208
31	62.46542	-189.995	-71.8005	284.2166	96.02566	18.6681	18.6681
32	70.67433	-187.097	-69.3063	284.0675	96.46583	18.75685	18.75685
33	78.68626	-183.871	-66.8318	283.7937	97.26826	18.91878	18.91878
34	86.44715	-180.352	-64.3905	283.3905	98.43692	19.15489	19.15489
35	93.90817	-176.582	-61.9955	282.8506	99.97771	19.4667	19.4667
36	101.0262	-172.609	-59.66	282.1645	101.8979	19.85614	19.85614
37	107.7639	-168.484	-57.3967	281.3205	104.2055	20.32541	20.32541
38	114.0907	-164.266	-55.2181	280.3045	106.9083	20.87689	20.87689

53,67 cont'd

	O	P	Q	R	S
1					
2					
3					
4					
5					
6					
7	Trans	r1x	r1y	Omega2	Omega3
8	angle				
9	45.00038	100	0	-.470633	-.470633
10	45.16074	99.61947	8.715574	-.39828	-.462292
11	45.63795	98.48078	17.36482	-.318791	-.445286
12	46.42083	96.59258	25.8819	-.235485	-.421555
13	47.49193	93.96926	34.20201	-.151543	-.393171
14	48.8291	90.63078	42.26183	-.069681	-.362075
15	50.40713	86.60254	50	.0080226	-.329899
16	52.19932	81.9152	57.35764	.0801607	-.297893
17	54.17878	76.60444	64.27876	.14593	-.266927
18	56.31936	70.71068	70.71068	.2050088	-.237538
19	58.59623	64.27876	76.60444	.257424	-.210002
20	60.98617	57.35764	81.9152	.3034301	-.184407
21	63.46771	50	86.60254	.3434119	-.160706
22	66.02101	42.26183	90.63078	.3778141	-.138776
23	68.62773	34.20201	93.96926	.4070918	-.11844
24	71.27084	25.8819	96.59258	.4316804	-.099501
25	73.93438	17.36482	98.48078	.4519773	-.081755
26	76.60325	8.715574	99.61947	.4683332	-.064997
27	79.26294	0	100	.4810491	-.049031
28	81.89941	-8.71557	99.61947	.4903768	-.03367
29	84.49883	-17.3648	98.48078	.4965219	-.018739
30	87.04744	-25.8819	96.59258	.4996479	-.004075
31	89.53143	-34.202	93.96926	.4998816	.0104733
32	91.93682	-42.2618	90.63078	.4973188	.0250439
33	94.24939	-50	86.60254	.4920305	.0397611
34	96.45464	-57.3576	81.9152	.4840701	.0547342
35	98.53779	-64.2788	76.60444	.4734813	.0700546
36	100.4839	-70.7107	70.71068	.4603062	.0857929
37	102.2779	-76.6044	64.27876	.4445945	.1019946
38	103.905	-81.9152	57.35764	.4264128	.1186762
39	105.3506	-86.6025	50	.4058535	.1358199
40	106.6009	-90.6308	42.26183	.3830428	.1533701

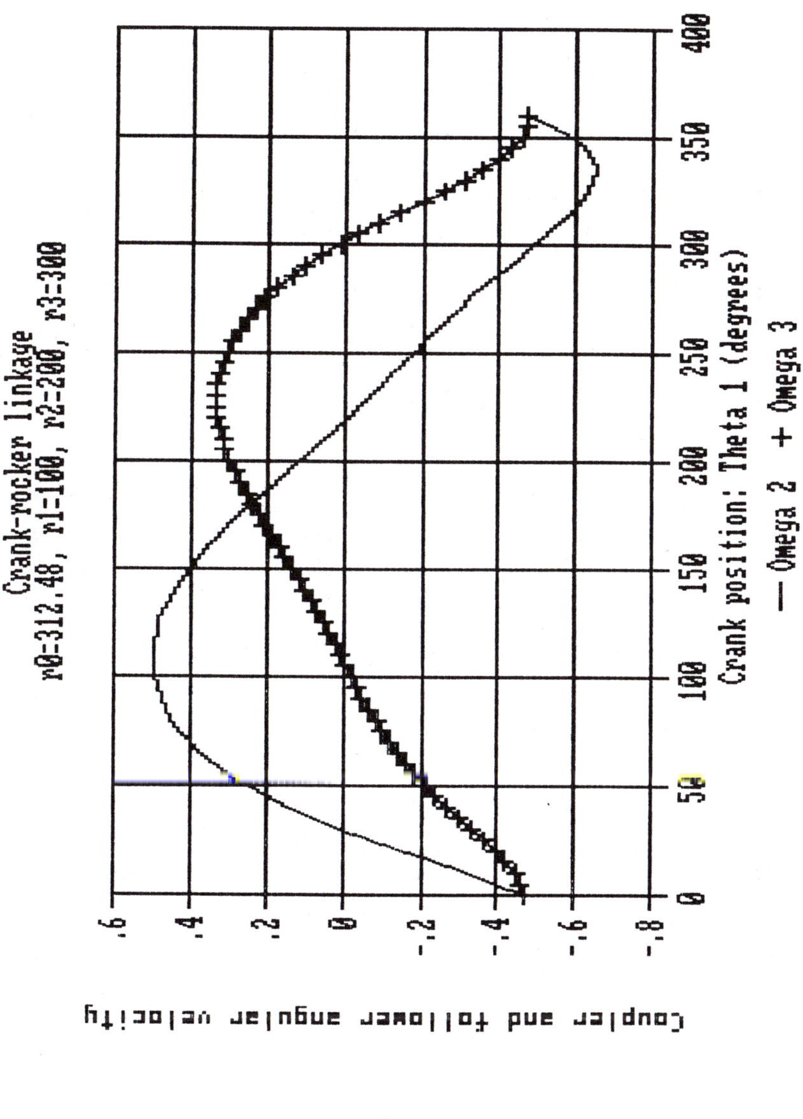
Crank-rocker linkage
r0=312.48, r1=100, r2=200, r3=300
Coupler and follower angular velocity
.6
.4
.2
0
-.2
-.4
-.6
-.8
0
50
100
150
200
250
300
350
400
Crank position: Theta 1 (degrees)
— Omega 2 + Omega 3

53,68

	A	B	C	D	E	F	G
1	Four-bar linkage.						
2	Position analysis based on the vector cross product method.						
3	If Rd-R2-R3 loop clockwise, config = 1; -1 if ccw.						
4	R0	deg	R1	R2	R3	config	
5	37	180	10	25	40	1	
6							
7	Theta1	Rdx	Rdy	Rd	Rdux	Rduy	a
8	deg						
9	0	-27.000	.000	27.000	-1.000	.000	31.556
10	5	-27.038	.872	27.052	-.999	.032	31.547
11	10	-27.152	1.736	27.207	-.998	.064	31.522
12	15	-27.341	2.588	27.463	-.996	.094	31.483
13	20	-27.603	3.420	27.814	-.992	.123	31.434
14	25	-27.937	4.226	28.255	-.989	.150	31.381
15	30	-28.340	5.000	28.777	-.985	.174	31.329
16	35	-28.808	5.736	29.374	-.981	.195	31.283
17	40	-29.340	6.428	30.035	-.977	.214	31.249
18	45	-29.929	7.071	30.753	-.973	.230	31.229
19	50	-30.572	7.660	31.517	-.970	.243	31.226
20	55	-31.264	8.192	32.320	-.967	.253	31.244
21	60	-32.000	8.660	33.151	-.965	.261	31.281
22	65	-32.774	9.063	34.004	-.964	.267	31.339
23	70	-33.580	9.397	34.870	-.963	.269	31.415
24	75	-34.412	9.659	35.742	-.963	.270	31.510
25	80	-35.264	9.848	36.613	-.963	.269	31.621
26	85	-36.128	9.962	37.477	-.964	.266	31.746
27	90	-37.000	10.000	38.328	-.965	.261	31.883
28	95	-37.872	9.962	39.160	-.967	.254	32.029
29	100	-38.736	9.848	39.969	-.969	.246	32.181
30	105	-39.588	9.659	40.750	-.971	.237	32.338
31	110	-40.420	9.397	41.498	-.974	.226	32.497
32	115	-41.226	9.063	42.211	-.977	.215	32.655
33	120	-42.000	8.660	42.884	-.979	.202	32.810
34	125	-42.736	8.192	43.514	-.982	.188	32.960
35	130	-43.428	7.660	44.098	-.985	.174	33.104
36	135	-44.071	7.071	44.635	-.987	.158	33.239
37	140	-44.660	6.428	45.121	-.990	.142	33.365
38	145	-45.192	5.736	45.554	-.992	.126	33.479
39	150	-45.660	5.000	45.933	-.994	.109	33.580
40	155	-46.063	4.226	46.257	-.996	.091	33.667
41	160	-46.397	3.420	46.523	-.997	.074	33.740
42	165	-46.659	2.588	46.731	-.998	.055	33.798
43	170	-46.848	1.736	46.880	-.999	.037	33.839
44	175	-46.962	.872	46.970	-1.000	.019	33.864
45	180	-47.000	.000	47.000	-1.000	.000	33.872
46	185	-46.962	-.872	46.970	-1.000	-.019	33.864
47	190	-46.848	-1.736	46.880	-.999	-.037	33.839
48	195	-46.659	-2.588	46.731	-.998	-.055	33.798
49	200	-46.397	-3.420	46.523	-.997	-.074	33.740
50	205	-46.063	-4.226	46.257	-.996	-.091	33.667
51	210	-45.660	-5.000	45.933	-.994	-.109	33.580
52	215	-45.192	-5.736	45.554	-.992	-.126	33.479
53	220	-44.660	-6.428	45.121	-.990	-.142	33.365
54	225	-44.071	-7.071	44.635	-.987	-.158	33.239
55	230	-43.428	-7.660	44.098	-.985	-.174	33.104
56	235	-42.736	-8.192	43.514	-.982	-.188	32.960

3.68 cont'd

	H	I	J	K	L	M	N
1							
2							
3							
4	Omega1	Alpha1	rBD	ThetaCBD			
5	100	0	15	20			
6							
7	r2x	r2y	theta2	r3x	r3y	theta3	theta3
8			deg			deg	deg
9	-4.556	24.581	100.499	31.556	-24.581	-37.918	322.082
10	-3.700	24.725	98.511	30.738	-25.596	-39.785	320.215
11	-2.734	24.850	96.278	29.886	-26.587	-41.657	318.343
12	-1.676	24.944	93.845	29.017	-27.532	-43.496	316.504
13	-.551	24.994	91.262	28.154	-28.414	-45.264	314.736
14	.619	24.992	88.582	27.318	-29.219	-46.925	313.075
15	1.808	24.935	85.852	26.532	-29.935	-48.449	311.551
16	2.995	24.820	83.120	25.814	-30.556	-49.809	310.191
17	4.159	24.652	80.424	25.181	-31.080	-50.986	309.014
18	5.284	24.435	77.797	24.645	-31.506	-51.967	308.033
19	6.358	24.178	75.266	24.214	-31.838	-52.746	307.254
20	7.371	23.889	72.851	23.893	-32.080	-53.322	306.678
21	8.318	23.576	70.566	23.682	-32.236	-53.697	306.303
22	9.194	23.248	68.422	23.580	-32.311	-53.879	306.121
23	9.999	22.913	66.424	23.581	-32.310	-53.877	306.123
24	10.733	22.579	64.576	23.679	-32.238	-53.703	306.297
25	11.397	22.251	62.880	23.867	-32.099	-53.368	306.632
26	11.993	21.936	61.334	24.136	-31.898	-52.887	307.113
27	12.524	21.637	59.938	24.476	-31.637	-52.272	307.728
28	12.992	21.359	58.689	24.880	-31.321	-51.538	308.462
29	13.401	21.105	57.586	25.336	-30.953	-50.699	309.301
30	13.752	20.878	56.627	25.836	-30.537	-49.767	310.233
31	14.049	20.679	55.808	26.371	-30.076	-48.755	311.245
32	14.293	20.511	55.129	26.933	-29.574	-47.676	312.324
33	14.487	20.375	54.586	27.513	-29.035	-46.542	313.458
34	14.631	20.271	54.179	28.105	-28.463	-45.363	314.637
35	14.727	20.202	53.907	28.700	-27.862	-44.151	315.849
36	14.777	20.166	53.767	29.294	-27.237	-42.915	317.085
37	14.779	20.164	53.760	29.881	-26.592	-41.666	318.334
38	14.736	20.196	53.884	30.456	-25.931	-40.412	319.588
39	14.646	20.261	54.138	31.014	-25.261	-39.162	320.838
40	14.510	20.358	54.522	31.553	-24.585	-37.924	322.076
41	14.328	20.487	55.033	32.069	-23.907	-36.704	323.296
42	14.099	20.645	55.671	32.561	-23.234	-35.510	324.490
43	13.822	20.831	56.434	33.026	-22.568	-34.346	325.654
44	13.499	21.042	57.320	33.463	-21.914	-33.220	326.780
45	13.128	21.276	58.325	33.872	-21.276	-32.134	327.866
46	[illegible]	[illegible]	[illegible]	[illegible]	-20.667	[illegible]	328.907
47	12.242	21.797	60.680	34.606	-20.061	-30.101	329.899
48	11.729	22.078	62.021	34.931	-19.490	-29.160	330.840
49	11.169	22.367	63.465	35.228	-18.946	-28.272	331.728
50	10.563	22.659	65.006	35.500	-18.433	-27.440	332.560
51	9.914	22.950	66.637	35.746	-17.950	-26.664	333.336
52	9.223	23.236	68.351	35.968	-17.501	-25.946	334.054
53	8.493	23.513	70.140	36.168	-17.085	-25.286	334.714
54	7.726	23.776	71.998	36.345	-16.705	-24.685	335.315
55	6.927	24.021	73.914	36.501	-16.361	-24.143	335.857
56	6.098	24.245	75.881	36.637	-16.053	-23.661	336.339
57	5.246	24.443	77.888	36.754	-15.783	-23.240	336.760
58	4.373	24.615	79.926	36.853	-15.551	-22.879	337.121
59	3.486	24.756	81.984	36.934	-15.359	-22.580	337.420
60	2.591	24.865	84.051	36.997	-15.206	-22.343	337.657

3,68 cont'd

	O	P	Q	R	S	T	U
1							
2							
3							
4							
5							
6							
7	Trans	r1x	r1y	Omega2	Omega3	rBDx	rBDy
8	angle						
9	41.583	10.000	.000	-37.037	-37.037	-7.613	12.925
10	41.704	9.962	.872	-42.355	-37.508	-7.160	13.181
11	42.066	9.848	1.736	-46.826	-37.236	-6.641	13.450
12	42.660	9.659	2.588	-50.327	-36.196	-6.064	13.720
13	43.474	9.397	3.420	-52.803	-34.410	-5.440	13.979
14	44.493	9.063	4.226	-54.259	-31.947	-4.780	14.218
15	45.699	8.660	5.000	-54.759	-28.909	-4.097	14.430
16	47.071	8.192	5.736	-54.406	-25.421	-3.405	14.608
17	48.590	7.660	6.428	-53.325	-21.615	-2.714	14.752
18	50.236	7.071	7.071	-51.653	-17.617	-2.035	14.861
19	51.988	6.428	7.660	-49.518	-13.544	-1.377	14.937
20	53.827	5.736	8.192	-47.040	-9.494	-.746	14.981
21	55.736	5.000	8.660	-44.318	-5.547	-.148	14.999
22	57.699	4.226	9.063	-41.438	-1.765	.413	14.994
23	59.699	3.420	9.397	-38.464	1.806	.936	14.971
24	61.721	2.588	9.659	-35.447	5.136	1.418	14.933
25	63.752	1.736	9.848	-32.421	8.206	1.859	14.884
26	65.780	.872	9.962	-29.413	11.004	2.260	14.829
27	67.790	.000	10.000	-26.438	13.527	2.621	14.769
28	69.772	-.872	9.962	-23.505	15.777	2.942	14.709
29	71.715	-1.736	9.848	-20.617	17.759	3.224	14.649
30	73.606	-2.588	9.659	-17.775	19.479	3.469	14.593
31	75.437	-3.420	9.397	-14.975	20.948	3.678	14.542
32	77.195	-4.226	9.063	-12.215	22.174	3.850	14.498
33	78.872	-5.000	8.660	-9.488	23.169	3.987	14.460
34	80.458	-5.736	8.192	-6.790	23.944	4.089	14.432
35	81.942	-6.428	7.660	-4.117	24.509	4.158	14.412
36	83.317	-7.071	7.071	-1.465	24.877	4.193	14.402
37	84.574	-7.660	6.428	1.168	25.058	4.195	14.401
38	85.704	-8.192	5.736	3.784	25.066	4.164	14.411
39	86.700	-8.660	5.000	6.380	24.911	4.100	14.429
40	87.555	-9.063	4.226	8.954	24.605	4.003	14.456
41	88.263	-9.397	3.420	11.502	24.163	3.874	14.491
42	88.819	-9.659	2.588	14.018	23.596	3.712	14.533
43	89.220	-9.848	1.736	16.492	22.917	3.518	14.582
44	89.461	-9.962	.872	18.915	22.140	3.293	14.634
45	89.542	-10.000	.000	21.277	21.277	3.036	14.690
46	89.461	-9.962	-.872	23.565	20.340	2.747	14.746
47	89.220	-9.848	-1.736	25.768	19.342	2.429	14.802
48	88.819	-9.659	-2.588	27.872	18.294	2.082	14.855
49	88.263	-9.397	-3.420	29.866	17.206	1.707	14.903
50	87.555	-9.063	-4.226	31.737	16.086	1.306	14.943
51	86.700	-8.660	-5.000	33.474	14.943	.880	14.974
52	85.704	-8.192	-5.736	35.065	13.783	.432	14.994
53	84.574	-7.660	-6.428	36.500	12.610	-.037	15.000
54	83.317	-7.071	-7.071	37.768	11.427	-.523	14.991
55	81.942	-6.428	-7.660	38.862	10.235	-1.024	14.965
56	80.458	-5.736	-8.192	39.770	9.036	-1.537	14.921
57	78.872	-5.000	-8.660	40.483	7.826	-2.059	14.858
58	77.195	-4.226	-9.063	40.992	6.603	-2.586	14.775
59	75.437	-3.420	-9.397	41.286	5.363	-3.115	14.673
60	73.606	-2.588	-9.659	41.353	4.099	-3.642	14.551

3.68 cont'd

	V	W	X	Y
1				
2				
3				
4				
5				
6				
7	vDx	vDy	vD	vD dir
8				deg
9	478.687	1281.959	1368.415	69.524
10	471.115	1299.453	1382.218	70.072
11	456.150	1295.771	1373.716	70.606
12	431.656	1271.108	1342.402	71.243
13	396.108	1226.917	1289.274	72.107
14	348.841	1165.657	1216.736	73.339
15	290.146	1090.394	1128.337	75.099
16	221.205	1004.395	1028.465	77.580
17	143.892	910.766	922.063	81.022
18	60.522	812.221	814.473	85.739
19	-26.407	710.963	711.453	92.127
20	-114.431	608.674	619.337	100.647
21	-201.281	506.571	545.095	111.670
22	-284.972	405.501	495.621	125.098
23	-363.853	306.035	475.443	139.933
24	-436.608	208.563	483.865	154.467
25	-502.239	113.367	514.875	167.280
26	-560.035	20.678	560.416	177.885
27	-609.529	-69.288	613.455	-173.515
28	-650.471	-156.305	668.987	-166.488
29	-682.782	-240.127	723.777	-160.624
30	-706.534	-320.486	775.824	-155.601
31	-721.918	-397.092	823.922	-151.187
32	-729.225	-469.641	867.371	-147.217
33	-728.825	-537.827	905.784	-143.575
34	-721.157	-601.344	938.979	-140.177
35	-706.708	-659.906	966.909	-136.961
36	-686.008	-713.250	989.612	-133.885
37	-659.613	-761.144	1007.188	-130.912
38	-628.100	-803.398	1019.783	-128.018
39	-592.054	-839.869	1027.574	-125.181
40	-552.062	-870.462	1030.765	-122.384
41	-508.703	-895.133	1029.583	-119.610
42	-462.541	-913.889	1024.274	-116.845
43	-414.120	-926.783	1015.097	-114.077
44	-363.959	-933.914	1002.328	-111.292
45	[illegible]	[illegible]	[illegible]	[illegible]
46	-260.339	-931.451	967.149	-105.616
47	-207.766	-922.210	945.324	-102.696
48	-155.219	-907.392	921.065	-99.702
49	-103.063	-888.707	894.663	-96.615
50	-51.635	-864.363	866.403	-93.417
51	-1.244	-836.567	836.568	-90.085
52	47.821	-804.013	805.434	-86.596
53	95.293	-767.386	773.280	-82.921
54	140.924	-726.857	740.392	-79.028
55	184.481	-682.581	707.071	-74.876
56	225.748	-634.699	673.651	-70.421
57	264.526	-583.337	640.513	-65.607
58	300.631	-528.607	608.115	-60.372
59	333.899	-470.605	577.025	-54.644
60	364.187	-409.415	547.953	-48.346
61	391.374	-345.103	521.794	-41.405
62	415.368	-277.722	499.660	-33.767

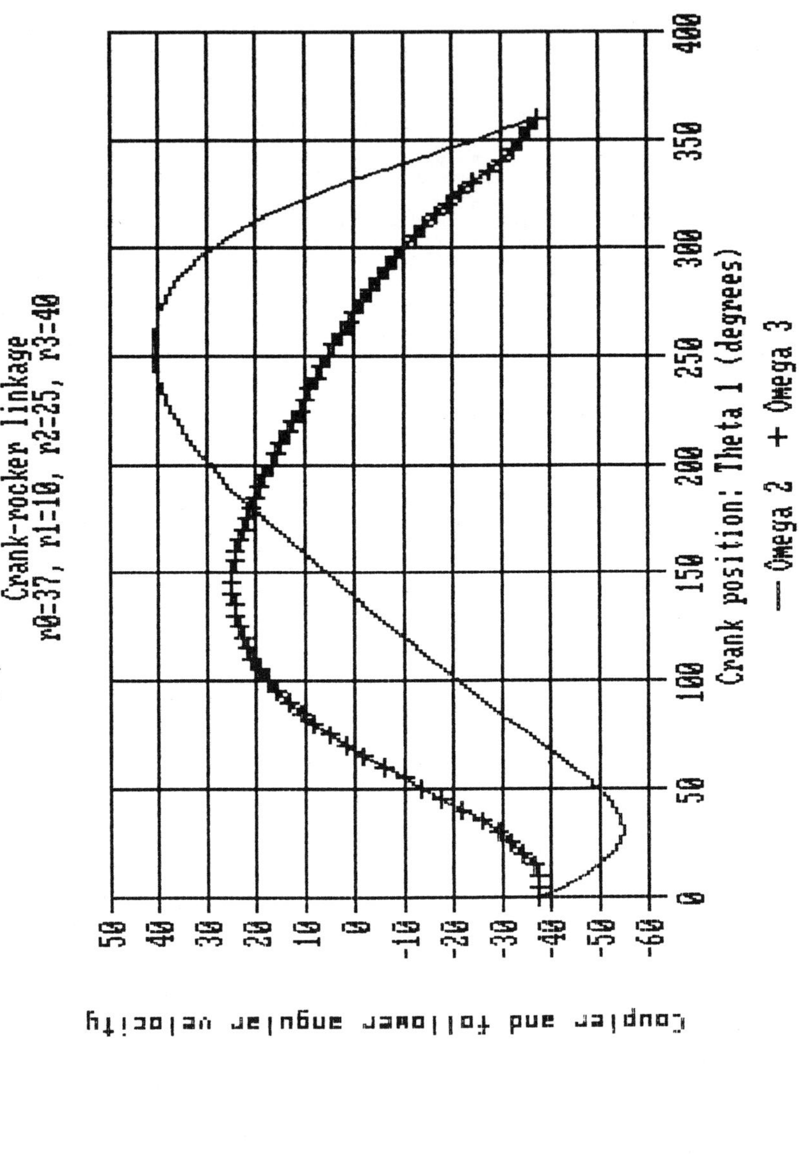
Crank-rocker linkage
r0=37, r1=10, r2=25, r3=40
Crank position: Theta 1 (degrees)
— Omega 2 + Omega 3
Coupler and follower angular velocity
50
40
30
20
10
0
-10
-20
-30
-40
-50
-60
0
50
100
150
200
250
300
350
400

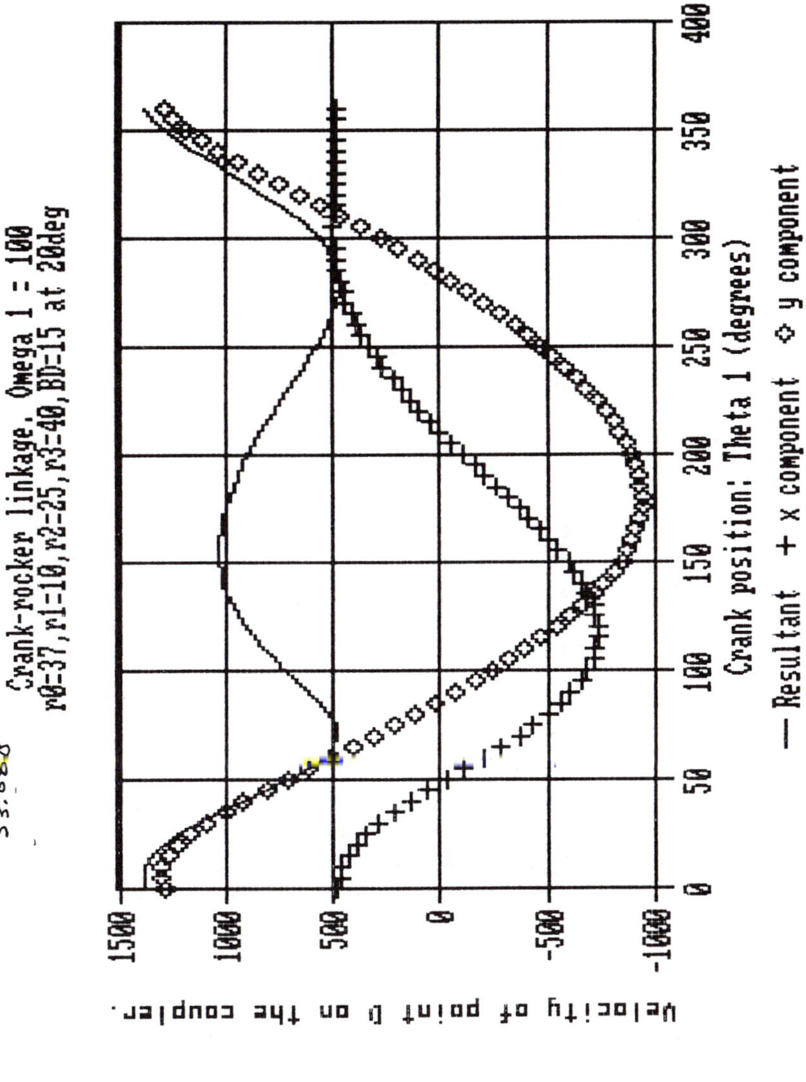
Crank-rocker linkage, Omega 1 = 100
r0=37,r1=10,r2=25,r3=40,BD=15 at 20deg
Velocity of point D on the coupler.
1500
1000
500
0
-500
-1000
0
50
100
150
200
250
300
350
400
Crank position: Theta 1 (degrees)
— Resultant + x component ◇ y component

S 3.69

	A	B	C	D	E	F	G
1	Four-bar linkage.						
2	Position analysis based on the vector cross product method.						
3	If Rd-R2-R3 loop clockwise, config = 1; -1 if ccw.						
4	R0	deg	R1	R2	R3	config	
5	38	180	13	27	41	1	
6							
7	Theta1	Rdx	Rdy	Rd	Rdux	Rduy	a
8	deg						
9	0	-25.000	.000	25.000	-1.000	.000	31.540
10	5	-25.049	1.133	25.075	-.999	.045	31.521
11	10	-25.197	2.257	25.298	-.996	.089	31.465
12	15	-25.443	3.365	25.664	-.991	.131	31.379
13	20	-25.784	4.446	26.165	-.985	.170	31.275
14	25	-26.218	5.494	26.787	-.979	.205	31.163
15	30	-26.742	6.500	27.520	-.972	.236	31.056
16	35	-27.351	7.456	28.349	-.965	.263	30.965
17	40	-28.041	8.356	29.260	-.958	.286	30.898
18	45	-28.808	9.192	30.239	-.953	.304	30.861
19	50	-29.644	9.959	31.272	-.948	.318	30.857
20	55	-30.544	10.649	32.347	-.944	.329	30.889
21	60	-31.500	11.258	33.451	-.942	.337	30.955
22	65	-32.506	11.782	34.575	-.940	.341	31.055
23	70	-33.554	12.216	35.708	-.940	.342	31.184
24	75	-34.635	12.557	36.841	-.940	.341	31.341
25	80	-35.743	12.803	37.966	-.941	.337	31.521
26	85	-36.867	12.951	39.075	-.943	.331	31.719
27	90	-38.000	13.000	40.162	-.946	.324	31.933
28	95	-39.133	12.951	41.220	-.949	.314	32.158
29	100	-40.257	12.803	42.244	-.953	.303	32.390
30	105	-41.365	12.557	43.229	-.957	.290	32.626
31	110	-42.446	12.216	44.169	-.961	.277	32.861
32	115	-43.494	11.782	45.062	-.965	.261	33.094
33	120	-44.500	11.258	45.902	-.969	.245	33.321
34	125	-45.456	10.649	46.687	-.974	.228	33.539
35	130	-46.356	9.959	47.414	-.978	.210	33.746
36	135	-47.192	9.192	48.079	-.982	.191	33.940
37	140	-47.959	8.356	48.681	-.985	.172	34.118

53.69 cont'd

	H	I	J	K	L	M	N
1							
2							
3							
4	Omega1	Alpha1	rBD	ThetaCBD			
5	50	0	12	20			
6							
7	r2x	r2y	theta2	r3x	r3y	theta3	theta3
8			deg			deg	deg
9	-6.540	26.196	104.018	31.540	-26.196	-39.712	320.288
10	-5.254	26.484	101.221	30.304	-27.617	-42.344	317.656
11	-3.796	26.732	98.082	28.994	-28.989	-44.996	315.004
12	-2.206	26.910	94.686	27.649	-30.274	-47.595	312.405
13	-.531	26.995	91.126	26.315	-31.441	-50.072	309.928
14	1.182	26.974	87.492	25.036	-32.468	-52.364	307.636
15	2.886	26.845	83.864	23.856	-33.345	-54.420	305.580
16	4.544	26.615	80.311	22.807	-34.071	-56.202	303.798
17	6.127	26.296	76.884	21.915	-34.652	-57.690	302.310
18	7.613	25.904	73.623	21.195	-35.097	-58.873	301.127
19	8.990	25.459	70.551	20.654	-35.418	-59.752	300.248
20	10.252	24.978	67.684	20.291	-35.627	-60.336	299.664
21	11.399	24.476	65.028	20.101	-35.734	-60.641	299.359
22	12.432	23.968	62.584	20.074	-35.750	-60.685	299.315
23	13.357	23.465	60.349	20.197	-35.681	-60.489	299.511
24	14.181	22.976	58.317	20.455	-35.533	-60.073	299.927
25	14.910	22.510	56.482	20.833	-35.313	-59.461	300.539
26	15.550	22.072	54.835	21.317	-35.023	-58.673	301.327
27	16.110	21.667	53.369	21.890	-34.667	-57.730	302.270
28	16.594	21.299	52.077	22.539	-34.249	-56.652	303.348
29	17.009	20.969	50.953	23.248	-33.771	-55.456	304.544
30	[illegible]	[illegible]	[illegible]	[illegible]	[illegible]	[illegible]	[illegible]
31	17.648	20.434	49.185	24.798	-32.650	-52.783	307.217
32	17.879	20.232	48.532	25.615	-32.014	-51.336	308.664
33	18.056	20.074	48.029	26.444	-31.333	-49.837	310.163
34	18.180	19.962	47.674	27.276	-30.611	-48.297	311.703
35	18.253	19.895	47.464	28.103	-29.854	-46.730	313.270
36	18.276	19.874	47.398	28.916	-29.066	-45.149	314.851
37	18.249	19.899	47.476	29.709	-28.255	-43.563	316.437
38	18.172	19.969	47.697	30.477	-27.426	-41.984	318.016
39	18.045	20.084	48.062	31.213	-26.584	-40.421	319.579

S 3.69 cont'd

	O	P	Q	R	S	T	U
1							
2							
3							
4							
5							
6							
7	Trans	r1x	r1y	Omega2	Omega3	rBDx	rBDy
8	angle						
9	36.270	13.000	.000	-26.000	-26.000	-6.713	9.946
10	36.435	12.951	1.133	-29.811	-26.537	-6.220	10.262
11	36.922	12.803	2.257	-32.826	-26.376	-5.649	10.587
12	37.718	12.557	3.365	-34.935	-25.495	-5.012	10.903
13	38.802	12.216	4.446	-36.118	-23.940	-4.325	11.193
14	40.144	11.782	5.494	-36.436	-21.810	-3.607	11.445
15	41.716	11.258	6.500	-36.006	-19.241	-2.875	11.650
16	43.487	10.649	7.456	-34.974	-16.377	-2.148	11.806
17	45.426	9.959	8.356	-33.492	-13.358	-1.438	11.913
18	47.505	9.192	9.192	-31.698	-10.300	-.758	11.976
19	49.697	8.356	9.959	-29.710	-7.297	-.115	11.999
20	51.980	7.456	10.649	-27.619	-4.419	.485	11.990
21	54.331	6.500	11.258	-25.496	-1.710	1.040	11.955
22	56.731	5.494	11.782	-23.387	.799	1.549	11.900
23	59.162	4.446	12.216	-21.324	3.095	2.012	11.830
24	61.609	3.365	12.557	-19.326	5.173	2.430	11.751
25	64.057	2.257	12.803	-17.401	7.035	2.805	11.668
26	66.492	1.133	12.951	-15.552	8.687	3.139	11.582
27	68.901	.000	13.000	-13.777	10.139	3.435	11.498
28	71.271	-1.133	12.951	-12.070	11.400	3.693	11.418
29	73.591	-2.257	12.803	-10.425	12.482	3.916	11.343
30	75.849	-3.365	12.557	-8.832	13.395	4.106	11.276
31	78.033	-4.446	12.216	-7.284	14.149	4.264	11.217
32	80.131	-5.494	11.782	-5.772	14.753	4.392	11.167
33	82.134	-6.500	11.258	-4.288	15.218	4.490	11.129
34	84.029	-7.456	10.649	-2.925	15.552	4.559	11.100
35	85.806	-8.356	9.959	-1.377	15.762	4.599	11.084
36	87.453	-9.192	9.192	.062	15.855	4.612	11.078
37	88.961	-9.959	8.356	1.496	15.841	4.597	11.085
38	90.319	-10.649	7.456	2.927	15.725	4.554	11.102
39	91.518	-11.258	6.500	4.356	15.516	4.483	11.131
40	92.548	-11.782	5.494	5.782	15.221	4.385	11.170

S 3.69 cont'd

	V	W	X	Y
1				
2				
3				
4				
5				
6				
7	vDx	vDy	vD	vD dir
8				deg
9	258.606	824.548	864.151	72.587
10	249.272	832.955	869.454	73.340
11	234.664	825.552	858.256	74.132
12	212.672	802.938	830.626	75.165
13	181.975	767.014	788.306	76.653
14	142.315	720.518	734.439	78.827
15	94.484	666.448	673.112	81.931
16	40.086	607.566	608.887	86.225
17	-18.810	546.101	546.425	91.973
18	-80.006	483.653	490.225	99.393
19	-141.429	421.240	444.349	108.559
20	-201.286	359.431	411.955	119.249
21	-258.117	298.484	394.610	130.852
22	-310.804	238.480	391.755	142.501
23	-358.534	179.416	400.920	153.416
24	-400.750	121.274	418.698	163.163
25	-437.101	64.061	441.771	171.662
26	-467.402	7.829	467.468	179.040
27	-491.593	-47.317	493.865	-174.502
28	-509.713	-101.225	519.667	-168.768
29	-521.878	-153.696	544.039	-163.590
[illegible]	[illegible]	[illegible]	[illegible]	[illegible]
31	-529.095	-253.375	586.635	-154.411
32	-524.638	-300.052	604.381	-150.234
33	-515.193	-344.253	619.624	-146.249
34	-501.086	-385.704	632.341	-142.413
35	-482.669	-424.144	642.548	-138.693
36	-460.312	-459.331	650.286	-135.061
37	-434.398	-491.051	655.616	-131.497
38	-405.323	-519.119	658.613	-127.982
39	-373.487	-543.387	659.365	-124.502
40	-339.291	-563.747	657.973	-121.042
41	-303.[illegible]	-580.129	654.552	-117.5[illegible]8

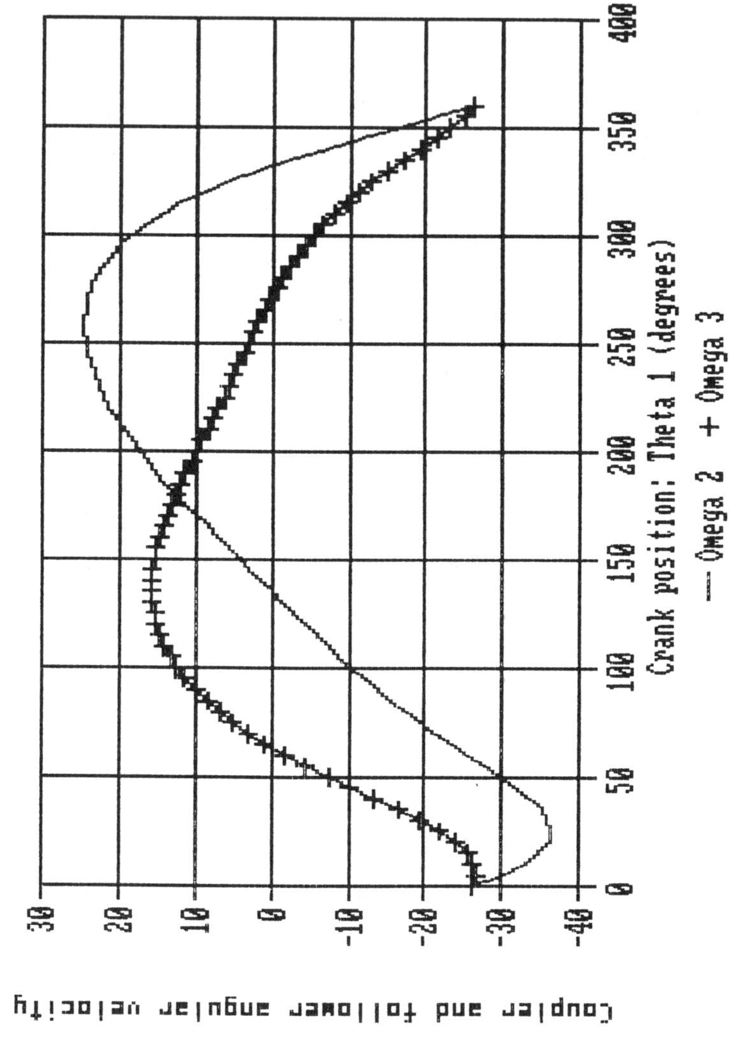
Crank-rocker linkage, Omega 1 = 50
r0=38, r1=13, r2=27, r3=41
Coupler and follower angular velocity
30
20
10
0
-10
-20
-30
-40
0
50
100
150
200
250
300
350
400
Crank position: Theta 1 (degrees)
— Omega 2 + Omega 3

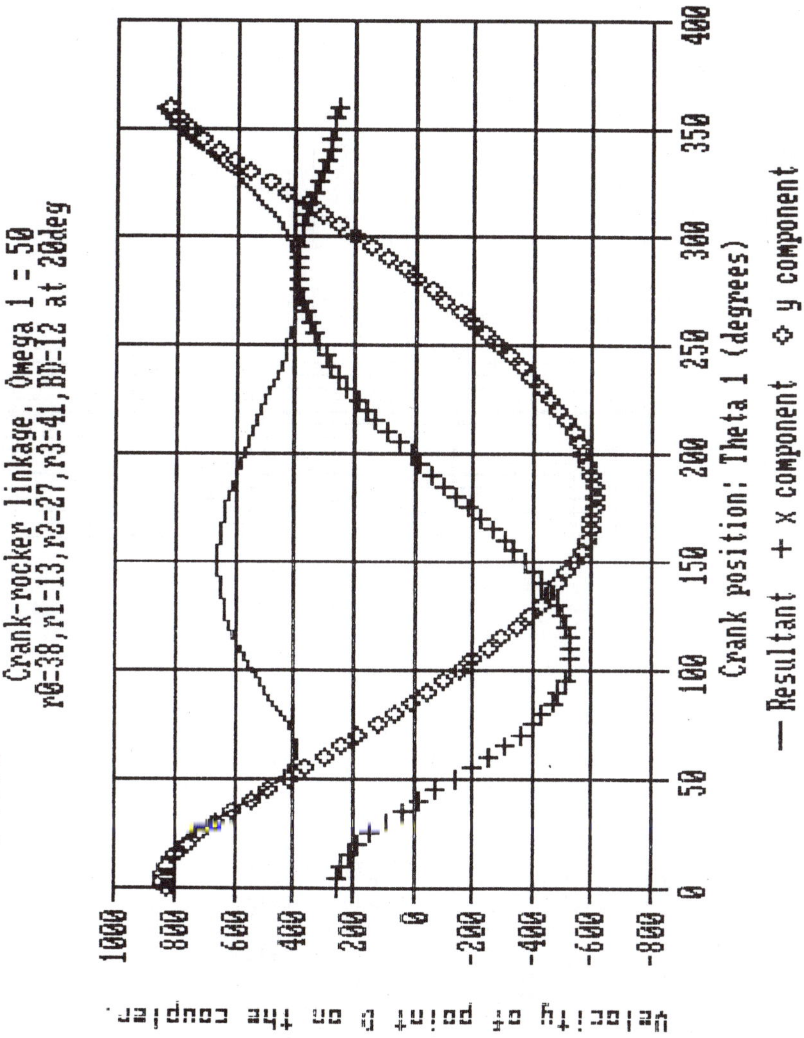

Crank-rocker linkage, Omega 1 = 50
r0=38,r1=13,r2=27,r3=41,BD=12 at 20deg
Crank position: Theta 1 (degrees)
— Resultant + x component ◇ y component
Velocity of point D on the coupler.

3.70

Slider crank linkage $\quad c_o := 1 \quad$ (clockwise configuration positive)

$$\theta_1 := 0, \frac{\pi}{9} .. 2\cdot\pi \qquad R_1 := 1 \qquad R_2 := 2.5 \qquad e := \begin{bmatrix} 0 \\ -0.4 \\ 0 \end{bmatrix} \qquad r_{0u} := \begin{bmatrix} -1 \\ 0 \\ 0 \end{bmatrix}$$

$$r_1\left[\theta_1\right] := \begin{bmatrix} R_1\cdot\cos\left[\theta_1\right] \\ R_1\cdot\sin\left[\theta_1\right] \\ 0 \end{bmatrix} \qquad i := \begin{bmatrix} 1 \\ 0 \\ 0 \end{bmatrix} \quad j := \begin{bmatrix} 0 \\ 1 \\ 0 \end{bmatrix} \quad k := \begin{bmatrix} 0 \\ 0 \\ 1 \end{bmatrix}$$

$$r_3\left[\theta_1\right] := r_1\left[\theta_1\right] + e$$

$$r_0\left[\theta_1\right] := \left[-r_3\left[\theta_1\right]\cdot r_{0u} + c_o\cdot\left[{R_2}^2 - \left[r_3\left[\theta_1\right]\cdot\left[r_{0u}\times k\right]\right]^2\right]^{.5}\right]\cdot r_{0u}$$

$$a\left[\theta_1\right] := -\left[r_3\left[\theta_1\right]\cdot\left[r_{0u}\times k\right]\right]\cdot\left[r_{0u}\times k\right]$$

$$r_2\left[\theta_1\right] := a\left[\theta_1\right] - c_o\cdot\left[{R_2}^2 - \left[r_3\left[\theta_1\right]\cdot\left[r_{0u}\times k\right]\right]^2\right]^{.5}\cdot r_{0u}$$

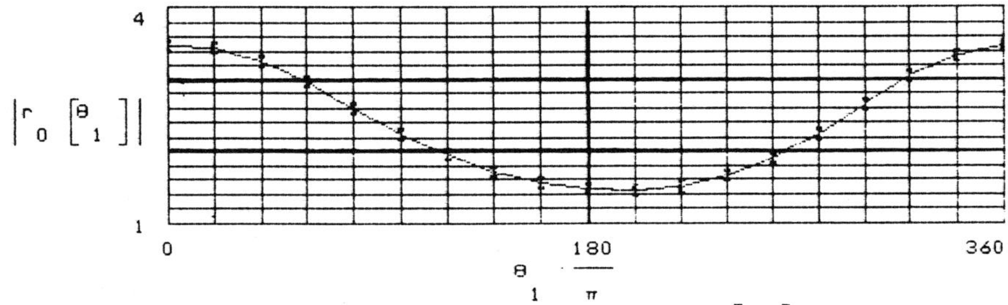

$$\omega_1 := 1$$

$$\omega_2\left[\theta_1\right] := -\omega_1\cdot\frac{r_1\left[\theta_1\right]\cdot i}{r_2\left[\theta_1\right]\cdot i}$$

$$v_c\left[\theta_1\right] := \omega_1\cdot k\times r_1\left[\theta_1\right] + \omega_2\left[\theta_1\right]\cdot k\times r_2\left[\theta_1\right]_0 \qquad V_c\left[\theta_1\right] := v_c\left[\theta_1\right]$$

$\theta_1\cdot\frac{180}{\pi}$	$\left|r_0\left[\theta_1\right]\right|$	$\omega_2\left[\theta_1\right]$	$V_c\left[\theta_1\right]$
0	3.468	-0.405	0.162
20	3.439	-0.376	-0.32
40	3.254	-0.308	-0.718
60	2.956	-0.204	-0.961
80	2.604	-0.071	-1.027
100	2.257	0.071	-0.943
120	1.956	0.204	-0.771
140	1.722	0.308	-0.568
160	1.56	0.376	-0.364
180	1.468	0.405	-0.162
200	1.448	0.394	0.05

3.70 cont'd

220	1.506	0.337	0.291
240	1.656	0.232	0.572
260	1.908	0.083	0.869
280	2.255	-0.083	1.1
300	2.656	-0.232	1.16
320	3.038	-0.337	0.994
340	3.327	-0.394	0.634
360	3.468	-0.405	0.162

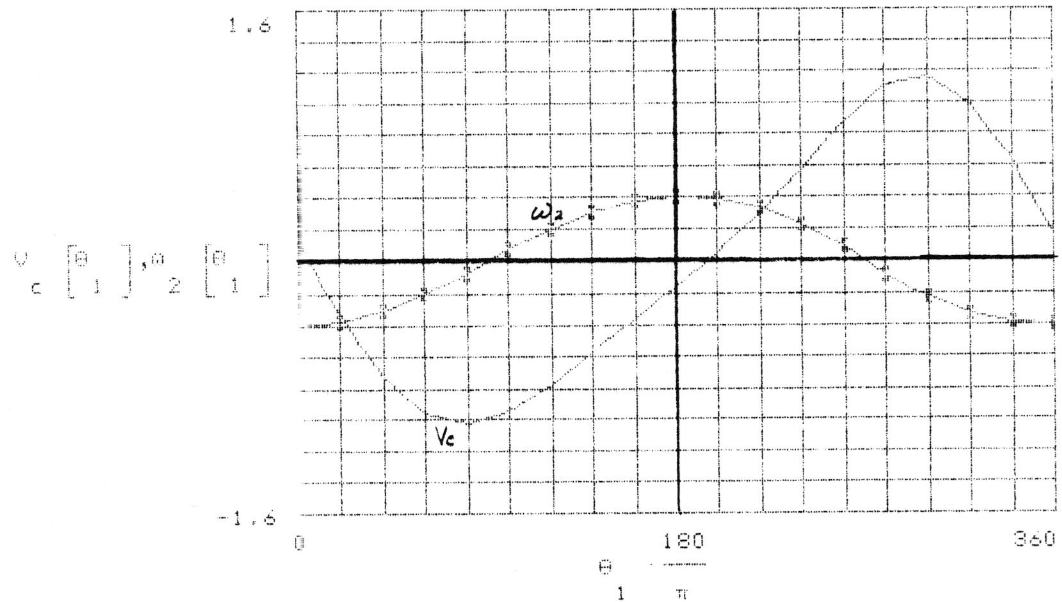

3.71

Slider crank linkage $\quad c_o := 1 \quad$ (clockwise configuration positive)

$$\theta_1 := 0, \frac{\pi}{9} .. 2\cdot\pi \qquad R_1 := 1 \qquad R_2 := 2 \qquad e := \begin{bmatrix} 0 \\ -0.5 \\ 0 \end{bmatrix} \qquad r_{0u} := \begin{bmatrix} -1 \\ 0 \\ 0 \end{bmatrix}$$

$$r_1\left[\theta_1\right] := \begin{bmatrix} R_1\cdot\cos\left[\theta_1\right] \\ R_1\cdot\sin\left[\theta_1\right] \\ 0 \end{bmatrix} \qquad i := \begin{bmatrix} 1 \\ 0 \\ 0 \end{bmatrix} \quad j := \begin{bmatrix} 0 \\ 1 \\ 0 \end{bmatrix} \quad k := \begin{bmatrix} 0 \\ 0 \\ 1 \end{bmatrix}$$

$$r_3\left[\theta_1\right] := r_1\left[\theta_1\right] + e$$

$$r_0\left[\theta_1\right] := \left[-r_3\left[\theta_1\right]\cdot r_{0u} + c_o\cdot\left[R_2^{\,2} - \left[r_3\left[\theta_1\right]\cdot\left[r_{0u} \times k\right]\right]^2\right]^{.5}\right]\cdot r_{0u}$$

$$a\left[\theta_1\right] := -\left[r_3\left[\theta_1\right]\cdot\left[r_{0u} \times k\right]\right]\cdot\left[r_{0u} \times k\right]$$

$$r_2\left[\theta_1\right] := a\left[\theta_1\right] - c_o\cdot\left[R_2^{\,2} - \left[r_3\left[\theta_1\right]\cdot\left[r_{0u} \times k\right]\right]^2\right]^{.5}\cdot r_{0u}$$

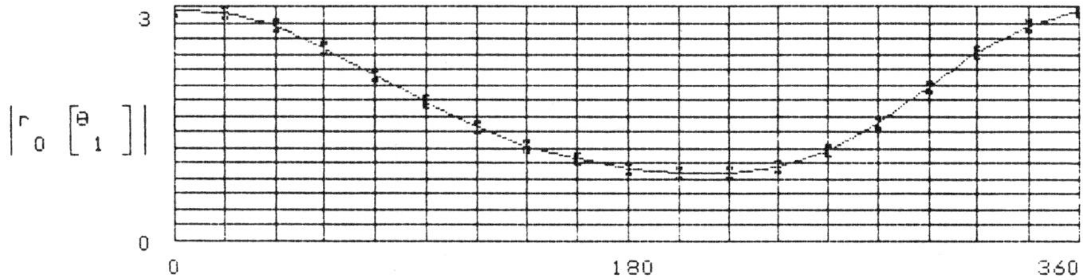

$$\omega_1 := 1 \qquad \omega_2\left[\theta_1\right] := -\omega_1\cdot\frac{r_1\left[\theta_1\right]\cdot i}{r_2\left[\theta_1\right]\cdot i}$$

$$v_c\left[\theta_1\right] := \omega_1\cdot k \times r_1\left[\theta_1\right] + \omega_2\left[\theta_1\right]\cdot k \times r_2\left[\theta_1\right]_0 \qquad V_c\left[\theta_1\right] := v_c\left[\theta_1\right]_0$$

$\theta_1\cdot\frac{180}{\pi}$	$\lvert r_0\left[\theta_1\right]\rvert$	$\omega_2\left[\theta_1\right]$	$V_c\left[\theta_1\right]$
0	2.936	-0.516	0.258
20	2.933	-0.471	-0.268
40	2.761	-0.384	-0.698
60	2.466	-0.254	-0.959
80	2.114	-0.089	-1.028
100	1.767	0.089	-0.941
120	1.466	0.254	-0.773
140	1.229	0.384	-0.588
160	1.054	0.471	-0.416
180	0.936	0.516	-0.258
200	0.874	0.518	-0.094

3.71 Cont'd

220	0.875	0.467	0.109
240	0.961	0.342	0.398
260	[illegible].166	0.13	0.792
280	[illegible].514	-0.13	1.177
300	[illegible].961	-0.342	1.334
320	[illegible].407	-0.467	1.176
340	[illegible].754	-0.518	0.778
360	[illegible].936	-0.516	0.258

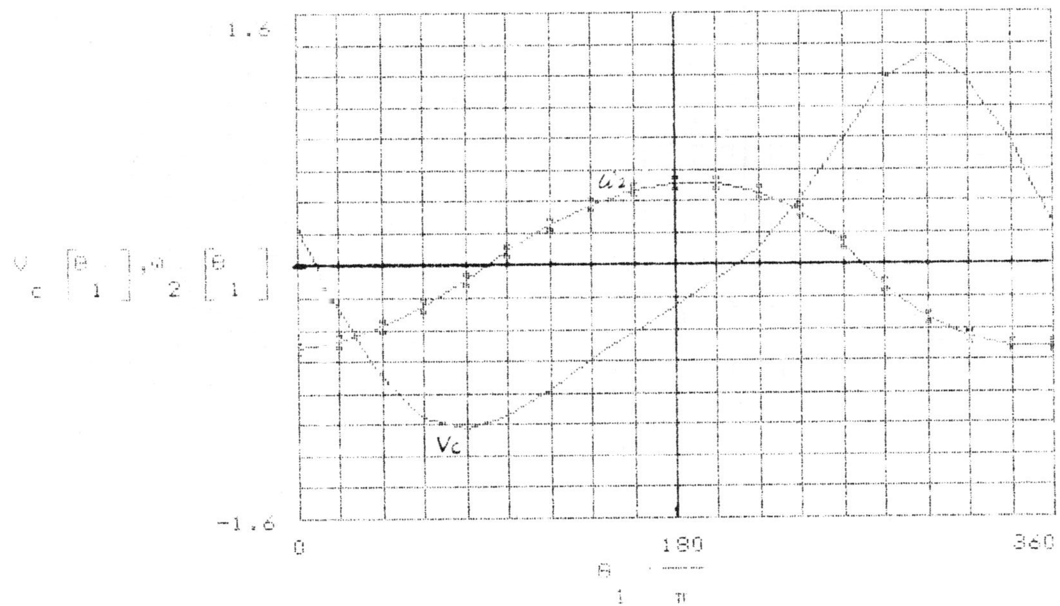

3.72

Slider crank linkage $\quad c_o := 1 \quad$ (clockwise configuration positive)

$$\theta_1 := 0, \frac{\pi}{9} .. 2\cdot\pi \qquad R_1 := 1 \qquad R_2 := 3 \qquad e := \begin{bmatrix} 0 \\ -0.7 \\ 0 \end{bmatrix} \qquad r_{0u} := \begin{bmatrix} -1 \\ 0 \\ 0 \end{bmatrix}$$

$$r_1\left[\theta_1\right] := \begin{bmatrix} R_1\cdot\cos\left[\theta_1\right] \\ R_1\cdot\sin\left[\theta_1\right] \\ 0 \end{bmatrix} \qquad i := \begin{bmatrix} 1 \\ 0 \\ 0 \end{bmatrix} \quad j := \begin{bmatrix} 0 \\ 1 \\ 0 \end{bmatrix} \quad k := \begin{bmatrix} 0 \\ 0 \\ 1 \end{bmatrix}$$

$$r_3\left[\theta_1\right] := r_1\left[\theta_1\right] + e$$

$$r_0\left[\theta_1\right] := \left[-r_3\left[\theta_1\right]\cdot r_{0u} + c_o\cdot\left[R_2^{\,2} - \left[r_3\left[\theta_1\right]\cdot\left[r_{0u}\times k\right]\right]^2\right]^{.5}\right]\cdot r_{0u}$$

$$a\left[\theta_1\right] := -\left[r_3\left[\theta_1\right]\cdot\left[r_{0u}\times k\right]\right]\cdot\left[r_{0u}\times k\right]$$

$$r_2\left[\theta_1\right] := a\left[\theta_1\right] - c_o\cdot\left[R_2^{\,2} - \left[r_3\left[\theta_1\right]\cdot\left[r_{0u}\times k\right]\right]^2\right]^{.5}\cdot r_{0u}$$

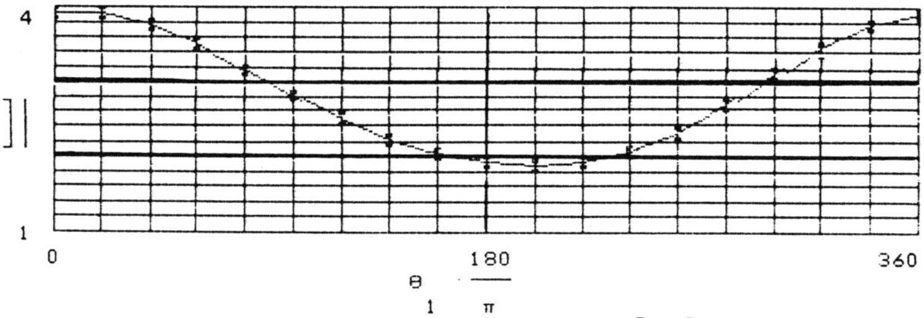

$$\omega_1 := 1 \qquad \omega_2\left[\theta_1\right] := -\omega_1\cdot\frac{r_1\left[\theta_1\right]\cdot i}{r_2\left[\theta_1\right]\cdot i}$$

$$v_c\left[\theta_1\right] := \omega_1\cdot k\times r_1\left[\theta_1\right] + \omega_2\left[\theta_1\right]\cdot k\times r_2\left[\theta_1\right], \qquad V_c\left[\theta_1\right] := v_c\left[\theta_1\right]\cdot i$$

$\theta_1\cdot\frac{180}{\pi}$	$\left|r_0\left[\theta_1\right]\right|$	$\omega_2\left[\theta_1\right]$	$V_c\left[\theta_1\right]$
0	3.917	-0.343	0.24
20	3.918	-0.315	-0.229
40	3.765	-0.255	-0.628
60	3.495	-0.167	-0.894
80	3.16	-0.058	-1.001
100	2.813	0.058	-0.968
120	2.495	0.167	-0.838
140	2.233	0.255	-0.657
160	2.039	0.315	-0.455
180	1.917	0.343	-0.24
200	1.874	0.334	-0.006

3.72 cont'd

220	1.917	0.286	0.259
240	2.059	0.195	0.56
260	2.309	0.07	0.867
280	2.656	-0.07	1.103
300	3.059	-0.195	1.172
320	3.449	-0.286	1.026
340	3.753	-0.334	0.69
360	3.917	-0.343	0.24

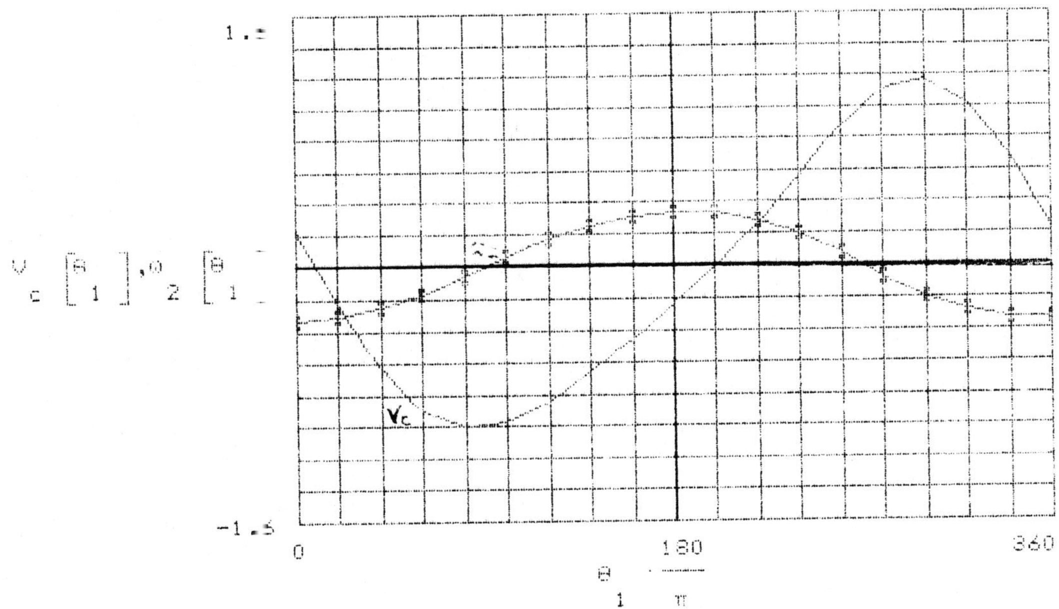

3.73 The vector components representing the link positions are determined as in chapter 2. Then equations 3.42 and 3.43 result in four simultaneous scalar equations which may be written in the matrix form

$A\ X = B$

where X is a column matrix of the four unknowns. Using mathematics software with matrix capability, the solution is given directly by

$X = A^{-1}\ B$

The equations and plotted results are shown in figure S3.73. The tabulated results are also given in that figure. For this linkage, the results are symmetric and antisymmetric about $\theta = \pi$ rad (180°).

3.74 The vector components representing the link positions are determined as in chapter 2. Then equations 3.42 and 3.43 result in four simultaneous scalar equations which may be written in the matrix form $A\ X = B$ where X is a column matrix of the four unknowns. Using mathematics software with matrix capability, the solution is given directly by $X = A^{-1}\ B$ The equations and plotted results are shown in figure S3.74. The tabulated results are also given in that figure. For this linkage, the results are symmetric and antisymmetric about $\theta = \pi$ rad (180°).

53.73

RSSR linkage. Link 1 rotates in xy plane; link 3 rotates in xz plane

$r_0 := 4 \quad r_1 := 1 \quad r_2 := 3.5 \quad r_3 := 2.5 \quad \theta := 0, \frac{\pi}{6} .. 2\cdot\pi$

$k := \begin{bmatrix} 0 \\ 0 \\ 1 \end{bmatrix} \quad \omega_1 := k \quad \alpha_1 := 0$

$c_x(\theta) := -r_0 + r_1\cdot\cos(\theta) \quad c_y(\theta) := r_1\cdot\sin(\theta) \quad r_{1x}(\theta) := r_1\cdot\cos(\theta)$

$r_{1y}(\theta) := r_1\cdot\sin(\theta)$

$$r_{2x}(\theta) := \frac{-r_2^{\,2} + c_y(\theta)^2 + r_3^{\,2} - c_x(\theta)^2}{2\cdot c_x(\theta)} \qquad r_{3x}(\theta) := -c_x(\theta) - r_{2x}(\theta)$$

$$r_{2y}(\theta) := -c_y(\theta) \qquad r_{2z}(\theta) := \left[r_3^{\,2} - r_{3x}(\theta)^2\right]^{.5} \qquad r_{3z}(\theta) := -r_{2z}(\theta)$$

$$\theta_3(\theta) := \text{angle}\left[r_{3z}(\theta), r_{3x}(\theta)\right]$$

$$A(\theta) := \begin{bmatrix} 0 & r_{2z}(\theta) & -r_{2y}(\theta) & r_{3z}(\theta) \\ -r_{2z}(\theta) & 0 & r_{2x}(\theta) & 0 \\ r_{2y}(\theta) & -r_{2x}(\theta) & 0 & -r_{3x}(\theta) \\ r_{2x}(\theta) & r_{2y}(\theta) & r_{2z}(\theta) & 0 \end{bmatrix} \qquad B(\theta) := \left|\omega_1\right|\cdot\begin{bmatrix} r_{1y}(\theta) \\ -r_{1x}(\theta) \\ 0 \\ 0 \end{bmatrix}$$

$$X(\theta) := A(\theta)^{-1}\cdot B(\theta)$$

$$I := \begin{bmatrix} 1 \\ 0 \\ 0 \\ 0 \end{bmatrix} \quad J := \begin{bmatrix} 0 \\ 1 \\ 0 \\ 0 \end{bmatrix} \quad K := \begin{bmatrix} 0 \\ 0 \\ 1 \\ 0 \end{bmatrix} \quad L := \begin{bmatrix} 0 \\ 0 \\ 0 \\ 1 \end{bmatrix}$$

$\omega_{2x}(\theta) := X(\theta)\cdot I \quad \omega_{2y}(\theta) := X(\theta)\cdot J \quad \omega_{2z}(\theta) := X(\theta)\cdot K$

$\omega_3(\theta) := X(\theta)\cdot L$

53.73 cont'd

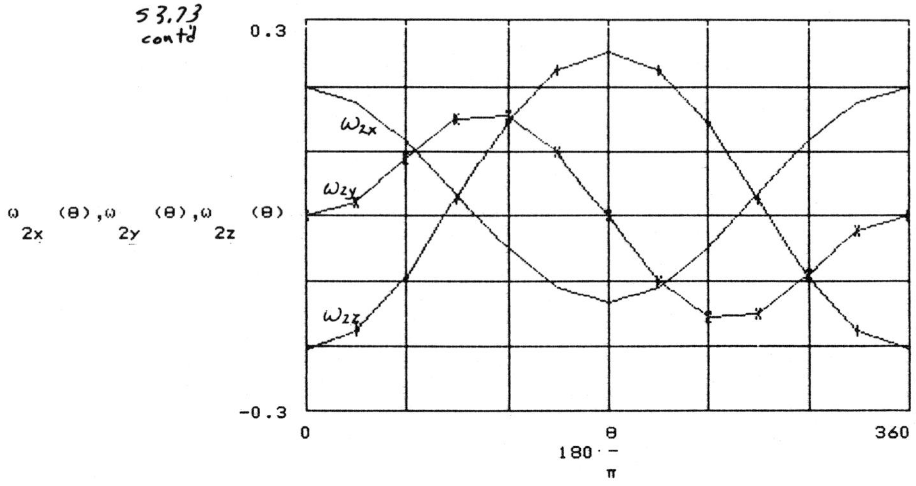

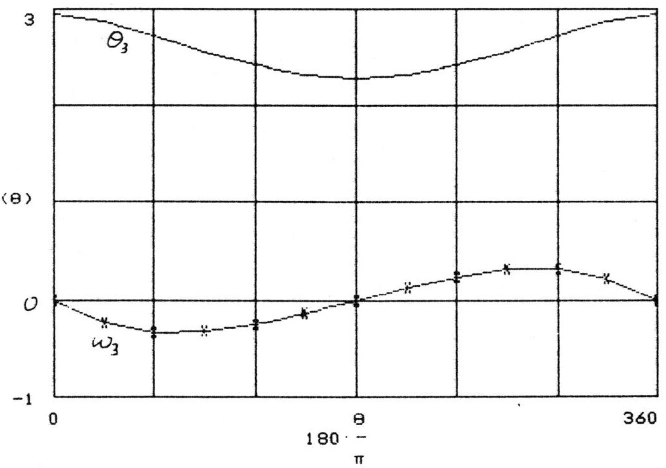

$$\omega_2(\theta) := \left[\omega_{2x}(\theta)^2 + \omega_{2y}(\theta)^2 + \omega_{2z}(\theta)^2\right]^{.5}$$

$180 \cdot \frac{\theta}{\pi}$	$\omega_2(\theta)$	$\theta_3(\theta)$	$\omega_3(\theta)$
0	0.286	2.94	0
30	0.251	2.879	-0.221
60	0.174	2.73	-0.324
90	0.158	2.559	-0.314

53.74

RSSR linkage. Link 1 rotates in xy plane; link 3 rotates in xz plane

$r_0 := 3.2 \quad r_1 := 1 \quad r_2 := 2.8 \quad r_3 := 2 \qquad \theta := 0, \frac{\pi}{6} .. 2\cdot\pi$

$k := \begin{bmatrix} 0 \\ 0 \\ 1 \end{bmatrix} \qquad \omega_1 := k \qquad \alpha_1 := 0$

$c_x(\theta) := -r_0 + r_1 \cdot \cos(\theta) \qquad c_y(\theta) := r_1 \cdot \sin(\theta) \qquad r_{1x}(\theta) := r_1 \cdot \cos(\theta)$

$r_{1y}(\theta) := r_1 \cdot \sin(\theta)$

$$r_{2x}(\theta) := \frac{-r_2^{\,2} + c_y(\theta)^2 + r_3^{\,2} - c_x(\theta)^2}{2\cdot c_x(\theta)} \qquad r_{3x}(\theta) := -c_x(\theta) - r_{2x}(\theta)$$

$r_{2y}(\theta) := -c_y(\theta) \qquad r_{2z}(\theta) := \left[r_3^{\,2} - r_{3x}(\theta)^2\right]^{.5} \qquad r_{3z}(\theta) := -r_{2z}(\theta)$

$\theta_3(\theta) := \operatorname{angle}\left[r_{3z}(\theta), r_{3x}(\theta)\right]$

$$A(\theta) := \begin{bmatrix} 0 & r_{2z}(\theta) & -r_{2y}(\theta) & r_{3z}(\theta) \\ -r_{2z}(\theta) & 0 & r_{2x}(\theta) & 0 \\ r_{2y}(\theta) & -r_{2x}(\theta) & 0 & -r_{3x}(\theta) \\ r_{2x}(\theta) & r_{2y}(\theta) & r_{2z}(\theta) & 0 \end{bmatrix} \qquad B(\theta) := \left|\omega_1\right| \cdot \begin{bmatrix} r_{1y}(\theta) \\ -r_{1x}(\theta) \\ 0 \\ 0 \end{bmatrix}$$

$X(\theta) := A(\theta)^{-1} \cdot B(\theta)$

$I := \begin{bmatrix} 1 \\ 0 \\ 0 \\ 0 \end{bmatrix} \quad J := \begin{bmatrix} 0 \\ 1 \\ 0 \\ 0 \end{bmatrix} \quad K := \begin{bmatrix} 0 \\ 0 \\ 1 \\ 0 \end{bmatrix} \quad L := \begin{bmatrix} 0 \\ 0 \\ 0 \\ 1 \end{bmatrix}$

$\omega_{2x}(\theta) := X(\theta)\cdot I \qquad \omega_{2y}(\theta) := X(\theta)\cdot J \qquad \omega_{2z}(\theta) := X(\theta)\cdot K$

$\omega_3(\theta) := X(\theta)\cdot L$

$$\omega_2(\theta) := \left[\omega_{2x}(\theta)^2 + \omega_{2y}(\theta)^2 + \omega_{2z}(\theta)^2\right]^{.5}$$

5.3.74 cont'd

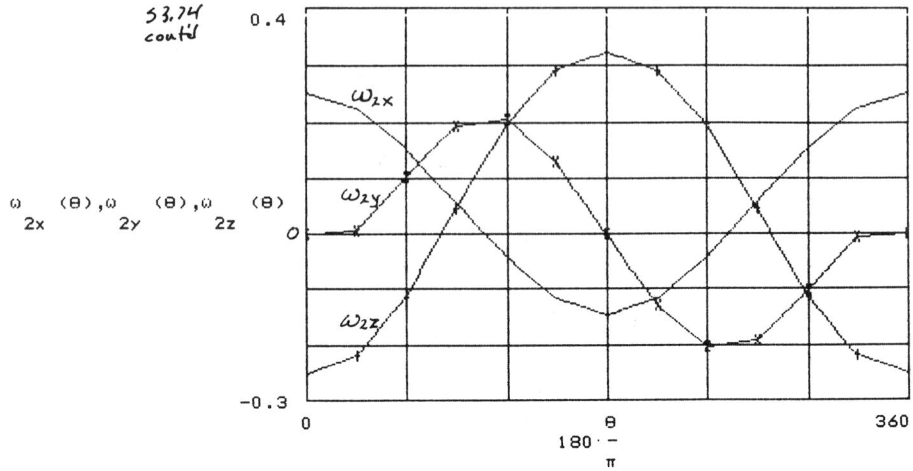

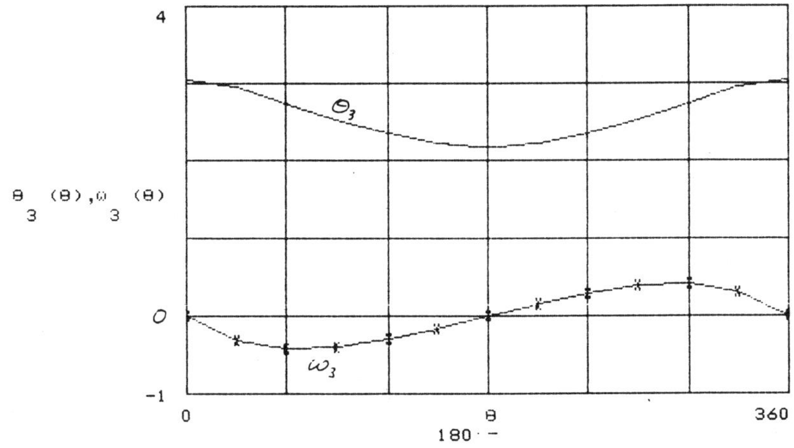

$180 \cdot \frac{\theta}{\pi}$	$\omega_2(\theta)$	$\theta_3(\theta)$	$\omega_3(\theta)$
0	0.357	3.028	0
30	0.314	2.941	-0.306
60	0.216	2.742	-0.422
90	0.207	2.525	-0.391
120	0.285	2.343	-0.296
150	0.34	2.221	-0.163
180	0.357	2.178	0
210	0.34	2.221	0.163
240	0.285	2.343	0.296

CHAPTER 4

4.1 The acceleration of a point in a rigid body rotating about a fixed point is

$a = \underset{\sim}{\alpha X R} + \underset{\sim}{\omega} X(\underset{\sim}{\omega X R})$ where

$$\underset{\sim}{\alpha X R} = \begin{vmatrix} \underset{\sim}{i} & \underset{\sim}{j} & \underset{\sim}{k} \\ -20 & -40 & 80 \\ 50 & 100 & -30 \end{vmatrix} = \underset{\sim}{i}6800 - \underset{\sim}{j}4600 + \underset{\sim}{k}0$$

$$\underset{\sim}{\omega X R} = \begin{vmatrix} \underset{\sim}{i} & \underset{\sim}{j} & \underset{\sim}{k} \\ 10 & -20 & 30 \\ 50 & 100 & -30 \end{vmatrix} = -\underset{\sim}{i}3600 + \underset{\sim}{j}1800 + \underset{\sim}{k}2000$$

$$\underset{\sim}{\omega} X(\underset{\sim}{\omega X R}) = \begin{vmatrix} \underset{\sim}{i} & \underset{\sim}{j} & \underset{\sim}{k} \\ 10 & -20 & 30 \\ -3600 & 1800 & 2000 \end{vmatrix}$$

$$= -\underset{\sim}{i}94000 - \underset{\sim}{j}128000 - \underset{\sim}{k}54000$$

$\underset{\sim}{a} = -\underset{\sim}{i}100{,}800 - \underset{\sim}{j}132{,}600 - \underset{\sim}{k}54{,}000$ (ans)

4.2 The acceleration of a point in a rigid body rotating about a fixed point is

$a = \underset{\sim}{\alpha X R} + \underset{\sim}{\omega} X(\underset{\sim}{\omega X R})$ where

$$\underset{\sim}{\alpha X R} = \begin{vmatrix} \underset{\sim}{i} & \underset{\sim}{j} & \underset{\sim}{k} \\ -20 & -40 & 80 \\ -40 & -80 & 100 \end{vmatrix}$$

$$= \underset{\sim}{i}2400 - \underset{\sim}{j}1200 + \underset{\sim}{k}0$$

$$\underset{\sim}{\omega X R} = \begin{vmatrix} \underset{\sim}{i} & \underset{\sim}{j} & \underset{\sim}{k} \\ 10 & -20 & 30 \\ -40 & -80 & 100 \end{vmatrix}$$

$$= \underset{\sim}{i}400 - \underset{\sim}{j}2200 - \underset{\sim}{k}1600$$

$$\underset{\sim}{\omega} X(\underset{\sim}{\omega X R}) = \begin{vmatrix} i & j & k \\ 10 & -20 & 30 \\ 400 & -2200 & -1600 \end{vmatrix}$$

$$= \underset{\sim}{i}9800 + \underset{\sim}{j}28000 - \underset{\sim}{k}14000$$

$\underset{\sim}{a} = \underset{\sim}{i}100{,}400 + \underset{\sim}{j}26{,}800 - \underset{\sim}{k}14{,}000$

4.3 $a_{(avg)} = \dfrac{140-120 \text{ in/sec}}{.01 \text{ sec}} = 2000 \text{ in/sec}^2$

4.4 $(1\ mi/hr)(5280\ ft/mi)(12\ in/ft)(1\ hr/3600\ sec) = 17.6\ in/sec$

$a_{(avg)} = (0 - 1000\ mi/hr)(17.6\ \frac{in/sec}{mi/hr}\ /\ (.05\ sec) = -352{,}000 in/sec^2$

4.5 a) $\alpha_{(avg)} = \frac{(1000 - 0\ rev/min)(2\pi\ rad/rev)}{(20\ sec)(60\ sec/min)} = 5.236\ rad/sec^2$

b) $-2.07\ rad/sec^2$

4.6 $\underset{\sim}{a}^n = \underset{\sim}{\omega} X (\underset{\sim}{\omega} X \underset{\sim}{R}) = V^2/R$

$= (50\ km/hr \times 1000\ m/km \times 1\ hr/3600\ s)^2/12m$

$\underset{\sim}{a}^n = 16.075\ m/s^2$ toward the center of the curve (see note, prob. 4.7).

4.7 $\underset{\sim}{a}^n = (50 \times 1000/3600)^2/10$

$= 19.29\ m/s^2$ toward the center of the curve. Note: if an automobile on an unbanked roadway exceeds a normal acceleration of 1/2 g ($4.9\ m/s^2$) or so, it may slide sideways or become unstable.

4.8 Tangential and normal accelerations are:

$\underset{\sim}{a}_B{}^t = \dot{\underset{\sim}{\omega}} X \underset{\sim}{R} = \alpha(OB) = 30 \times 125 = 3750\ mm/s^2$

$\perp$ OB (upward and to the left)

$\underset{\sim}{a}_B{}^n = \underset{\sim}{\omega} X (\underset{\sim}{\omega} X \underset{\sim}{R}) = \omega^2(OB) = 24^2 \times 125 = 72000 mm/s^2$

$||$OB toward O

$a_B = 72098\ mm/s^2$ downward and 3° to the left of OB. ($\gamma = 93^o$)

4.9 $\underset{\sim}{a}_B{}^t = \alpha(OB) = 100 \times 125 = 12500\ mm/s^2$

$\perp$ OB (downward and to the right)

$\underset{\sim}{a}_B{}^n = 72000\ mm/s^2$ $||$OB toward O

$\underset{\sim}{a}_B = 73077\ mm/s^2$ approximately downward in the direction of BO rotated 9.85^o cc. ($\gamma = 80.15^o$)

4.10 The location of P is given by:

$\underset{\sim}{R} = \underset{\sim}{R}_o + \underset{\sim}{r} = \underset{\sim}{i}40 + \underset{\sim}{i}20 + \underset{\sim}{j}60 = \underset{\sim}{i}60 + \underset{\sim}{j}60$ (mm)

Velocity is given by:

$$\dot{\underset{\sim}{R}} = \dot{\underset{\sim}{R}}_o + \dot{\underset{\sim}{r}}_r + \underset{\sim}{\omega} X \underset{\sim}{r} = \underset{\sim}{i}10 + \underset{\sim}{i}5 + \underset{\sim}{j}15 + \begin{vmatrix} \underset{\sim}{i} & \underset{\sim}{j} & \underset{\sim}{k} \\ 0 & 0 & 150 \\ 20 & 60 & 0 \end{vmatrix}$$

$= -\underset{\sim}{i}8985 + \underset{\sim}{j}3015$ (mm/s)

4.10 (Cont'd.)

The acceleration of P is given by:

$$\ddot{R} = \ddot{R}_o + \dot{\omega} X r + \omega X(\omega X r) + \ddot{r}_r + 2\omega X \dot{r}_r$$

where: $$\dot{\omega} X r = \begin{vmatrix} i & j & k \\ 0 & 0 & 80 \\ 20 & 60 & 0 \end{vmatrix} = -i4800 + j1600$$

For planar motion:

$$\omega X(\omega X r) = -\omega^2 r = -150^2(i20 + j60)$$

$$= -i450{,}000 - j135{,}0000$$

$$2\omega X \dot{r}_r = 2 \begin{vmatrix} i & j & k \\ 0 & 0 & 150 \\ 5 & 15 & 0 \end{vmatrix} = -i4500 + j1500$$

Thus,

$$\ddot{R} = -i20 - i4800 + j1600 - i450000 - j1350000 + i10$$

$$+ j30 - i4500 + j1500$$

$$\ddot{R} = -\ i459{,}310 - j1{,}346{,}870$$

4.11 Using results from the above problem

$$R = R_o + r = i20 + j60$$

$$\dot{R} = \dot{R}_o + \dot{r}_r + \omega X r \qquad \text{where } \omega X r = \begin{vmatrix} i & j & k \\ 0 & 0 & 150 \\ 20 & 60 & 0 \end{vmatrix}$$

$$= k10 + i5 + j15 \qquad\qquad = -i9000 + j3000$$

$$-i9000 + j3000$$

$$\dot{R} = -i8995 + j3015 + k10$$

$$\ddot{R} = \ddot{R}_o + \dot{\omega} X r + \omega X(\omega X r) + \ddot{r}_r + 2\omega X \dot{r}_r$$

$$= k20 \quad i1000 + j1600 \quad i450000 - j1350000$$

$$+ i10 + j30 - i4500 + j1500$$

$$\ddot{R} = -i459{,}290 - j1{,}346{,}870 + k20$$

4.12 $a_D = i36{,}863 - j73{,}838$ mm/s^2. See microcomputer results.

4.13 See microcomputer results. In this case, point D is the same as point C.

$\underset{\sim}{a}_C = \underset{\sim}{i}31,381 - \underset{\sim}{j}48,131$ mm/s^2.

$r_3 = 20\angle{-84.67^\circ} = \underset{\sim}{i}1.86 - \underset{\sim}{j}19.91$

Alternatively, $\underset{\sim}{a}_C = \underset{\sim}{\alpha}_3 X(-\underset{\sim}{r}_3) + \underset{\sim}{\omega}_3 X(\underset{\sim}{\omega}_3 X(-\underset{\sim}{r}_3))$

$$\underset{\sim}{a}_C = \begin{vmatrix} \underset{\sim}{i} & \underset{\sim}{j} & \underset{\sim}{k} \\ 0 & 0 & \alpha_3 \\ -r_{3x} & -r_{3y} & 0 \end{vmatrix} - \omega_3^2(-\underset{\sim}{r}_3)$$

$$= \alpha_3(\underset{\sim}{i}r_{3y} - \underset{\sim}{j}r_{3x}) + \omega_3^2(\underset{\sim}{i}r_{3x} + \underset{\sim}{j}r_{3y})$$

$$= -1338.7(-\underset{\sim}{i}19.91 - \underset{\sim}{j}1.86) + 50.417^2(\underset{\sim}{i}1.86 - \underset{\sim}{j}19.91)$$

$a_C = \underset{\sim}{i}31,381 - \underset{\sim}{j}48,119$

4.14 See microcomputer results where $\omega_1 = 1$.

At $\theta_1 = 15^\circ$, $\theta_2 = 22.56^\circ$, $\theta_3 = -126.81^\circ$,

$\omega_3/\omega_1 = -0.1290$, $\alpha_3/\omega_1^2 = 1.357$ etc.

4.15 See microcomputer results. This is a drag link mechanism.

At $\theta_1 = 15^\circ$, $\theta_2 = -77.6^\circ$, $\theta_3 = 159.6^\circ$,

$\omega_3/\omega_1 = 1.388$, $\alpha_3/\omega_1^2 = -0.1482$ etc.

4.16 and 4.17 See program listings and plots. T1 represents crank angle and A Represents slider acceleration $\div\ \omega_1^2R$

4.18 See sketch. $\alpha_2 = 32$ rad/sec^2 cc.

4.19 See sketch. $\alpha_2 = 46$ rad/sec^2 cc.

4.20 See sketch. $\alpha_2 = 9$ rad/sec^2 cw.

4.21 $a_D = 23,000$ in/sec^2 as shown. $\alpha_2 = 2230$ rad/sec^2 cc.

4.22 $a_D = 25,600$ in/sec^2 as shown. $\alpha_2 = 900$ rad/sec^2 cc.

4.23 $a_D = 5040$ in/sec^2 as shown. $\alpha_2 = 147$ and $\alpha_3 = 267$ rad/sec^2, both cc.

4.24 $a_D = 4900$ in/sec^2 as shown. $\alpha_2 = 110$ rad/sec^2 cc and $\alpha_3 = 110$ rad/sec^2 cw.

4.25 See sketch. $\alpha_2 = 138$ and $\alpha_3 = 440$ rad/sec^2, both cc.

4.26 See sketch. $\alpha_2 = 20$ rad/sec^2 cc and $\alpha_3 = 155$ rad/sec^2 cw.

4.27 See sketch. $\alpha_1 = 63$ rad/sec^2 cc.

4.28 See sketch. $\alpha_1 = 500$ rad/sec^2 cc.

4.12 THETA 2= 18.754902 DEGREES
THETA 3= -84.6683618 DEGREES

VELOCITY ANALYSIS BY
CROSS PRODUCT METHOD

OMEGA 2= 12.2596363 RADIANS/SECOND

OMEGA 3= 50.4174164 RADIANS/SECOND

THE VELOCITY OF POINT D IS

-981.141586 (I)
-356.593345 (J)

OR IN POLAR FORM

1043.93373 <-160.026457

ALPHA 2= 1211.87016 RADIANS/SECOND**2

ALPHA 3= -1338.66707 RADIANS/SECOND**2

THE ACCELERATION OF POINT D IS

36862.6039 (I)
-73838.0156 (J)

OR IN POLAR FORM

82528.2021 <-63.4699738

4.13 THETA 2= 18.754902 DEGREES
THETA 3= -84.6683618 DEGREES

VELOCITY ANALYSIS BY
CROSS PRODUCT METHOD

OMEGA 2= 12.2596363 RADIANS/SECOND

OMEGA 3= 50.4174164 RADIANS/SECOND

THE VELOCITY OF POINT D IS

-1003.98575 (I)
-93.6961347 (J)

OR IN POLAR FORM

1008.34833 <-174.668362

4.13 continued

```
ALPHA 2= 1211.87016 RADIANS/SECOND**2


ALPHA 3= -1338.66707 RADIANS/SECOND**2

THE ACCELERATION OF POINT D IS

31381.4247 (I)
-48130.5779 (J)

OR IN POLAR FORM

57457.3437 <-56.8954093
```

4.14

```
THETA 1            THETA 2             THETA 3          (DEGREES)
     OMEGA 1            OMEGA 2             OMEGA 3          (RADIANS/S)
          ALPHA 1            ALPHA 2             ALPHA 3
                                                     (RADIANS/S**2)


15                 22.555547           -126.805616

     1                  -.346659199         -.129004052

          0                  .624840614          1.3571598



30                 18.510494           -126.333008

     1                  -.199182268         .172963175

          0                  .481104135          .929036904



45                 16.3527593          -122.192105

     1                  -.0956718567        .362078105

          0                  .318036507          .538576193



60                 15.464503           -115.889868

     1                  -.0272811268        .467170259

          0                  .214162014          .284439413
```

4.14 continued

75		15.4329473	-108.439853
	1	.0206473961	.5192379
	0	.158230836	.124951785
90		16.0321846	-100.48259
	1	.0580920953	.537039568
	0	.131474337	.0170304975
105		17.153182	-92.4501361
	1	.0909504302	.530389392
	0	.121640837	-.064881246
120		18.754902	-84.6683617
	1	.122596363	.504174164
	0	.121187016	-.133866706
135		20.8337591	-77.4098833
	1	.154733796	.460943534
	0	.124518881	-.195275422
150		23.4011034	-70.9159101
	1	.187661192	.402556602
	0	.12631445	-.249381403
165		26.4621727	-65.3979959
	1	.22025634	.331237272
	0	.121071313	-.293515182

4.14 Angular acceleration of link 3

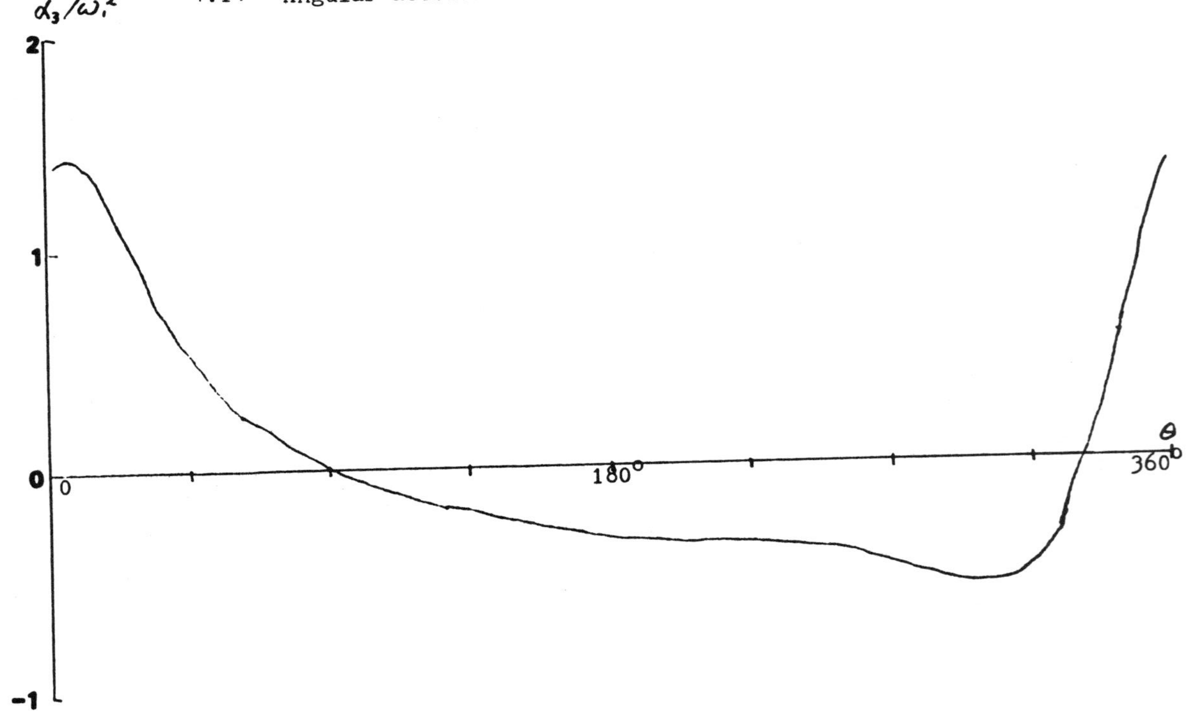

4.15

```
THETA 1             THETA 2             THETA 3         (DEGREES)
       OMEGA 1             OMEGA 2             OMEGA 3          (RADIANS/S)
              ALPHA 1             ALPHA 2             ALPHA 3
                                                          (RADIANS/S**2)


15                  -77.5687127       159.568883

       1                   1.20779224         1.38753281

              0                   -.798105576        -.148174722


30                  -61.0231934       -179.992968

       1                   .999929852         1.33331068

              0                   -.761987431        -.246085715


45                  -47.4336722       -160.490606

       1                   .818522144         1.26681225

              0                   -.614246372        -.251496205


60                  -36.2504233       -141.964285

       1                   .679962725         1.20475777

              0                   -.445616594        -.220112144


75                  -26.8239726       -124.301912

       1                   .583417237         1.15170725

              0                   -.296233473        -.186351765


90                  -18.5689333       -107.375185

       1                   .522711533         1.1061715

              0                   -.171005358        -.163911618
```

4.15 continued

105		-10.9904442	-91.0973068
1		.492545804	1.06450766
0	-.157168655	-.0609666982	
120		-3.65230717	-75.4423505
1		.49053736	1.02235338
0		.0461276098	-.167872841
135		3.86951055	-60.4525389
1		.517362865	.975040282
0		.160558696	-.196407546
150		12.0249349	-46.2384798
1		.575399302	.918344743
0		.2833648	-.237891957
165		21.2898822	-32.9572658
1		.664828558	.850866043
0		.39497234	-.274410812
180		32.0891837	-20.7419172
1		.77777779	.777777772
0		.456396414	-.275645355
195		-315.343757	-9.59090542
1		.897348753	.711311236
0		.445868882	-.22351425

4.16

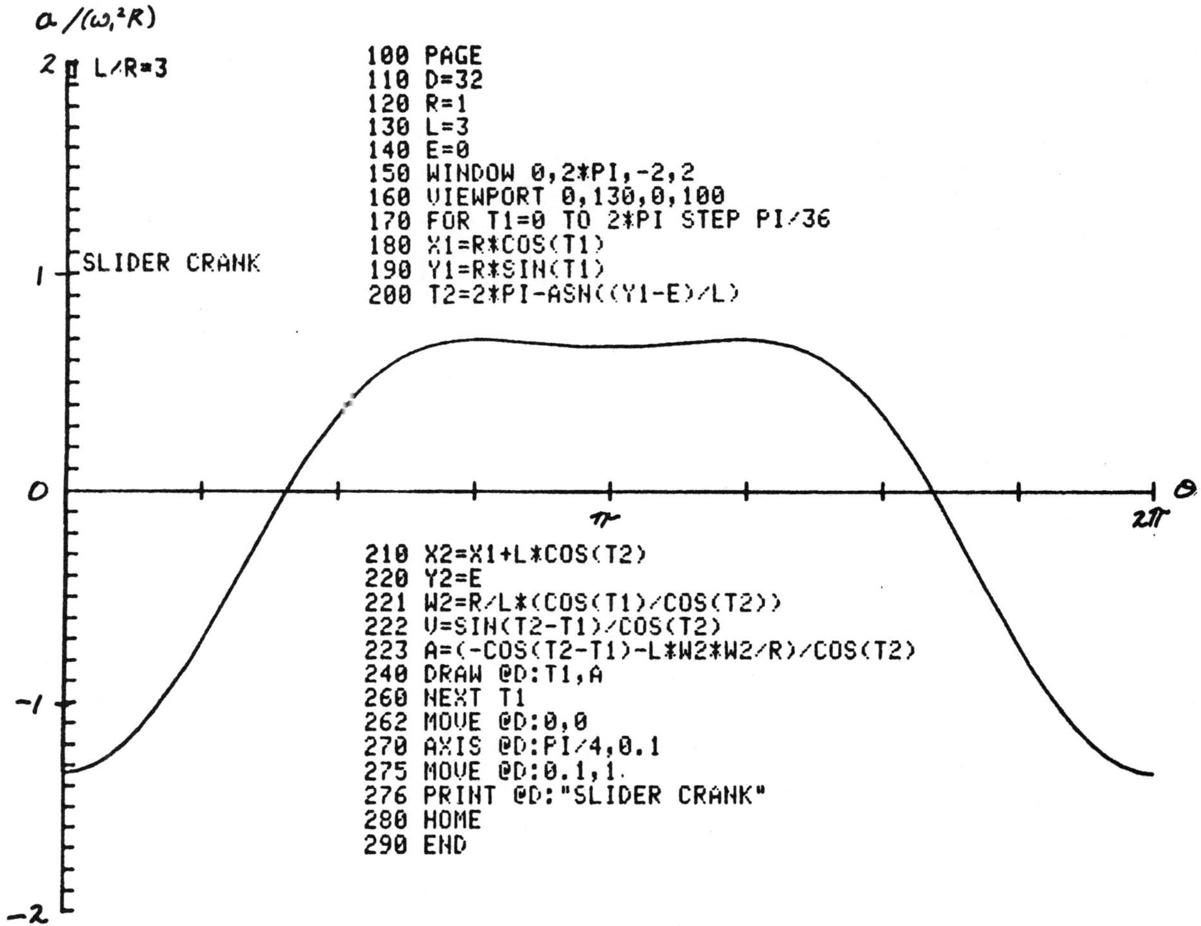


```
100 PAGE
110 D=32
120 R=1
130 L=3
140 E=0
150 WINDOW 0,2*PI,-2,2
160 VIEWPORT 0,130,0,100
170 FOR T1=0 TO 2*PI STEP PI/36
180 X1=R*COS(T1)
190 Y1=R*SIN(T1)
200 T2=2*PI-ASN((Y1-E)/L)
210 X2=X1+L*COS(T2)
220 Y2=E
221 W2=R/L*(COS(T1)/COS(T2))
222 V=SIN(T2-T1)/COS(T2)
223 A=(-COS(T2-T1)-L*W2*W2/R)/COS(T2)
240 DRAW @D:T1,A
260 NEXT T1
262 MOVE @D:0,0
270 AXIS @D:PI/4,0.1
275 MOVE @D:0.1,1
276 PRINT @D:"SLIDER CRANK"
280 HOME
290 END
```

4.17

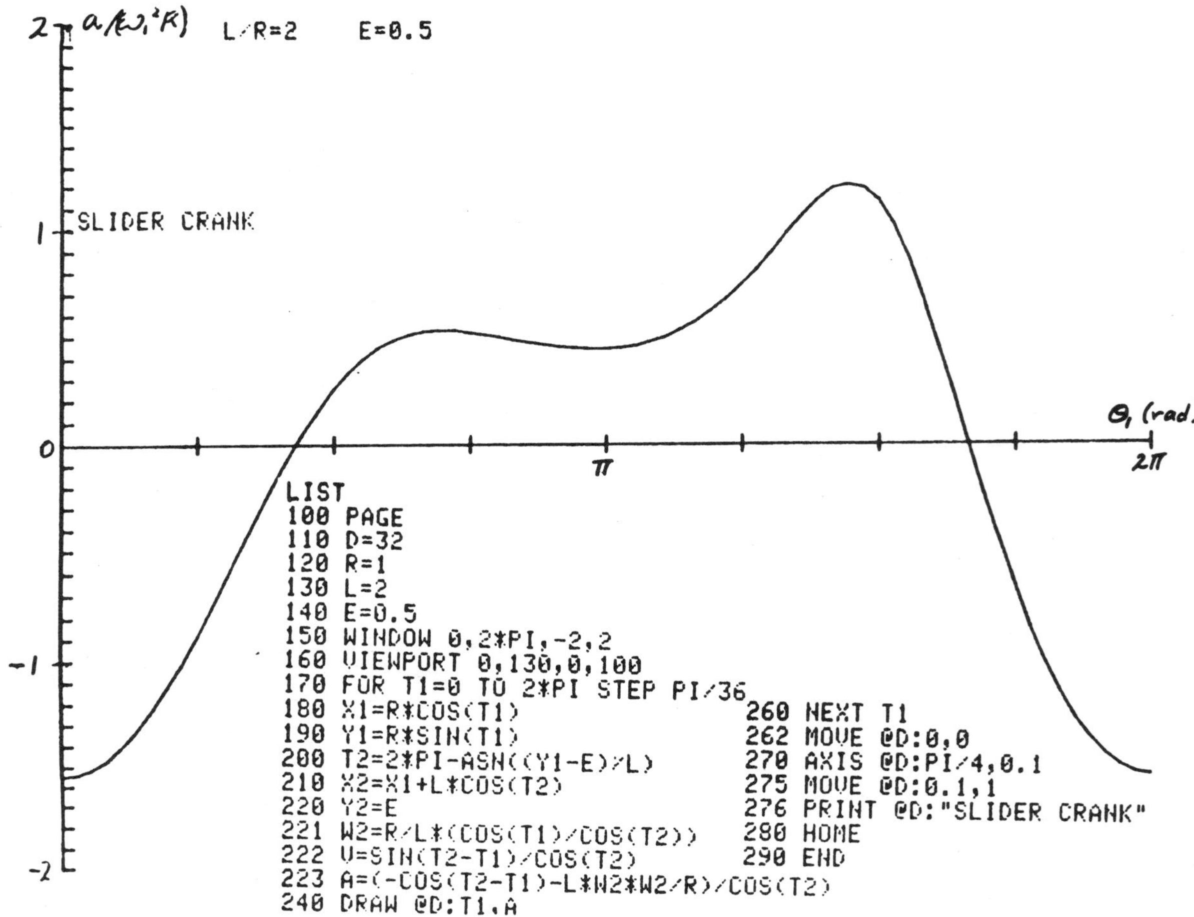


```
LIST
100 PAGE
110 D=32
120 R=1
130 L=2
140 E=0.5
150 WINDOW 0,2*PI,-2,2
160 VIEWPORT 0,130,0,100
170 FOR T1=0 TO 2*PI STEP PI/36
180 X1=R*COS(T1)
190 Y1=R*SIN(T1)
200 T2=2*PI-ASN((Y1-E)/L)
210 X2=X1+L*COS(T2)
220 Y2=E
221 W2=R/L*(COS(T1)/COS(T2))
222 V=SIN(T2-T1)/COS(T2)
223 A=(-COS(T2-T1)-L*W2*W2/R)/COS(T2)
240 DRAW @D:T1,A
260 NEXT T1
262 MOVE @D:0,0
270 AXIS @D:PI/4,0.1
275 MOVE @D:0.1,1
276 PRINT @D:"SLIDER CRANK"
280 HOME
290 END
```

4.18

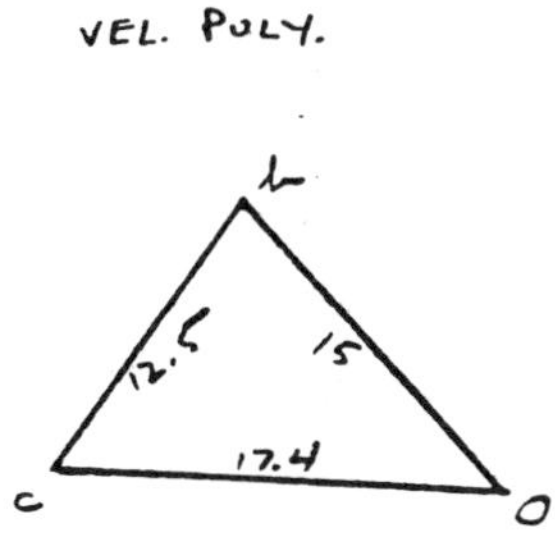

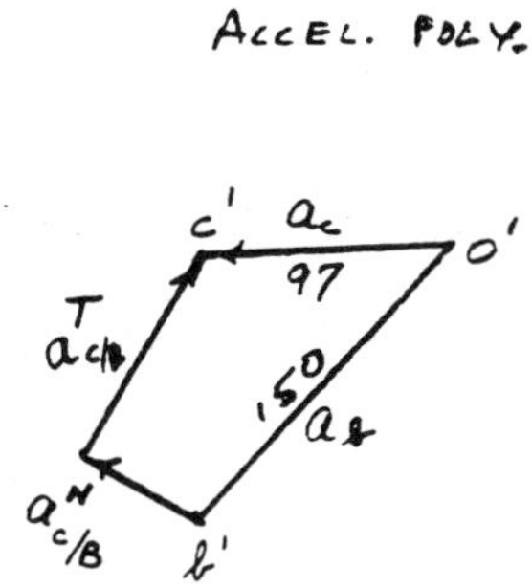

$a^N_B = \omega_1{}^2 O_1B = 10^2(1.5) = 150$ in/sec

$a^N_{C/B} = \frac{bc^2}{BC} = \frac{12.5^2}{3} = 51$ in/sec

$\alpha_2 = \frac{a^t_{C/B}}{CB} = \frac{96}{3} = 32 \frac{\text{rad}}{\text{sec}^2}$ ↻

4.19

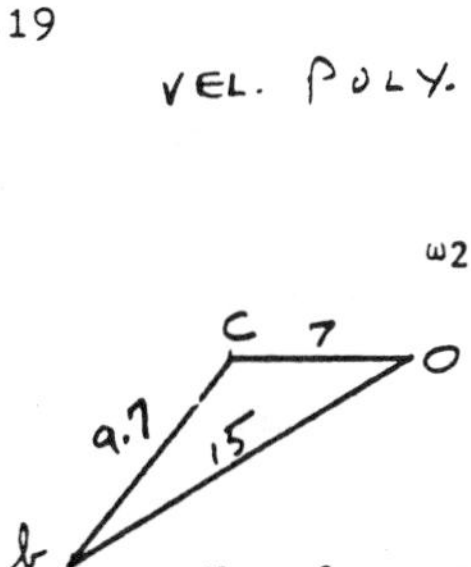

$\omega_2 = \frac{bc}{BC} = \frac{9.7}{3}$

$= 3.23$ ↻

$a^N_B = \omega_1{}^2 O_1B = 150$

$a^N_{C/B} = \frac{bc^2}{BC} = 31$

$\alpha_2 = \frac{a^t_{C/B}}{CB} = 45.7$ ↻ rad/sec²

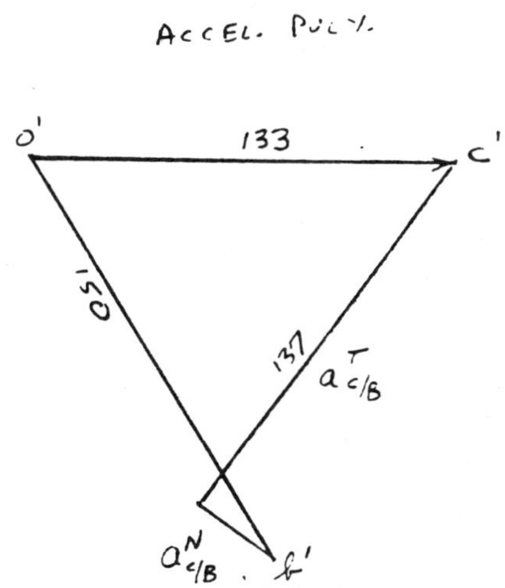

4.20

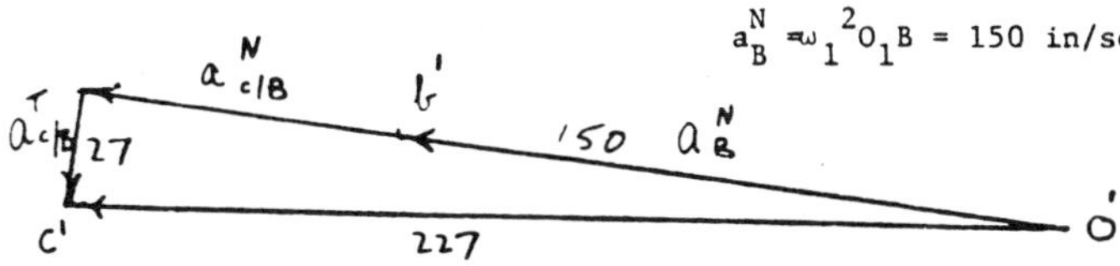

$a^N_B = \omega_1{}^2 O_1B = 150$ in/sec²

$a^N_{C/B} = \frac{bc^2}{BC} = \frac{15^2}{3} = 75$

$\alpha_2 = \frac{a^T_{C/B}}{CB} = \frac{27}{3} = 9$ CW

4.21

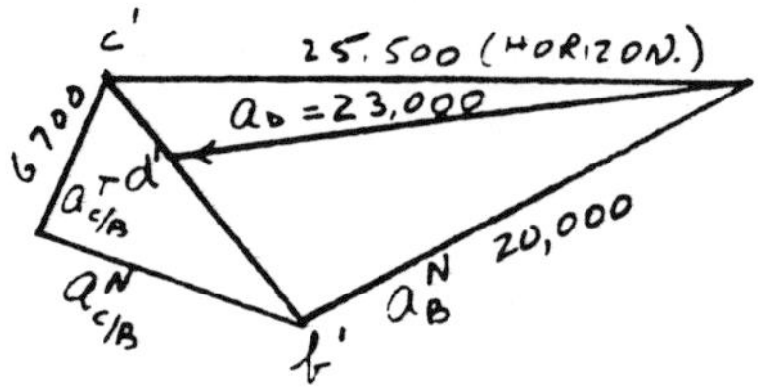

$\alpha_1 = 0$

$a^N_B = \omega_1^{\,2} O_1B = 100^2(2) = 20{,}000$

$a^N_{C/B} = \dfrac{bc^2}{BC} = \dfrac{180^2}{3} = 10{,}800$

$\dfrac{b^1d^1}{b^1c^1} = \dfrac{BD}{BC} \quad b^1d^1 = 128\frac{2}{3} = 85.5$

$\alpha_2 = a^T_{C/B}/CB = \dfrac{6700}{3} = 2230 \dfrac{rad}{sec^2}$ ↶

4.22

$\alpha_1 = 2000$

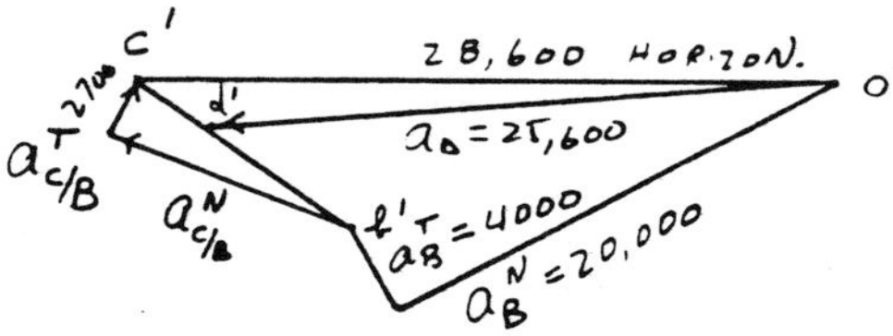

$a^T_B = \alpha_1 O_1B = 2000(2) = 4000 \text{ in/sec}^2$

$\alpha_2 = \dfrac{a^T_{C/B}}{CB} = \dfrac{2700}{3} = 900 \dfrac{rad}{sec^2}$ ↶

$\dfrac{b^1d^1}{b^1c^1} = \dfrac{BD}{BC} \quad b^1d^1 = 11{,}100\frac{2}{3} = 7{,}400$

4.23

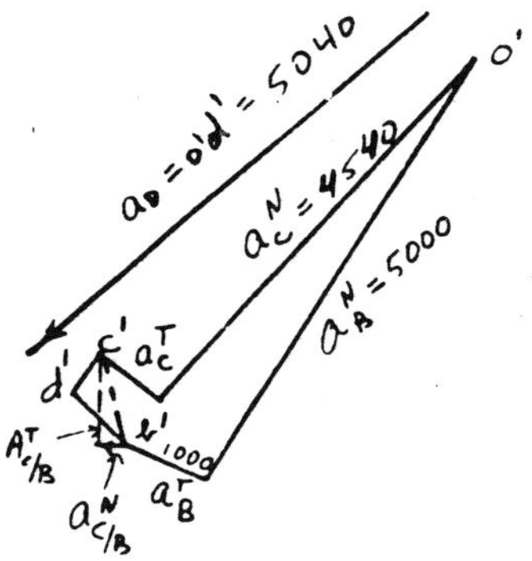

$a^N_B = \omega_1^{\,2} O_1B = 50^2(20) = 5{,}000 \text{ in/sec}^2$

$a^N_{C/B} = \dfrac{bc^2}{BC} = \dfrac{28^2}{6} = 130 \quad a^N_C = \dfrac{oc^2}{O_3C} = \dfrac{117^2}{3} = 4540$

$\alpha_2 = \dfrac{a^T CB}{CB} = \dfrac{880}{6} = 147 \dfrac{rad}{sec^2}$ ↶

$\alpha_3 = \dfrac{a^T_C}{O_3C} = \dfrac{800}{3} = 267 \dfrac{rad}{sec^2}$ ↶

4.24

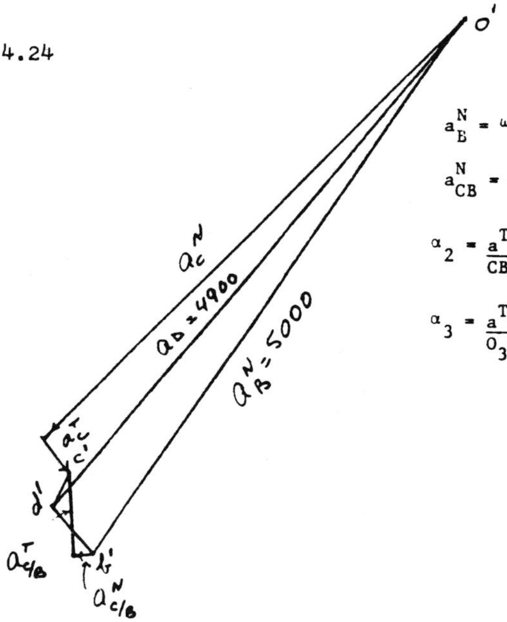

$$a_B^N = \omega_1^2 O_1B = 5000 \text{ in/sec}^2$$

$$a_{CB}^N = \frac{bc^2}{BC} = 130$$

$$\alpha_2 = \frac{a^T{}_{CB}}{CB} = \frac{660}{6} = 110 \frac{\text{rad}}{\text{sec}^2} \circlearrowleft$$

$$\alpha_3 = \frac{a^T{}_C}{O_3C} = \frac{330}{3} = 110 \frac{\text{rad}}{\text{sec}^2} \circlearrowright$$

4.25

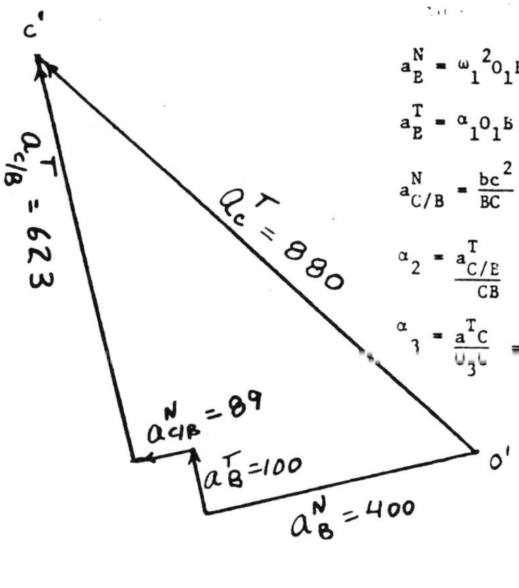

$$a_B^N = \omega_1^2 O_1B = 400 \text{ in/sec}^2$$

$$a_B^T = \alpha_1 O_1B = 100$$

$$a_{C/B}^N = \frac{bc^2}{BC} = 89 \qquad a_C^N = \frac{oc^2}{O_3C} = 0$$

$$\alpha_2 = \frac{a_{C/B}^T}{CB} = 138.5 \frac{\text{rad}}{\text{sec}^2} \circlearrowleft$$

$$\alpha_3 = \frac{a^T{}_C}{O_3C} = 440 \frac{\text{rad}}{\text{sec}^2} \circlearrowleft$$

4.26 $a_B^N = \omega_1^2 O_1B = 20^2(1) = 400$

$a_B^T = \alpha_1 O_1B = 100(1) = 100$

$a_{C/B}^N = \frac{bc^2}{BC} = \frac{20^2}{4.5} = 89 \quad a_C^N = 0$

$\alpha_2 = \frac{a_{C/B}^T}{BC} = \frac{90}{4.5} = 20$↻

$\alpha_3 = \frac{a_C^T}{O_3C} = \frac{310}{2} = 155$↺

See 2.9b

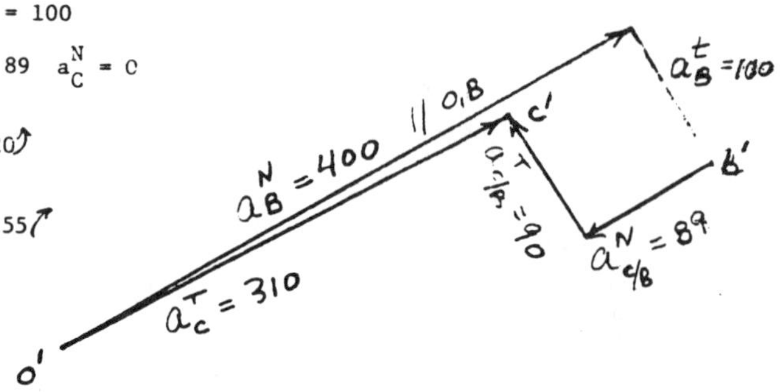

4.27

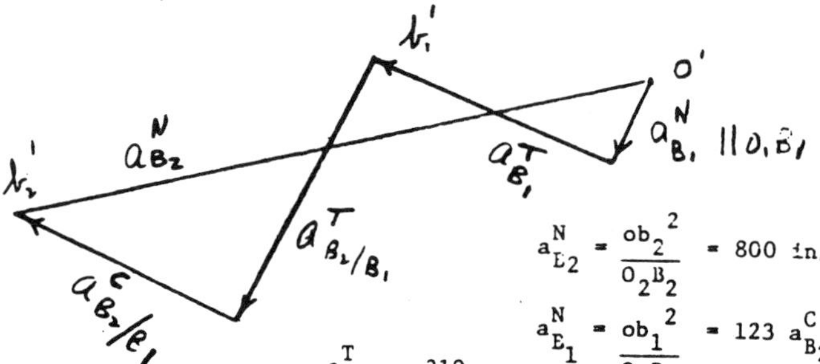

$a_{B2}^N = \frac{ob_2^2}{O_2B_2} = 800 \text{ in/sec}^2$

$a_{B_1}^N = \frac{ob_1^2}{O_1B_1} = 123 \quad a_{B_2B_1}^C = 2b_1b_2\omega_1 = 314$

$\alpha_1 = \frac{a_{B_1}^T}{O_1B_1} = \frac{310}{4.9} = 63 \frac{\text{rad}}{\text{sec}^2}$↻

4.28

$a_{B_2}^N = \omega_2^2 O_2B_2 = 800 \text{ in/sec}^2$

$a_{B_1}^N = \frac{ob_1^2}{O_1B_1} = 221 \quad \alpha_1 = \frac{a_{B_1}^T}{O_1B_1} = 500 \frac{\text{rad}}{\text{sec}^2}$↺

$a_{B_2}^N = a_{B_1}^N + a_{B_1}^r + a_{B2B_1}^T + a_{B2B_1}^C$

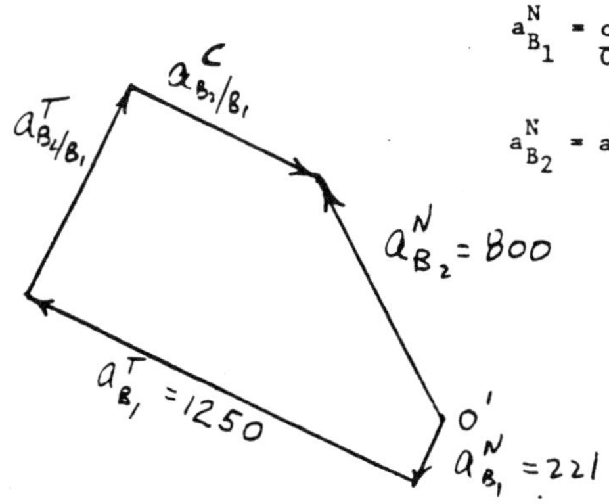

.29 a) See figure. The in line slider crank linkage is represented by the vector equation

$$\tilde{r}_o + \tilde{r}_1 + \tilde{r}_2 = 0 \quad \text{(a)}$$

Position

$$\tilde{i}r_o + \tilde{i}r_{1x} + \tilde{j}r_{1y} + \tilde{i}r_{2x} + \tilde{j}r_{2y} = 0 \quad \text{(b)}$$

From the $\tilde{j}$ components:

$$r_{2y} = -r_{1y} \qquad \sin\theta_2 = r_{2y}/r_2 \quad \text{(c)}$$

Using the $\tilde{i}$ components:

$$r_o = -r_{1x} - r_{2x} = -r_1 \cos\theta_1 - r_2 \cos\theta_2 \quad \text{(d)}$$

Velocity

Differentiating eq (a) with respect to time:

$$\dot{\tilde{r}}_o + \tilde{\omega}_1 X \tilde{r}_1 + \tilde{\omega}_2 X \tilde{r}_2 = 0 \quad \text{(e)}$$

from which

$$\tilde{i}\dot{r}_o + \begin{vmatrix} \tilde{i} & \tilde{j} & \tilde{k} \\ 0 & 0 & \omega_1 \\ r_{1x} & r_{1y} & 0 \end{vmatrix} + \begin{vmatrix} \tilde{i} & \tilde{j} & \tilde{k} \\ 0 & 0 & \omega_2 \\ r_{2x} & r_{2y} & 0 \end{vmatrix} = 0 \quad \text{(f)}$$

$$\tilde{i}\dot{r}_o - \tilde{i}\omega_1 r_{1y} + \tilde{j}\omega_1 r_{1x} - \tilde{i}\omega_2 r_{2y} + \tilde{j}\omega_2 r_{2x} = 0 \quad \text{(g)}$$

Using the $\tilde{j}$ components:

$$\omega_1 r_{1x} + \omega_2 r_{2x} = 0 \qquad \omega_2 = -\omega_1 r_{1x}/r_{2x} \quad \text{(h)}$$

From the $\tilde{i}$ components:

$$\dot{r}_o = \omega_1 r_{1y} + \omega_2 r_{2y} \quad \text{(i)}$$

Acceleration

Differentiating eq (e) with respect to time:

$$\ddot{r}_o - \tilde{i}\alpha_1 r_{1y} + \tilde{j}\alpha_1 r_{1x} - \tilde{i}\omega_1^2 r_{1x} - \tilde{j}\omega_1^2 r_{1y}$$
$$- \tilde{i}\alpha_2 r_{2y} + \tilde{j}\alpha_2 r_{2x} - \tilde{i}\omega_2^2 r_{2x} - \tilde{j}\omega_2^2 r_{2y} = 0 \quad \text{(j)}$$

Using the $\tilde{j}$ components:

$$\alpha_1 r_{1x} - \omega_1^2 r_{1y} + \alpha_2 r_{2x} - \omega_2^2 r_{2y} = 0$$

$$\alpha_2 = (-\alpha_1 r_{1x} + \omega_1^2 r_{1y} + \omega_2^2 r_{2y})/r_{2x} \quad \text{(k)}$$

From the $\tilde{i}$ components:

$$\ddot{r}_o = \alpha_1 r_{1y} + \omega_1^2 r_{1x} + \alpha_2 r_{2y} + \omega_2^2 r_{2x} \quad \text{(l)}$$

Position

For $\theta_1 = 30^o$, the position equations yield

$$r_{2y} = -r_{1y} = -r_1 \cos\theta_1 = -2 \cos 30^o = -1$$

$\sin\theta_2 = r_{2y}/r_2 = -1/3 \quad \theta_2 = -19.47^o$ (or 199.47^o for the other assembly configuration)

$$r_o = -r_1\cos\theta_1 - r_2\cos\theta_2 = -2\cos 30^o - 3\cos(-19.47^o) = -4.56 \text{ in}$$

4.29 a) (Cont'd.)

(i.e. $\underset{\sim}{r}_o = -\underset{\sim}{i}4.56$)

Velocity

For $\omega_1 = 100$ rad/s cc (+ for cc)

$\omega_2 = -\omega_1 r_{1x}/r_{2x} = -\omega_1 r_1 \cos\theta_1/(r_2 \cos\theta_2)$

$= -100 \times 2\cos 30^\circ/(3\times\cos(-19.47^\circ)) = -61.24$ rad/s

$\dot{r}_o = \omega_1 r_{1y} + \omega_2 r_{2y} = 100\times 2\sin 30^\circ - 61.24\times 3\sin(-19.47^\circ)$

$= 161.24$ (to the left).

Acceleration

For $\alpha_1 = 0$

$\alpha_2 = (-\alpha_1 r_{1x} + \omega_1^2 r_{1y} + \omega_2^2 r_{2y})/r_{2x}$

$= [100^2\times 2\ \sin 30^\circ + 61.24^2\times 3\ \sin(-19.47^\circ)]/[3\cos(-19.47^\circ)]$

$= 2209.7$ rad/s^2 cc

$\ddot{r}_o = \alpha_1 r_{1y} + \omega_1^2 r_{1x} + \alpha_2 r_{2y} + \omega_2^2 r_{2x}$

$= 100^2\times 2\cos 30^\circ + 2209.7\times 3\sin(-19.47^\circ) + 61.24^2\times 3\cos(-19.47^\circ)$

$= 25{,}717$ in/s^2 to the left

b) For $\theta_1 = 0^\circ$, $\theta_2 = 0^\circ$, $\underset{\sim}{r}_o = -\underset{\sim}{i}5$

$\omega_2 = -\omega_1 r_{1x}/r_{2x} = -100\times 2/3 = -66.67$ rad/s

$\alpha_2 = (-\alpha_1 r_1 \cos\theta_1 + \omega_1^2 r_1 \sin\theta_1 + \omega_2^2 r_2 \sin\theta_2)/(r_2 \cos\theta_2)$

$= 0$

$\ddot{r}_o = \alpha_1 r_1 \sin\theta_1 + \omega_1^2 r_1 \cos\theta_1 + \alpha_2 r_2 \sin\theta_2 + \omega_2^2 r_2 \cos\theta_2$

$= \omega_1^2 r_1 + \omega_2^2 r_2 = 100^2\times 2 + 66.67^2\times 3 = 33{,}333$ in/s^2 to left.

For $\theta_1 = 180^\circ$, $\theta_2 = 0$, $\underset{\sim}{r}_o = -\underset{\sim}{i}1$, $\alpha_2 = 0$

$\ddot{r}_o = 100^2\times 2\times(-1) + 66.67^2\times 3 = -6667$ (6667 in/s^2 to right)

Comparison of results

Approximate equation:

i) for $\theta_1 = 30^\circ$

$a_C = R\omega^2(\cos\theta + \frac{R}{L}\cos 2\theta) = 2(100)^2(\cos 30^\circ + \frac{2}{3}\cos 60^\circ)$

$= 24{,}000$ in/sec^2 to the left.

Compare with prob. 4.21: $a_C = 25{,}500$ (graphically).

ii) See sketch. For $\theta = 0^\circ$,

$a_C = R\omega^2(1 + \frac{R}{L}) = 33{,}000$ in/sec^2 to left.

For 180°, $a_C = R\omega^2(1 - \frac{R}{L}) = 6670$ to right. (Values of A_C using the "approximate equation" are exact at limiting positions.)

4.30 Same as 4.29b. See sketch. α_1 does not affect a_C since tangintial accelerations are equal and opposite.

4.29

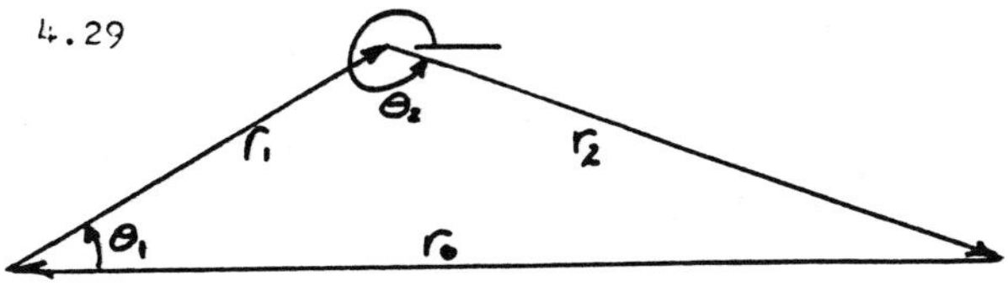

4.30

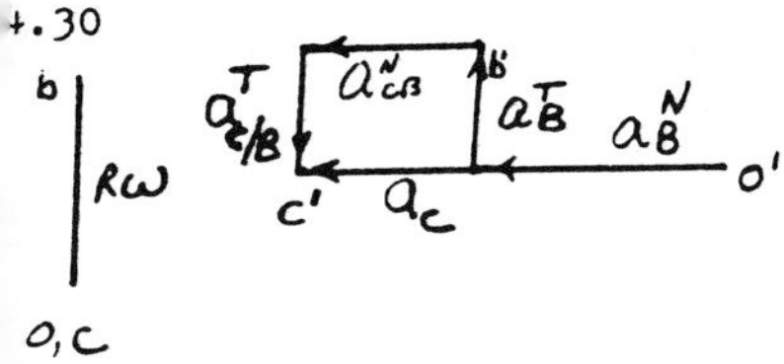

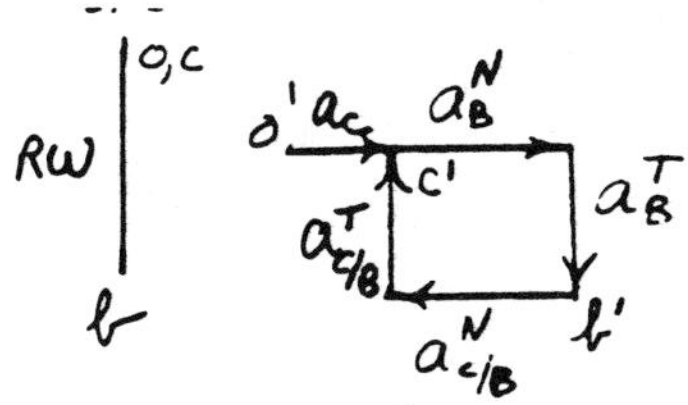

a) $bc = \omega R \quad a^N_{CB} = \frac{bc^2}{BC} = \frac{(\omega R)^2}{L}$

$a_c = \omega^2 R(1 + \frac{R}{L}) \leftarrow$

b) $a_c = \omega^2 R(1 - \frac{R}{L}) \rightarrow$

4.31

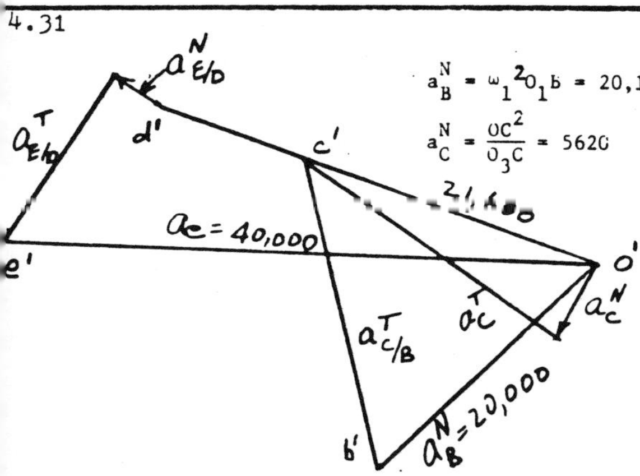

$a^N_B = \omega_1^2 O_1 B = 20,100 \quad a^N_{C/B} = \frac{bc^2}{BC} = 585$

$a^N_C = \frac{oc^2}{O_3C} = 5620 \quad \frac{o^1d^1}{o^1c^1} = \frac{O_3D}{O_3C}$

$o^1d^1 = 21,000\frac{6}{4} = 32,400$

$a^N_{E/D} = \frac{de^2}{DE} = 3310$

4.31 $a_E = 40{,}000\ in/sec^2$ as shown.

4.32 $a_E = 45{,}000\ in/sec^2$ as shown.

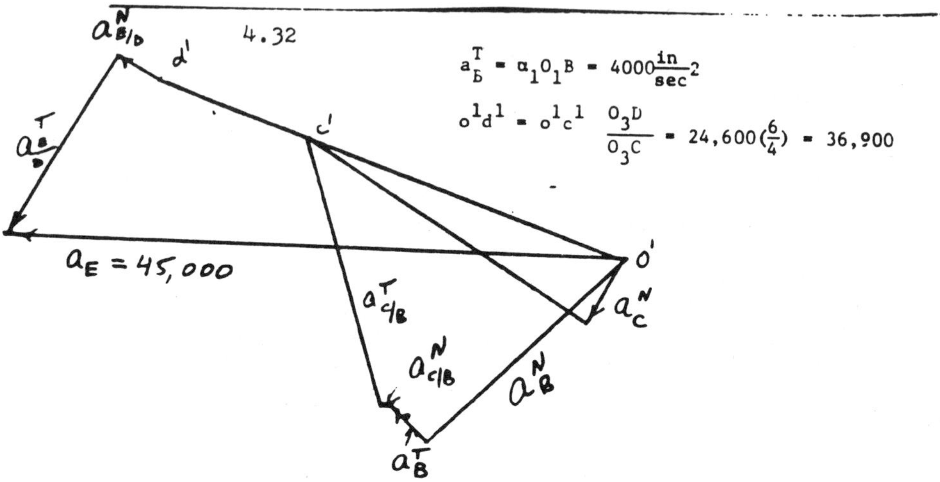

4.33 Cam follower acceleration $a_{C2} = 70.7\ in/sec^2$ as shown.

4.33

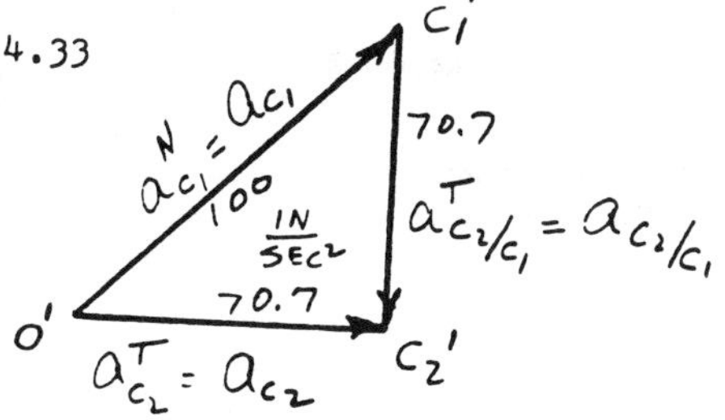

$a^N_{c_1} = 0_1c\,\omega_1^2 = 100 \text{ in/s}^2$

$a^T_{c_1} = 0$

4.34

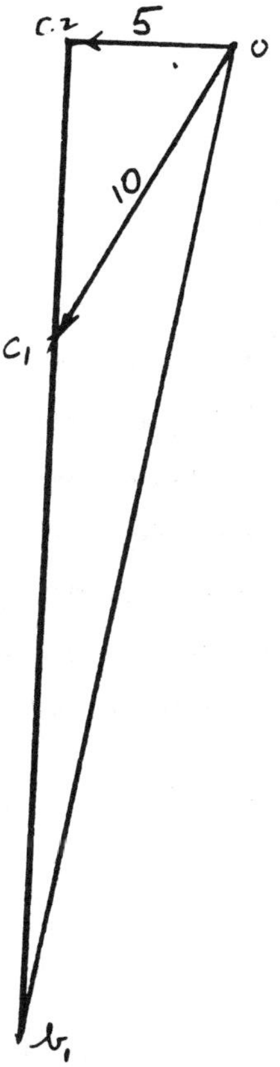

VEL. POLY.

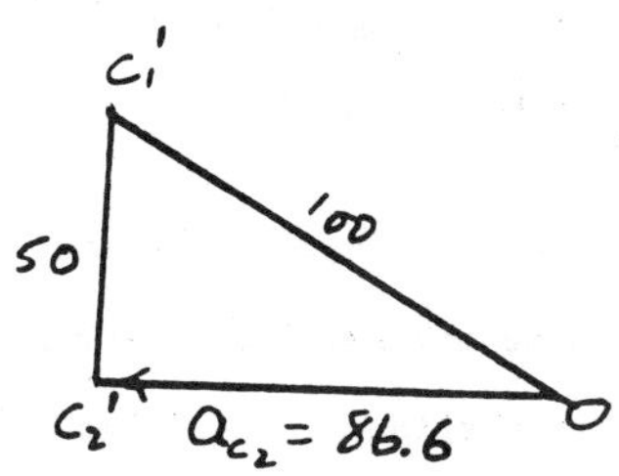

ACCEL. POLY.

4.34 Cam follower acceleration a_{C2} = 70.7 in/sec^2 as shown.

4.35 a_D = 700; a_E = 1100; a_F = 600 in/sec^2 as shown.
(based on O_3C = 2 in)

4.36 α_2 = 90 rad/sec^2 CC. See sketch.

4.37 a_D = 700 in/sec^2 as shown.

4.38 a_D = 4020 in/sec^2 as shown.

4.39 α_2 = 210 rad/sec^2 CW.

4.40 α_2 = 3160 rad/sec^2 CW; a_D = 12,000 in/sec^2 as shown.

4.41 $a_C = \omega^2 R\ (1 + \frac{R}{L})$ from which $\omega^2 = \frac{50{,}000}{4(1 + \frac{4}{10})} = 8930.$
Crank speed $= \frac{60}{2\pi}\ \omega = 903$ RPM.

4.42 a_E = 230 in/sec^2 as shown. The problem is straightforward as stated since we may solve the four bar linkage O_1BCO_3 and then find e' by proportions. Beginning with link 5, we cannot make use of a four bar linkage. A trial and error method may be employed, assuming successive values of α_1 and working through the problem to obtain α_5. When the value of α_5 thus obtained is close enough to the given value, our trial solution is satisfactory.

4.43 See sketch. α_1 = 5780 and α_3 = 5150 rad/sec^2, both CW.

4.35

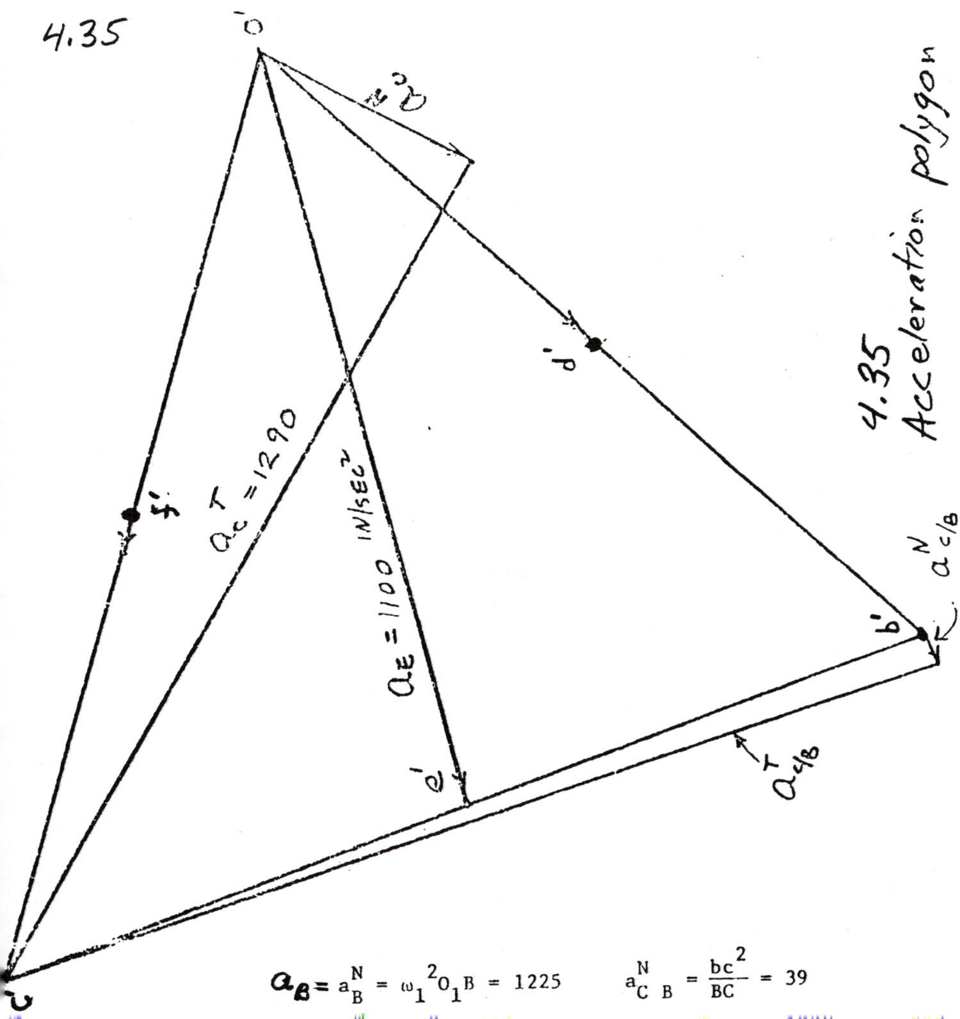

$a_B = a_B^N = \omega_1^2 O_1B = 1225$ $\qquad a_{C\,B}^N = \frac{bc^2}{BC} = 39$

$a_C^N = \frac{oc^2}{O_3C} = 333$ $\qquad \alpha_2 = \frac{a_{C\,B}^T}{CB} = \frac{1330}{3.25} = 409 \frac{\text{rad}}{\text{sec}^2}$

$\alpha_3 = \frac{a_C^T}{O_3C} = \frac{1290}{2} = 645 \circlearrowleft$

$a_D = a_B/2 = 612.5$

$a_F = 680$

$a_E = 1100$

4.36

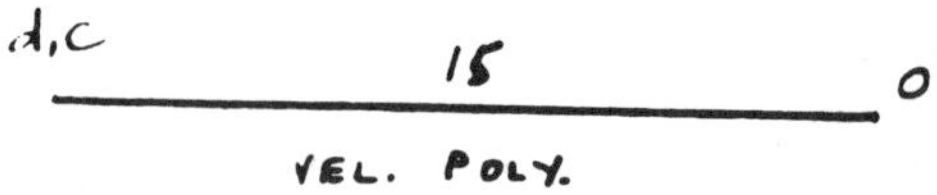

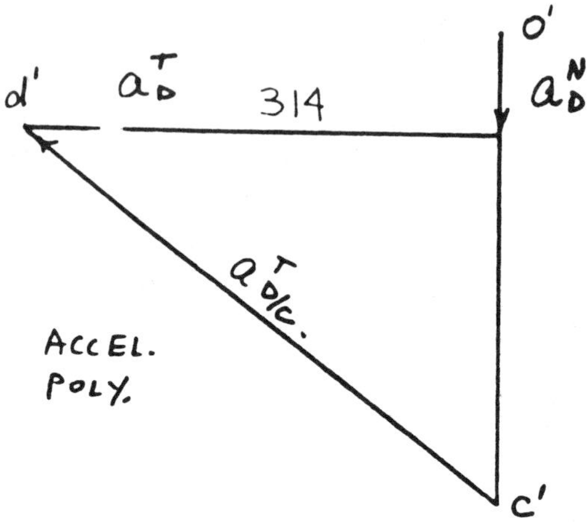

$a_C^N = \omega_1^2\ O_1C = 300\ in/s^2$

$a_{d/c}^N = 0$

$a_D^N = \frac{od^2}{O_2D} = 64.3$

$\alpha_2 = \frac{a_D^T}{O_2D} = \frac{314}{3.5} = 90$ rad/s² CCW

$\omega_2 = \frac{od}{O_2D} = \frac{15}{3.5} = 4.29$ rad/s

4.37

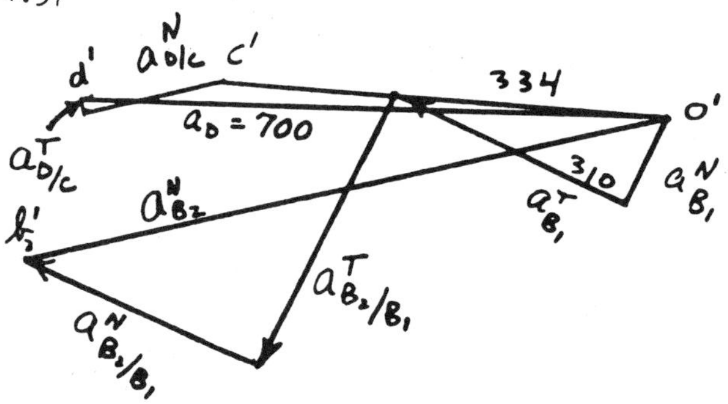

$a_c = 545$

$\frac{o^1c^1}{o^1b^1} = \frac{O_1C}{O_1B}$

$\frac{o^1c^1}{334} = \frac{8}{4.9}$

$a_{D/C}^N = \frac{cd^2}{CD} = \frac{16^2}{1.5} = 171$

$a_D = 700\ in/sec^2$

$\alpha_1 = \frac{a^t_{B_1}}{O_1B_1} = \frac{310}{4.9} = 63.3$ ↺

4.38

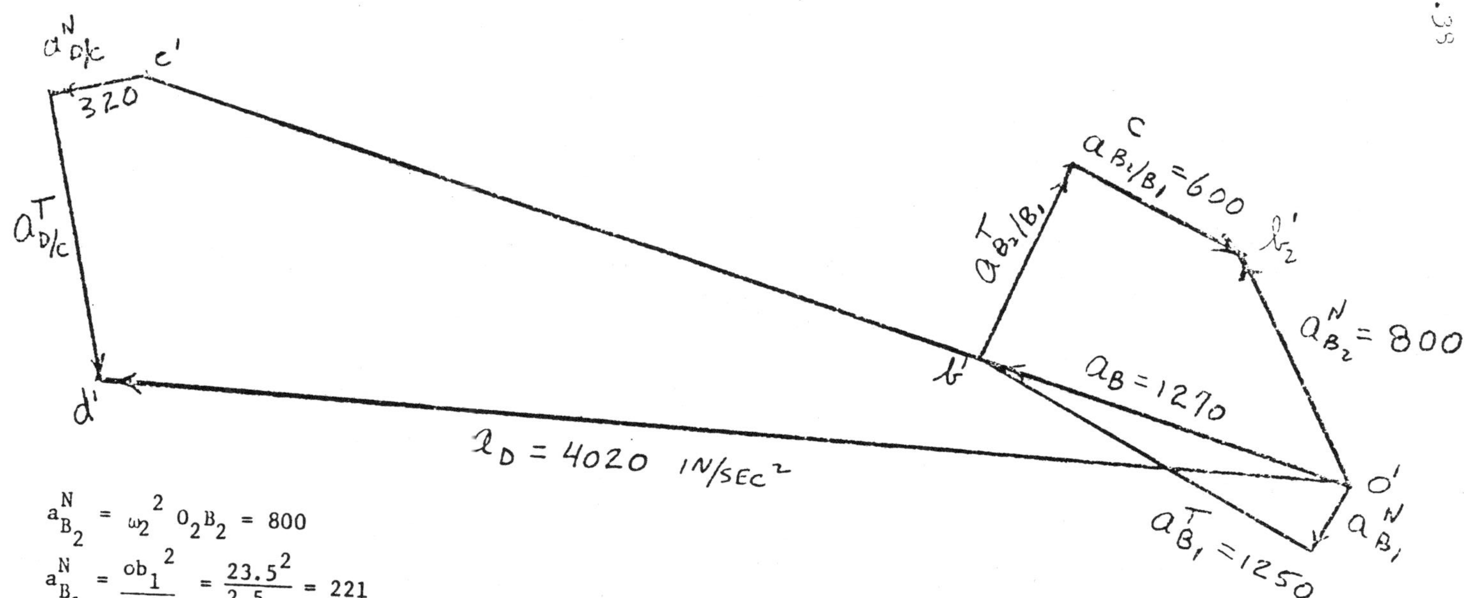

$$a^N_{B_2} = \omega_2^2 \, O_2B_2 = 800$$

$$a^N_{B_1} = \frac{ob_1^2}{O_1B_1} = \frac{23.5^2}{2.5} = 221$$

$$\frac{o^1c^1}{o^1b^1_1} = \frac{O_1C}{O_1B} \qquad o^1c^1 = 1270\,\frac{8}{2.5} = 4060, \quad a^C_{B2/B1} = 2\omega_1 b_1 b_2 = 600 \searrow$$

$$a^N_{D/C} = \frac{cd^2}{CD} = \frac{31^2}{3} \doteq 320 \qquad a^N_{B_2} = a^N_{B_1} \mathbin{+\!\!\!\rightarrow} a^T_{B_1} \mathbin{+\!\!\!\rightarrow} a^T_{B2/B1} \mathbin{+\!\!\!\rightarrow} a^c_{B2/B1}$$

$$\alpha_1 = \frac{a^t_{B_1}}{O_1B_1} = \frac{1250}{2.5} = 500 \text{ rad/sec}^2$$

4.39

$O_1B_1 = 2.2$ in.

$ob_1 = \omega_1 O_1 B_1 = 66$ in/sec

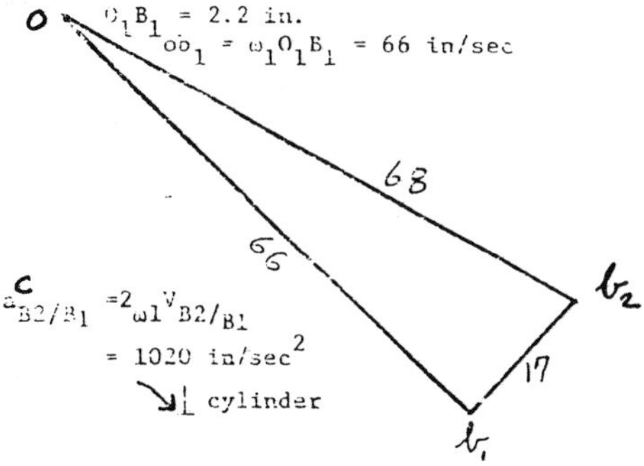

$a^C_{B2/B1} = 2\omega_1 V_{B2/B1}$

$= 1020$ in/sec^2

↘ ⊥ cylinder

$\alpha_2 = a^t_{B2}/O_2B_2$

$= 210 \frac{rad}{sec^2}$ ↻

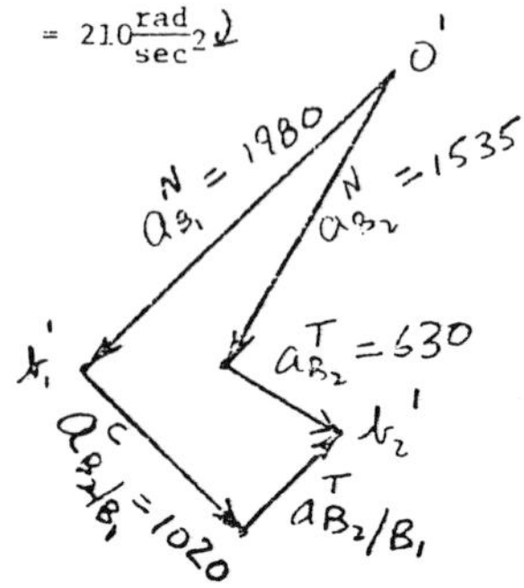

ACCEL. POLY.

$$\overbrace{a^N_{B_2} \leftrightarrow a^T_{B_2}}^{o^1 b^1_2} = \overbrace{a^N_{B_1} \leftrightarrow a^T_{B2/B1} \leftrightarrow a^C_{B2/B.}}^{o^1 b^1_2}$$

$\frac{ob_2^2}{O_2B_2}$	?	$\frac{ob_1^2}{O_1B_1}$	?	$2\omega_1 b_1 b_2$
1533		1980		1020

4.40

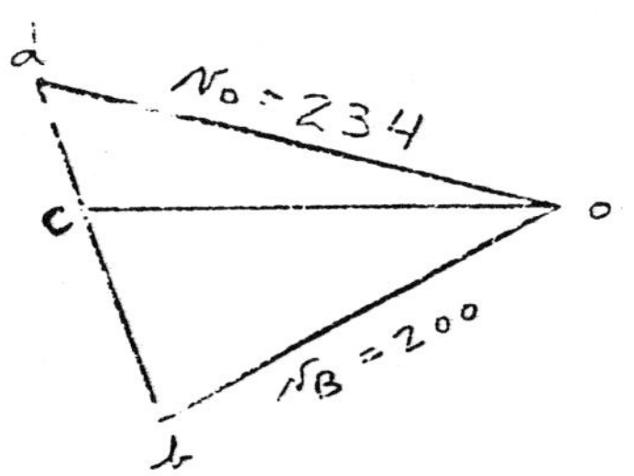

VEL. POLY.

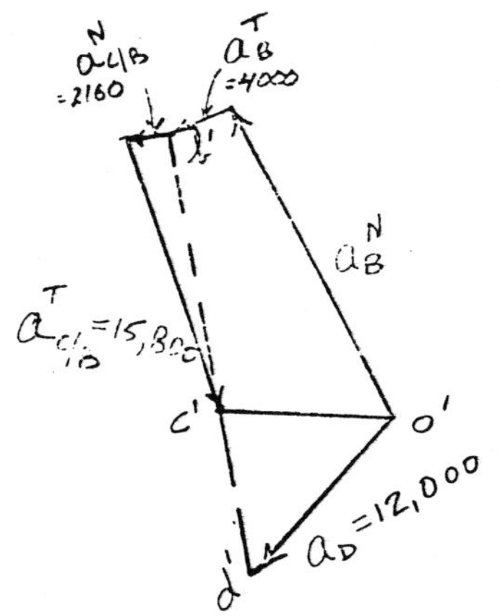

ACCEL. POLY.

$\frac{bd}{bc} = \frac{BD}{BC} = \frac{8}{5}$

$\omega_2 = \frac{bc}{BC} \; \frac{104}{5} = 20.8 \frac{rad}{sec}$ ↻

$a^N_B = \omega_1^2 O_1 B = 20,000$ in $\quad a^T_B = \alpha_1 O_1 B = 4000$

$a^N_{C/B} = \frac{bc^2}{BC} = 2160 \quad \frac{b^1 d^1}{b^1 c^1} = \frac{BD}{BC} \quad b^1 d^1 = 25,600$

$\alpha_2 = \frac{a^T_{C/B}}{C} = 3160$ rad/sec ↻

4.42 (see 3.51)

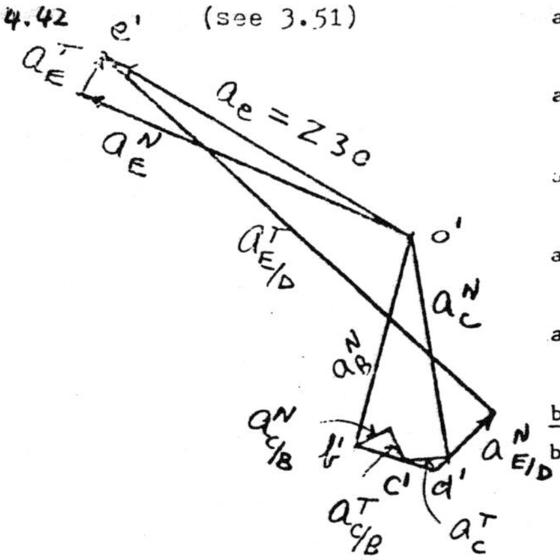

$$a_B^N = \frac{ob^2}{O_1B} = \frac{20.4^2}{3} = 139$$

$$a_C^N = \frac{oc^2}{O_3C} = \frac{16.7^2}{2} = 140$$

$$a_{C/B}^N = \frac{bc^2}{BC} = \frac{8.2^2}{2.25} = 30$$

$$a_{E/D}^N = \frac{de^2}{DE} = \frac{11.9^2}{2.75} = 51$$

$$a_E^N = \frac{oe^2}{O_5E} = \frac{15^2}{1} = 225$$

$$\frac{b^1d^1}{b^1c^1} = \frac{BD}{BC} \quad b^1d^1 = 35\ \frac{3.25}{2.25} = 51$$

4.43

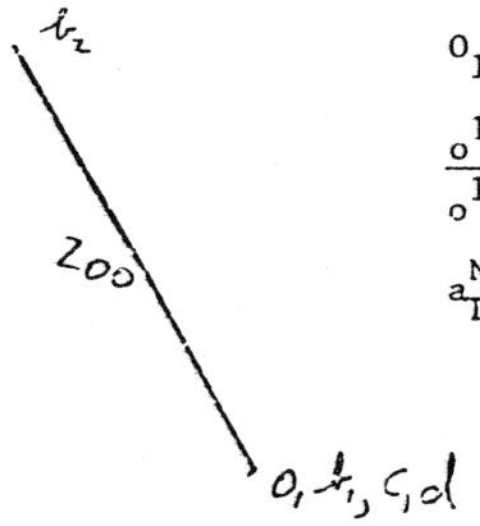

VEL. POLY.

$$O_1B_1 = O_1O_2 \sin 60^\circ = 3.46 \text{ in.}$$

$$\frac{o^1c^1}{o^1b^1_1} = \frac{OC}{OB_1} \qquad o^1c^1 = 20{,}000\ \frac{7}{3.46} = 40{,}400$$

$$a_{D/C}^N = 0$$

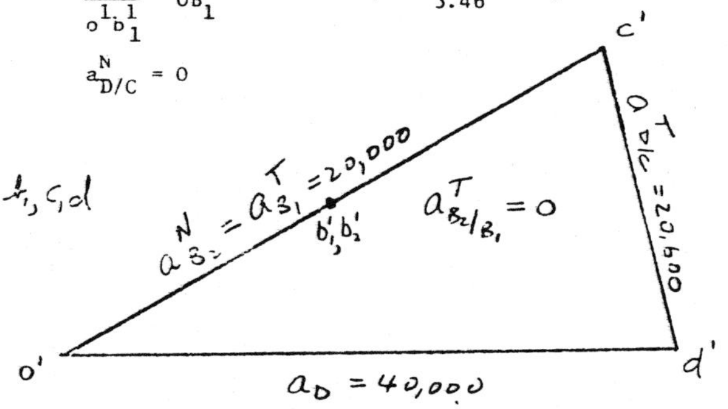

ACCEL. POLY.

$$a_{B2}^N + a_{B_2}^T = a_{B_1}^N + a_{B_1}^t + a_{B2/B1}^t + a_{B2/B1}^C$$

$$\omega_2^2\, O_2B_2 \quad 0 \quad 0\ \nearrow? \quad \nwarrow \quad ? \quad 0$$

$$\alpha_1 = \frac{a_{B1}^T}{O_1B_1} = 5780 \circlearrowright$$

$$\alpha_3 = \frac{a_{D/C}^T}{DC} = 5150 \circlearrowright$$

4.44

Velocity

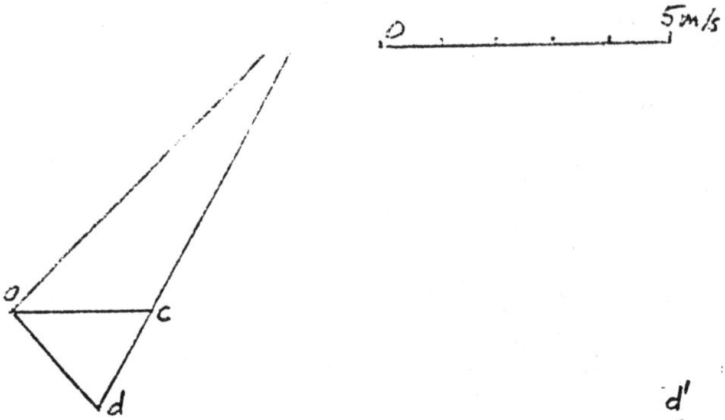

Acceleration

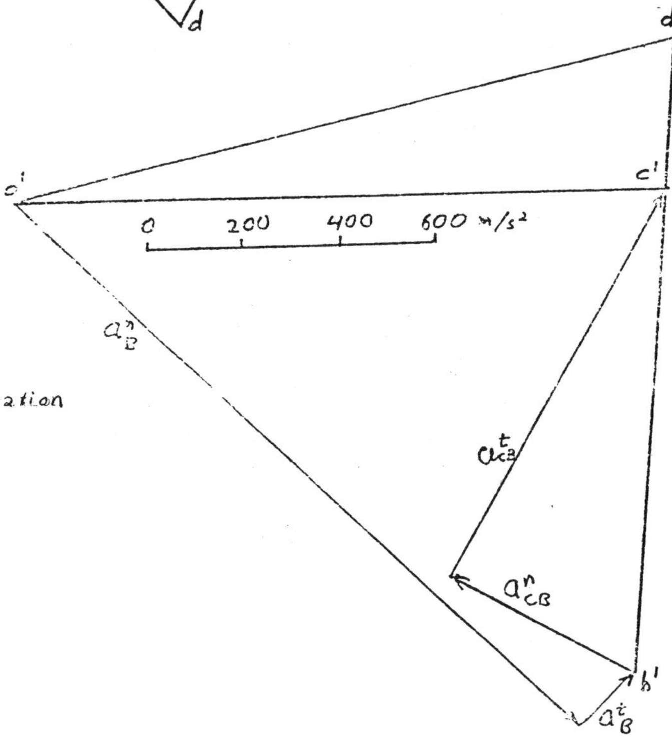

4.44 Velocity

(See figure). Referring to velocity polygon obcd,

$ob = \omega_1 O_1B = 200 \times 0.040 = 8$ m/s

$oc = 2.5$ and $bc = 6.5$ m/s (measured).

$bd = bc \times BD/BC = 6.5 \times 130/99 = 8.5$ m/s

Acceleration (See figure)

$\underset{\sim}{o'c'} = \underset{\sim}{o'b'} + \underset{\sim}{b'c'}$

$\underset{\sim}{a}_C^n + \underset{\sim}{a}_C^t = \underset{\sim}{a}_B^n + \underset{\sim}{a}_B^t + \underset{\sim}{a}_{CB}^n + \underset{\sim}{a}_{CB}^t$

$\underset{\sim}{a}_C^n = 0$ (straight path)

$\underset{\sim}{a}_B^n = \omega_1^2 O_1B = 200^2 \times 0.040 = 1600$ m/s$^2 \perp O_1B$ toward O_1

$o'b' = a_B = \underset{\sim}{a}_B^n + \underset{\sim}{a}_B^t = 1608$ m/s

$\underset{\sim}{a}_B^t = \alpha_1 O_1B = 4000 \times 0.04 = 160$ m/s$^2 \perp O_1B$

$\underset{\sim}{a}_{CB}^n = bc^2/BC = 6.5^2/0.099 = 427$ m/s$^2 \parallel BC$

Acceleration polygon o'b'c' is completed to obtain

$o'c' = a_C = a_C^t = 1350$ and $a_{CB}^t = 925$ m/s^2

Acceleration image

$b'd' = b'c' \times BD/BC = 1030 \times 130/0.099 = 1352$ m/s

$a_D = o'd' = 1425$ m/s

Angular acceleration

For $\alpha_1 = 4000$ rad/s^2 cw (given)

$\alpha_2 = a_{CB}^t/CB = 925/0.099 = 9340$ rad/s^2, cc

4.45

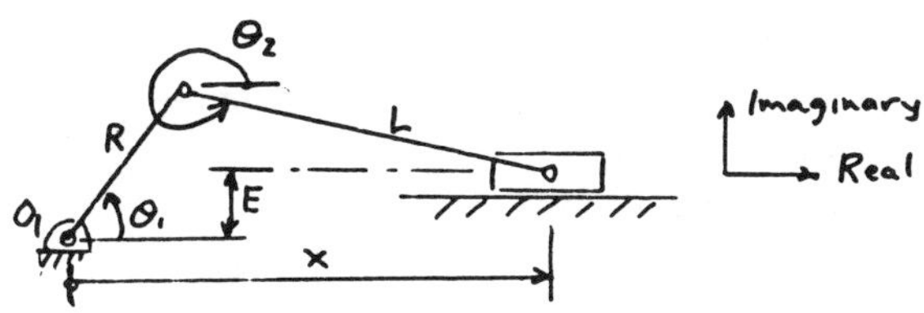

See figure. Using complex number notation, slider location is given by:

$x = Re^{j\theta_1} + Le^{j\theta_2}$

Velocity $V = \dfrac{dx}{dt} = j\omega_1 Re^{j\theta_1} + j\omega_2 Le^{j\theta_2}$

Acceleration $a = \dfrac{dv}{dt}$

$= -\omega_1^2 Re^{j\theta_1} + j\alpha_1 Re^{j\theta_1} - \omega_2^2 Le^{j\theta_2} + j\alpha_2 Le^{j\theta_2}$

where $\theta_2 = \arcsin \dfrac{E - R_1 \sin\theta_1}{L}$

(For the configuration shown, $-90 < \theta_2 < 90^o$ if $R+E<L$)

4.45 (Cont'd.)

ω_2 is found by noting that the imaginary part of V is 0

$\text{Im } V = j[\omega_1 R\cos\theta_1 + \omega_2 L\cos\theta_2] = 0$ from which

$$\omega_2 = -\frac{\omega_1 R\cos\theta_1}{L\cos\theta_2}$$

Similarly,

$\text{Im } a = j[-\omega_1^2 R\sin\theta_1 + \alpha_1 R\cos\theta_1 - \omega_2^2 L\sin\theta_2 + \alpha_2 L\cos\theta_2] = 0$

from which $\alpha_2 = \dfrac{\omega_1^2 R\sin\theta_1 - \alpha_1 R\cos\theta_1 + \omega_2^2 L\sin\theta_2}{L\cos\theta_2}$

Slider acceleration is the real part of

a(Im a = 0):

$a = -\omega_1^2 R\cos\theta_1 - \alpha_1 R\sin\theta_1 - \omega_2^2 L\cos\theta_2 - \alpha_2 L\sin\theta_2$

or, substituting for α_2:

$a = -\omega_1^2 R\cos\theta_1 - \alpha_1 R\sin\theta_1 - \omega_2^2\cos\theta_1$

$\quad - \sin\theta_2\left[\dfrac{\omega_1^2 R\sin\theta_1 - \alpha_1 R\cos\theta_1 + \omega_2^2 L\sin\theta_2}{\cos\theta_2}\right]$

Combining terms using trigonometric identities:

$\cos(\theta_1-\theta_2) = \cos\theta_1\cos\theta_2 + \sin\theta_1\sin\theta_2$

$\sin(\theta_1-\theta_2) = \sin\theta_1\cos\theta_2 - \cos\theta_1\sin\theta_2$

$$a = \frac{-\omega_1^2 R\cos(\theta_1-\theta_2) - \alpha_1 R\sin(\theta_1-\theta_2) - \omega_2^2 L}{\cos\theta_2} \quad \text{(ans)}$$

where θ_1, ω_1 and α_1 are given and θ_2 and ω_2 are found as above.

4.46 a) The approximate slider acceleration is given by:

$$a = R\omega^2(\cos\theta + \frac{R}{L}\cos 2\theta) \quad \text{for constant } \omega,$$

$$\frac{da}{dt} = R\omega^3\{-\sin\theta - 2\frac{R}{L}\sin 2\theta\}$$

$$\frac{da}{dt} = 0 \text{ yields } \sin\theta + 2\frac{R}{L}\sin 2\theta = 0$$

For L/R = 2, $\sin\theta + \sin 2\theta = 0$

The roots are θ = 0, 120*, 180, and 240°* from which

b)

θ	$a/(R\omega^2) = \cos\theta + \frac{1}{2}\cos 2\theta$
0°	1.5 (pos. toward crankshaft)
120	-0.75*
180	-0.5
240	-0.75*

$a_{max} = 1.5\ R\omega^2$ (ans)

*Approximate values since the equation is approximate. See plot of $a/(R\omega^2)$ vs. θ based on exact equations.

4.46

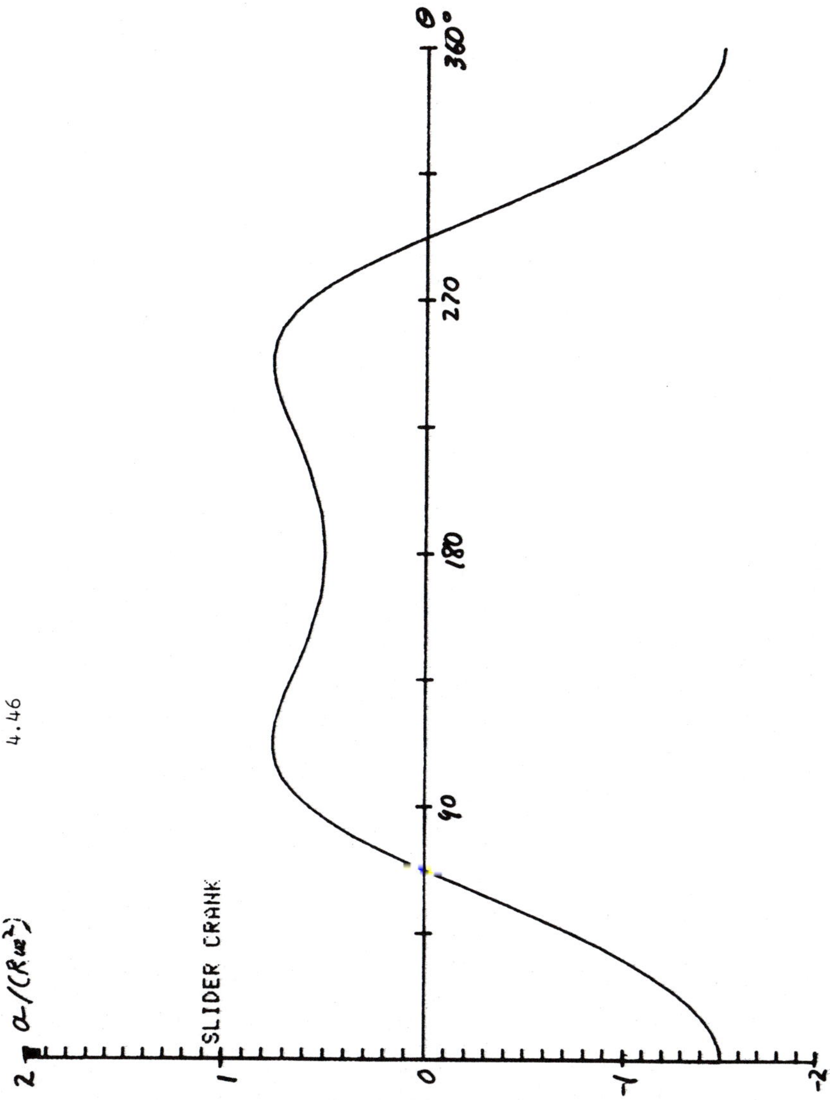

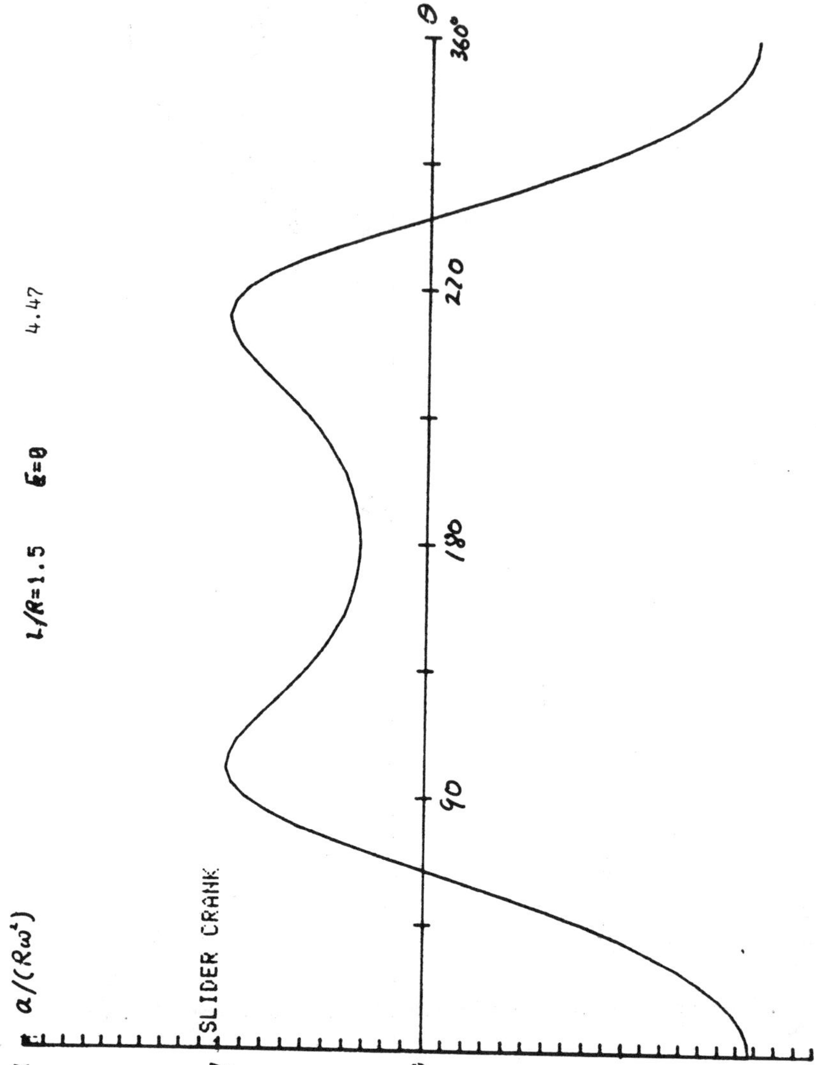
a/(Rω²)
SLIDER CRANK
L/R=1.5
ε=0
4.47
2
1
0
-1
-2
90
180
220
360°
θ

4.48

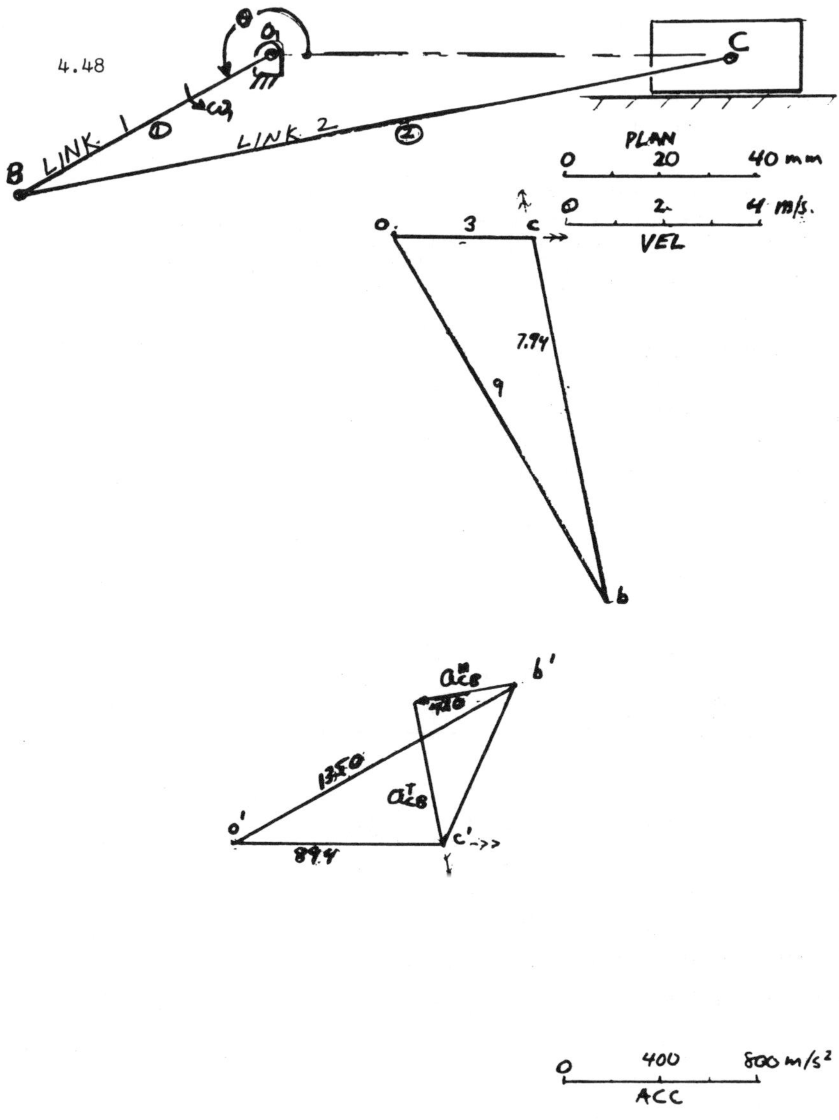

LINK 1
LINK 2
B
C
PLAN
0 20 40 mm
0 2 4 m/s.
VEL
o
c
b
3
7.94
9
b'
o'
c'
1350
894
0 400 800 m/s²
ACC

4.47 a) See above problem. For L/R = 1.5, extremum values of acceleration are found by solving

$$f(\theta) = \sin\theta + \frac{2}{1.5}\sin 2\theta = 0$$

The roots may be found using a program for zeros of functions $[f(\theta) = 0]$. The roots are θ = 0, 112.03*, 180 and 247.97°* from which

b)

θ	$a/R\omega^2 = \cos\theta + \frac{1}{1.5}\cos 2\theta$
0°	1.667
112.03	-0.854*
180	-0.3333
247.97	-0.854*

$a_{max} = 1.6667\ R\omega^2$ (ans)

*Use of the approximate acceleration equation will result in substantial errors except at θ=0 and 180°. For the in-line slider crank linkage, $a_{max}=(1+R/L)R\omega^2$ and occurs at θ=0. See plot of $a/(R\omega^2)$ based on exact equations.

4.48 See figure.

$ob = \omega_1 O_1B = 150 \times 0.06 = 9$ m/s.

$oc = 3$ m/s $bc = 7.94$ m/s (measured)

$o'b' = a_B{}^n = \omega_1{}^2 O_1B = 150^2 \times 0.06 = 1350$ m/s^2

$a_{CB}{}^n = bc^2/BC = 7.94^2/0.15 = 420$ m/s^2

$o'c' = a_C = 894$ m/s^2 (measured).

4.49 See figure. Let r_o lie along the real axis as shown.

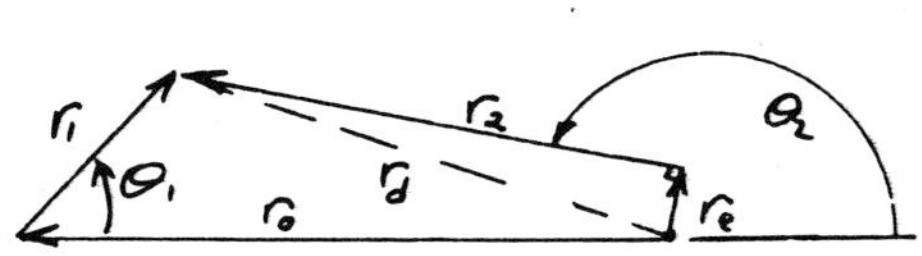

Then, $-r_o + r_1e^{j\theta_1} = r_ee^{j(\theta_2-\pi/2)} + r_2e^{j\theta_2}$ (1)

where eccentricity (identified as r_e instead of e) is measured perpendicular to r_2. We may write

$-r_o + r_1e^{j\theta_1} = (-jr_e + r_2)e^{j\theta_2}$ (2)

The square of the diagonal is given by

$r_d{}^2 = r_e{}^2 + r_2{}^2 = r_o{}^2 + r_1{}^2 - 2r_or_1\cos\theta_1$ (3)

4.49 (Cont'd.)

Given r_o, r_1, r_e and θ_1, we may find r_2 and θ_2:

$$r_2 = \sqrt{r_o^2 - r_1^2 - r_e^2 - 2r_or_1\cos\theta_1} \quad \text{(ans)} \qquad (4)$$

$$\theta_2 = 180^o - \tan^{-1}[r_1\sin\theta_1/(r_o - r_1\cos\theta_1)] + \tan^{-1}[r_e/r_2] \text{(ans)} \quad (5)$$

The proper angle quadrant must be checked for certain configurations. Differentiating eq. 2 with respect to time:

$$j\omega_1 r_1 e^{j\theta_1} = [j\omega_2(-jr_e + r_2) + \dot{r}_2]e^{j\theta_2} \qquad (6)$$

Multiplying eq. 6 by $e^{-j\theta_2}$:

$$j\omega_1 r_1 e^{j(\theta_1-\theta_2)} = j\omega_2(-jr_e + r_2) + \dot{r}_2 \qquad (7)$$

The imaginary components of eq. 7 yield:

$$\omega_1 r_1 \cos(\theta_1-\theta_2) = \omega_2 r_2 \quad \text{from which}$$

$$\omega_2 = \omega_1 r_1 \cos(\theta_1-\theta_2)/r_2 \quad \text{(ans)} \qquad (8)$$

Using the real components

$$-\omega_1 r_1 \sin(\theta_1-\theta_2) = \omega_2 r_e + \dot{r}_2$$

from which

$$\dot{r}_2 = -\omega_2 r_e - \omega_1 r_1 \sin(\theta_1-\theta_2) \quad \text{(ans)} \qquad (9)$$

Differentiating eq. 6:

$$(j\alpha_1 - \omega_1^2)r_1 e^{j\theta_1} = \{j\alpha_2(-jr_e + r_2) + j\omega_2\dot{r}_2 + \ddot{r}_2$$
$$+ j\omega_2[j\omega_2(-jr_e + r_2) + \dot{r}_2]\}e^{j\theta_2}$$

Multiplying by $e^{-j\theta_2}$:

$$(j\alpha_1 - \omega_1^2)r_1 e^{j(\theta_1-\theta_2)} = \alpha_2(r_e + jr_2) + j\omega_2\dot{r}_2 + \ddot{r}_2$$
$$+ \omega_2^2(jr_e - r_2) + j\omega_2\dot{r}_2 \qquad (10)$$

The real components of eq. 10 yield

$$-\alpha_1 r_1 \sin(\theta_1-\theta_2)$$
$$-\omega_1^2 r_1 \cos(\theta_1-\theta_2) = \alpha_2 r_e + \ddot{r}_2 - \omega_2^2 r_2 \quad \text{from which}$$

$$\ddot{r}_2 = \omega_2^2 r_2 - \omega_1^2 r_1 \cos(\theta_1-\theta_2) - \alpha_2 r_e - \alpha_1 r_1 \sin(\theta_1-\theta_2) \text{(ans)} \qquad (11)$$

where α_2 is given by equating the imaginary components:

$$\alpha_1 r_1 \cos(\theta_1-\theta_2) - \omega_1^2 r_1 \sin(\theta_1-\theta_2) = \alpha_2 r_2 + 2\omega_2\dot{r}_2 + \omega_2^2 r_e \text{ from which}$$

$$\alpha_2 = \frac{\alpha_1 r_1 \cos(\theta_1-\theta_2) - \omega_1^2 r_1 \sin(\theta_1-\theta_2) - 2\omega_2\dot{r}_2 - \omega_2^2 r_e}{r_2} \quad \text{(ans)} \qquad (12)$$

4.50 See sketch. Let r_o lie in the negative direction on the real axis. Then,

$$-r_o + r_1 e^{j\theta_1} = r_2 e^{j\theta_2} \qquad (1)$$

$$\theta_1 + \phi + 180^o - \theta_2 = 180^o \quad \text{from which}$$

$$\theta_2 = \theta_1 + \phi \quad \text{(ans)} \qquad (2)$$

4.50 (Cont'd.)

Using the law of sines:

$r_1 = r_o\sin(180^o-\theta_2)/\sin\phi = r_o\sin\theta_2/\sin\phi$ (ans) (3)

$r_2 = r_o\sin\theta_1/\sin\phi$ (ans) (4)

Differentiating eq. 1,

$$(jr_1\omega + \dot{r}_1)e^{j\theta_1} = (jr_2\omega + \dot{r}_2)e^{j\theta_2} \quad (5)$$

where, $\omega_2 = \omega_1 = \omega$ by differentiation of eq. 2 where $\phi = \theta_2 - \theta_1$ (a constant)

Multiplying eq. 5 by $e^{-j\theta_2}$

$(jr_1\omega + \dot{r}_1)e^{j(\theta_1-\theta_2)} = jr_2\omega + \dot{r}_2$ or

$$(jr_1\omega + \dot{r}_1)e^{-j\phi} = jr_2\omega + \dot{r}_2 \quad (6)$$

The imaginary components of eq. 6 yield

$r_1\omega\cos(-\phi) + \dot{r}_1\sin(-\phi) = r_2\omega$

(recalling $\phi = \theta_2 - \theta_1$)

Thus:

$$\dot{r}_1 = \frac{\omega(r_1\cos\phi - r_2)}{\sin\phi} \quad (7)$$

Multiplying eq. 5 by $e^{-j\theta_1}$

$jr_1\omega + \dot{r}_1 = (jr_2\omega + \dot{r}_2)e^{j\phi}$

The imaginary components yield

$r_1\omega = r_2\omega\cos\phi + \dot{r}_2\sin\phi$ from which

$$\dot{r}_2 = \frac{\omega(r_1 - r_2\cos\phi)}{\sin\phi} \quad (8)$$

A second differentiation of eq. 2 yields

$\alpha_2 = \alpha_1 = \alpha$ for constant ϕ

Differentiating eq. 5:

$$(j\dot{r}_1\omega + jr_1\alpha + \ddot{r}_1 - \omega^2 r_1 + j\omega\dot{r}_1)e^{j\theta_1} = (j\dot{r}_2\omega + jr_2\alpha + \ddot{r}_2 - \omega^2 r_2 + j\omega\dot{r}_2)e^{j\theta_2} \quad (9)$$

Multiplying by $e^{-j\theta_2}$, where $\phi = \theta_2 - \theta_1$ yields

$(2j\dot{r}_1\omega+jr_1\alpha+\ddot{r}_1-\omega^2r_1)e^{-j\phi}=2j\dot{r}_2\omega+jr_2\alpha+\ddot{r}_2-\omega^2r_2$ from which

$(2j\dot{r}_1\omega+jr_1\alpha+\ddot{r}_1-\omega^2r_1)(\cos\phi-j\sin\phi)=2j\dot{r}_2\omega+jr_2\alpha+\ddot{r}_2-\omega^2r_2$

Equating the imaginary components:

$2\dot{r}_1\omega\cos\phi+r_1\alpha\cos\phi+\omega^2r_1\sin\phi-\ddot{r}_1\sin\phi=2\dot{r}_2\omega+r_2\alpha$ from which

$$\ddot{r}_1 = \frac{2\dot{r}_1\omega\cos\phi+r_1\alpha\cos\phi+\omega^2r_1\sin\phi-2\dot{r}_2\omega-r_2\alpha}{\sin\phi} \quad (10)$$

Similarly, the imaginary parts are equated after multiplying eq. 9 by $e^{-j\theta_1}$ resulting in

4.50 (Cont'd.)

$$2j\dot{r}_1\omega + jr_1\alpha + \ddot{r}_1 - \omega^2 r_1 = (2j\dot{r}_2\omega + jr_2\alpha + \ddot{r}_2 - \omega^2 r_2)e^{j\phi}$$

from which

$$\ddot{r}_2 = \frac{2\dot{r}_1\omega + r_1\alpha - 2\dot{r}_2\omega\cos\phi - r_2\alpha\cos\phi + r_2\omega^2\sin\phi}{\sin\phi} \qquad (11)$$

4.51 See sketch.

As above, $\theta_2 = \theta_1 + \phi$ and

$r_1 = r_o\sin\theta_2/\sin\phi$

$r_2 = r_o\sin\theta_1/\sin\phi$

$\omega = \dot{\theta}_1 = \dot{\theta}_2 \quad \alpha = \ddot{\theta}_1 = \ddot{\theta}_2$

$r_2 = r_o + r_1$

Let B_1 lie on link 1, B_2 on link 2 and B on the sliding member. Then,

$$\underset{\sim}{V}_B = \underset{\sim}{\omega}\times\underset{\sim}{r}_1 + \dot{\underset{\sim}{r}}_{1r} = \underset{\sim}{\omega}\times\underset{\sim}{r}_2 + \dot{\underset{\sim}{r}}_{2r} \qquad (1)$$

where

$$\omega\times r_1 = \omega r_1 \begin{vmatrix} \underset{\sim}{i} & \underset{\sim}{j} & \underset{\sim}{k} \\ 0 & 0 & 1 \\ \cos\theta_1 & \sin\theta_1 & 0 \end{vmatrix} = \omega r_1(-\underset{\sim}{i}\sin\theta_1 + \underset{\sim}{j}\cos\theta)$$

$\dot{\underset{\sim}{r}}_{1r} = \dot{r}_1\underset{\sim}{r}_1{}^u = \dot{r}_1(\underset{\sim}{i}\cos\theta_1 + \underset{\sim}{j}\sin\theta_1)$ etc.

Thus, from eq. 1,

$$\omega r_1(-\underset{\sim}{i}\sin\theta_1 + \underset{\sim}{j}\cos\theta_1) + \dot{r}_1(\underset{\sim}{i}\cos\theta_1 + \underset{\sim}{j}\sin\theta_1) =$$

$$\omega r_2(-\underset{\sim}{i}\sin\theta_2 + \underset{\sim}{j}\cos\theta_2) + \dot{r}_2(\underset{\sim}{i}\cos\theta_2 + \underset{\sim}{j}\sin\theta_2) \qquad (2)$$

Equating the $\underset{\sim}{i}$ components:

$\omega(-r_1\sin\theta_1 + r_2\sin\theta_2) = -\dot{r}_1\cos\theta_1 + \dot{r}_2\cos\theta_2$

and the $\underset{\sim}{j}$ components

$\omega(r_1\cos\theta_1 - r_2\cos\theta_2) = -\dot{r}_1\sin\theta_1 + \dot{r}_2\sin\theta_2$

In matrix form:

$$\begin{bmatrix} -\cos\theta_1 & \cos\theta_2 \\ -\sin\theta_1 & \sin\theta_2 \end{bmatrix} \begin{bmatrix} \dot{r}_1 \\ \dot{r}_2 \end{bmatrix} = \omega \begin{bmatrix} -r_1\sin\theta_1 + r_2\sin\theta_2 \\ r_1\cos\theta_1 - r_2\cos\theta_2 \end{bmatrix}$$

from which

$$\dot{r}_1 = \frac{\omega}{D}\begin{vmatrix} -r_1\sin\theta_1 + r_2\sin\theta_2 & \cos\theta_2 \\ r_1\cos\theta_1 - r_2\cos\theta_2 & \sin\theta_2 \end{vmatrix}$$

4.51 (Cont'd.)

where $D = -\cos\theta_1\sin\theta_2 + \sin\theta_1\cos\theta_2 = \sin(\theta_1 - \theta_2)$

$= \sin(-\phi) = -\sin\phi$

Thus

$$\dot{r}_1 = \omega(r_2\sin^2\theta_2 - r_1\sin\theta_1\sin\theta_2 - r_1\cos\theta_1\cos\theta_2$$

$$+ r_2\cos^2\theta_2)/(-\sin\phi) = \frac{\omega(r_2 - r_1\cos(\theta_1 - \theta_2))}{-\sin\phi}$$

$$\dot{r}_1 = \frac{\omega(r_1\cos\phi - r_2)}{\sin\phi} \tag{3}$$

Similarly

$$\dot{r}_2 = \frac{\omega(r_1 - r_2\cos\phi)}{\sin\phi} \tag{4}$$

Acceleration:

$$\underset{\sim}{a}_B = \underset{\sim}{\omega}\times(\underset{\sim}{\omega}\times\underset{\sim}{r}_1) + 2\underset{\sim}{\omega}\times\dot{\underset{\sim}{r}}_{1r} + \ddot{\underset{\sim}{r}}_{1r} + \underset{\sim}{\alpha}\times\underset{\sim}{r}_1$$

$$= \underset{\sim}{\omega}\times(\underset{\sim}{\omega}\times\underset{\sim}{r}_2) + 2\underset{\sim}{\omega}\times\dot{\underset{\sim}{r}}_{2r} + \ddot{\underset{\sim}{r}}_{2r} + \underset{\sim}{\alpha}\times\underset{\sim}{r}_2 \tag{5}$$

where

$$\underset{\sim}{\omega}\times(\underset{\sim}{\omega}\times\underset{\sim}{r}_1) = \omega^2 r_1(-\underset{\sim}{i}\cos\theta_1 - \underset{\sim}{j}\sin\theta_1)$$

$$2\underset{\sim}{\omega}\times\dot{\underset{\sim}{r}}_{1r} = 2\omega\dot{r}_1\begin{vmatrix} \underset{\sim}{i} & \underset{\sim}{j} & \underset{\sim}{k} \\ 0 & 0 & 1 \\ \cos\theta_1 & \sin\theta_1 & 0 \end{vmatrix} = 2\omega\dot{r}_1(-\underset{\sim}{i}\sin\theta_1 + \underset{\sim}{j}\cos\theta_1)$$

$$\ddot{r}_{1r} = \ddot{r}_1(\underset{\sim}{i}\cos\theta_1 + \underset{\sim}{j}\sin\theta_1)$$

$$\underset{\sim}{\alpha}\times\underset{\sim}{r}_1 = \alpha r_1(-\underset{\sim}{i}\sin\theta_1 + \underset{\sim}{j}\cos\theta_1) \qquad \text{etc.}$$

Substituting in eq. 5 and equating the $\underset{\sim}{i}$ components:

$$-\omega^2 r_1\cos\theta_1 - 2\omega\dot{r}_1\sin\theta_1 + \ddot{r}_1\cos\theta_1 - \alpha r_1\sin\theta_1 =$$

$$-\omega^2 r_2\cos\theta_2 - 2\omega\dot{r}_2\sin\theta_2 + \ddot{r}_2\cos\theta_2 - \alpha r_2\sin\theta_2 \tag{6}$$

Similarly for the $\underset{\sim}{j}$ components

$$-\omega^2 r_1\sin\theta_1 + 2\omega\dot{r}_1\cos\theta_1 + \ddot{r}_1\sin\theta_1 + \alpha r_1\cos\theta_1 =$$

$$-\omega^2 r_2\sin\theta_2 + 2\omega\dot{r}_2\cos\theta_2 + \ddot{r}_2\sin\theta_2 + \alpha r_2\cos\theta_2 \tag{7}$$

Writing eqs. 6 and 7 in matrix form:

$$\begin{bmatrix} \cos\theta_1 & -\cos\theta_2 \\ \sin\theta_1 & -\sin\theta_2 \end{bmatrix} \begin{bmatrix} \ddot{r}_1 \\ \ddot{r}_2 \end{bmatrix} = \begin{bmatrix} a \\ b \end{bmatrix}$$

4.50 and
4.51

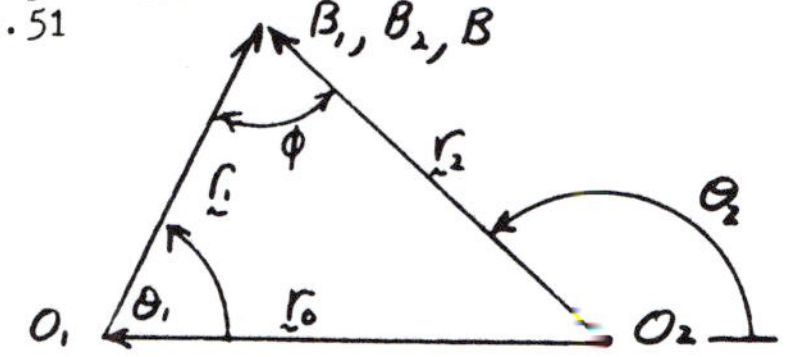

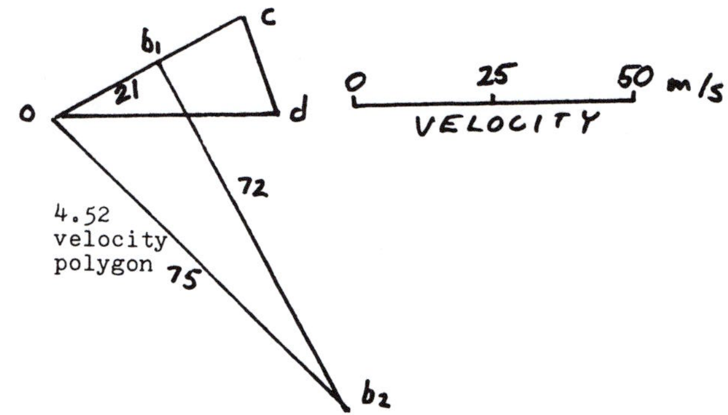

4.52
velocity
polygon

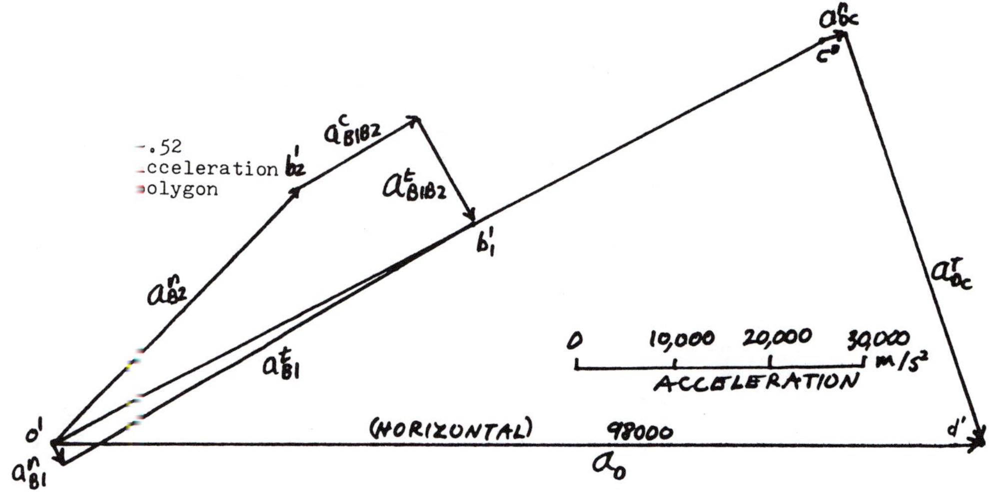

.52
cceleration
olygon

4.51 (Cont'd.)

where $a = \omega^2(r_1\cos\theta_1 - r_2\cos\theta_2) + 2\omega(\dot{r}_1\sin\theta_1 - \dot{r}_2\sin\theta_2)$

$+ \alpha(r_1\sin\theta_1 - r_2\sin\theta_2)$

$b = \omega^2(r_1\sin\theta_1 - r_2\sin\theta_2) + 2\omega(-\dot{r}_1\cos\theta_1 + \dot{r}_2\cos\theta_2)$

$+ \alpha(-r_1\cos\theta_1 + r_2\cos\theta_2)$

$$\ddot{r}_1 = \frac{-a\sin\theta_2 + b\cos\theta_2}{D}$$

where $D = -\cos\theta_1\sin\theta_2 + \cos\theta_2\sin\theta_1$

$= \sin(\theta_1 - \theta_2) = -\sin\phi$

$\ddot{r}_1 = [a\sin\theta_2 - b\cos\theta_2]/\sin\phi$

$\ddot{r}_1 = [\omega^2 r_1(\cos\theta_1\sin\theta_2 - \cos\theta_2\sin\theta_1)$

$+ \omega^2 r_2(-\cos\theta_2\sin\theta_2 + \sin\theta_2\cos\theta_2)$

$+ 2\omega\dot{r}_1(\sin\theta_1\sin\theta_2 + \cos\theta_1\cos\theta_2)$

$+ 2\omega\dot{r}_2(-\sin^2\theta_2 - \cos^2\theta_2)$

$+ \alpha r_1(\sin\theta_1\sin\theta_2 + \cos\theta_1\cos\theta_2)$

$+ \alpha r_2(-\sin^2\theta_2 - \cos^2\theta_2)]/\sin\phi$

$\ddot{r}_1 = [\omega^2 r_1\sin\phi + 2\omega\dot{r}_1\cos\phi - 2\omega\dot{r}_2 + \alpha r_1\cos\phi - \alpha r_2]/\sin\phi$

$\ddot{r}_2 = [b\cos\theta_1 - a\sin\theta_1]/D = [-b\cos\theta_1 + a\sin\theta_1]/\sin\phi$

$\ddot{r}_2 = [\omega^2 r_1(-\sin\theta_1\cos\theta_1 + \cos\theta_1\sin\theta_1) + \omega^2 r_2(\cos\theta_1\sin\theta_2 - \cos\theta_2\sin\theta_1)$

$+ 2\omega\dot{r}_1(\cos^2\theta_1 + \sin^2\theta_1) + 2\omega\dot{r}_2(-\cos\theta_2\cos\theta_1 - \sin\theta_2\sin\theta_1)$

$+ \alpha r_1(\cos^2\theta_1 + \sin^2\theta_1) + \alpha r_2(-\cos\theta_2\cos\theta_1 - \sin\theta_2\sin\theta_1)]/\sin\phi$

$\ddot{r}_2 = [\omega^2 r_2\sin\phi + 2\omega\dot{r}_1 - 2\omega\dot{r}_2\cos\phi + \alpha r_1 + \alpha r_2\cos\phi]/\sin\phi$

4.52 (a) See figure. See also solution 3.58.

$\underset{\sim}{a}^n_{B1} + \underset{\sim}{a}^t_{B1} = \underset{\sim}{a}_{B2}^n + \underset{\sim}{a}_{B2}^t + \underset{\sim}{a}^c_{B1B2} + \underset{\sim}{a}^t_{B1B2}$

$a^n_{B1} = ob_1^2/O_1B_1 = 21^2/0.221 = 1995 \text{ m/s}^2$

$a^n_{B2} = \omega_2^2 O_2B_2 = 500^2 \times 0.150 = 37500 \text{ m/s}^2$

$a^t_{B2} = \alpha_2 O_2B_2 = 0$

$a^c_{B1B2} = 2\omega_1 \times V_{B1B2} = 2 \times 95 \times 72 = 13{,}700 \text{ m/s}^2$

$o^1b_1^1 = 50500 \text{ m/s}^2$

4.52 (Cont'd.)

b) $\frac{o'c'}{o'b_1'} = \frac{O_1C}{O_1B}$ $\quad o'c' = 91{,}400\ m/s^2$

c) $a^n_{DC} = \frac{cd^2}{CD} = 19^2/0.160 = 2250\ m/s^2$

d) $a_D = 98{,}000\ m/s^2$ to right

$\alpha_1 = a^t_{B1}/O_1B_1 = 50{,}000/0.221 = 226{,}000\ rad/s^2$ cw

4.53 See figure for $\theta_1 = 45^o$.

<u>Velocity</u>

$ob = \omega_1 O_1B = 100x0.08 = 8\ m/s \perp O_1B$ (downward and to right)

$bc = 6\ m/s$ and $oc = 7.6\ m/s$ (to right) from velocity polygon.

<u>Acceleration</u>

$\underset{\sim}{a}_C = \underset{\sim}{a}_B^n + \underset{\sim}{a}_B^t + \underset{\sim}{a}^n_{CB} + \underset{\sim}{a}^t_{CB}$

$a_B^n = \omega_1^2 O_1B = 100^2x0.08 = 800\ m/s^2$ parallel to BO_1 toward O_1.

$a_B^t = \alpha_1 O_1B = 0$

$a^n_{CB} = bc^2/BC = 6^2/0.3 = 120\ m/s^2$ parallel to CB toward B.

$\alpha_2 = a^t_{CB}/CB = 560\ m/s^2/0.3\ m = 1870\ m/s^2$ cc.

See plot of a_C vs θ_1 for other values of θ_1.

4.54 See velocity polygon for counterclockwise rotation. $V_B = ob$ is upward and to left. The acceleration analysis is the same as the above problem.

4.55 See figure for $\theta_1 = 45^o$.

<u>Velocity</u>

$ob_1 = \omega_1 O_1B_1 = 10x40 = 400\ mm/s$

$ob_2 = 329$ $\quad b_1b_2 = 215\ mm/s$ (from polygon)

$oc = ob_2 \times (O_2C/O_2B_2) = 329x150/131 = 385\ mm/s$

$\omega_2 = ob_2/O_2B_2 = 329/131 = 2.57\ rad/s$ ccw.

<u>Acceleration</u>

$\underset{\sim}{a}^n_{B1} + \underset{\sim}{a}^t_{B1} = \underset{\sim}{a}^n_{B2} + \underset{\sim}{a}^t_{B2} + \underset{\sim}{a}^t_{B1B2} + \underset{\sim}{a}^c_{B1B2}$

$a^n_{B1} = \omega_1^2 O_1B = 10^2x40 = 4000\ mm/s^2$. $\quad a^t_{B1} = \alpha_1 O_1B_1 = 0$.

$a^n_{B2} = ob_2^2/O_2B_2 = 335^2/131 = 863\ mm/s^2$

$a^c_{B1B2} = 2\omega_2 V_{B1B2} = 2x2.57x215 = 1105\ mm/s^2$

$a^t_{B2} = 1050$ $\quad a_{B2} = o'b_2' = 1400\ mm/s^2$ (from polygon)

$o'c' = o'b_2' \times O_2C/O_2B_2 = 1400x150/131 = 1600\ mm/s^2$

$\alpha_2 = a^t_{B2}/O_2B_2 = 1050/131 = 8\ rad/s^2$ cw

For other values of θ_1, see plot of α_2 vs θ_1.

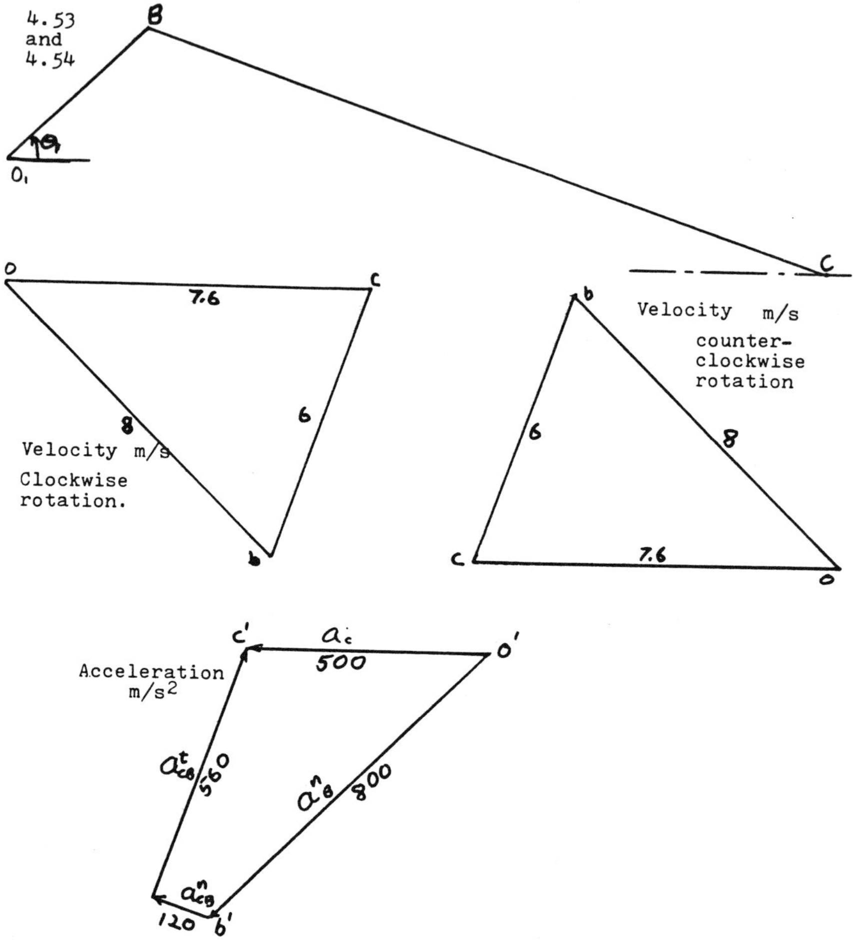
4.53
and
4.54
B
O_1
C
o
c
7.6
8
6
b
Velocity m/s
Clockwise
rotation.
b
Velocity m/s
counter-
clockwise
rotation
6
8
c
7.6
o
c'
a_c
500
o'
Acceleration
m/s²
a^t_{cB}
560
a^n_B
800
a^n_{cB}
120
b'

4.53 and 4.54

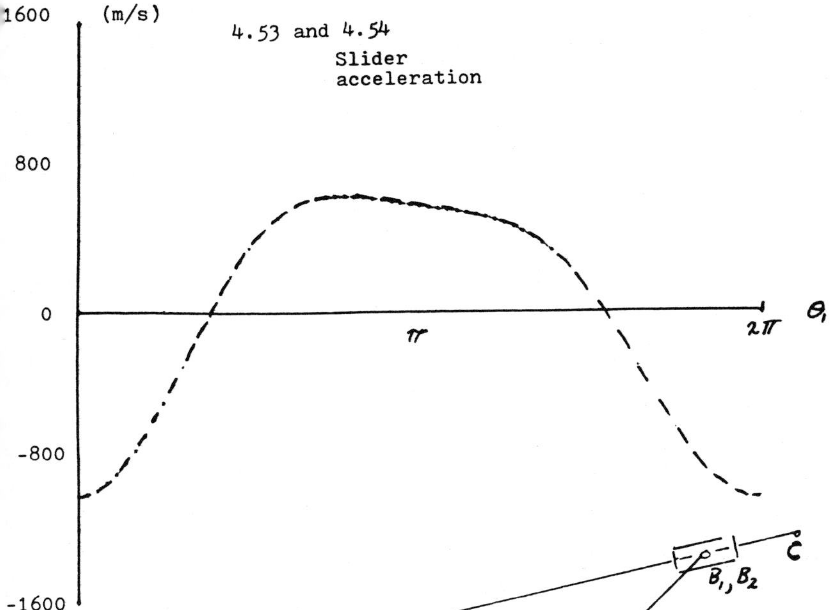

4.55

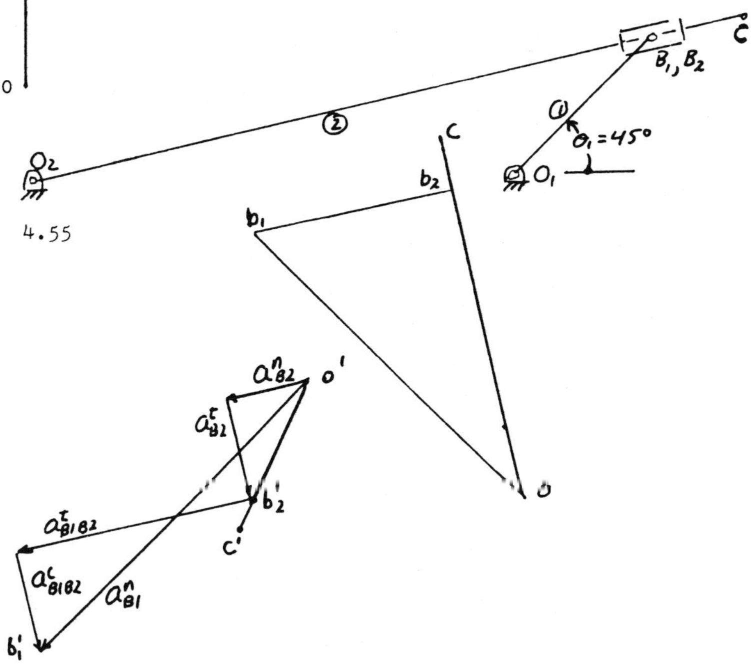

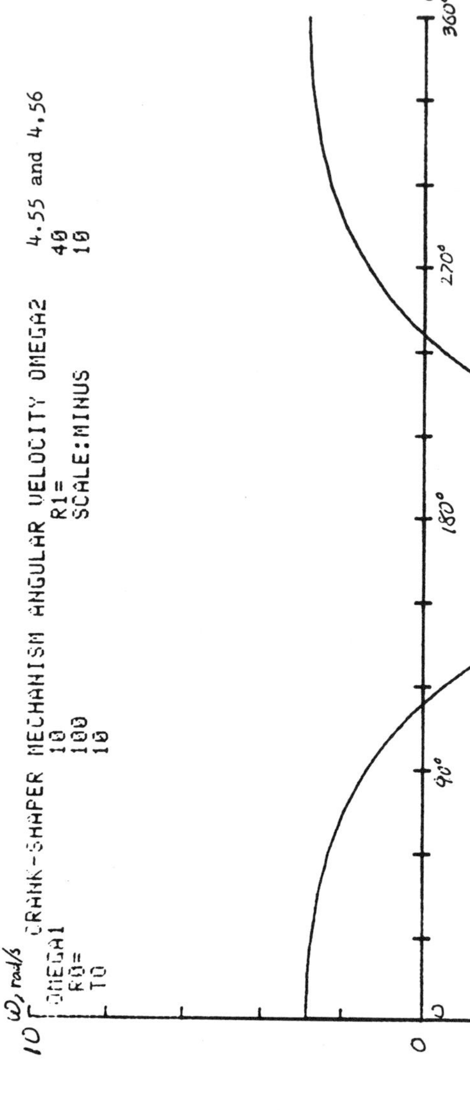
CRANK-SHAPER MECHANISM ANGULAR VELOCITY OMEGA2 4.55 and 4.56
OMEGA1 10
R0= 100
T0 10
R1= 40
SCALE:MINUS 10
ω, rad/s
10
0
-10
90°
180°
270°
360°
θ1

4.55 and 4.56

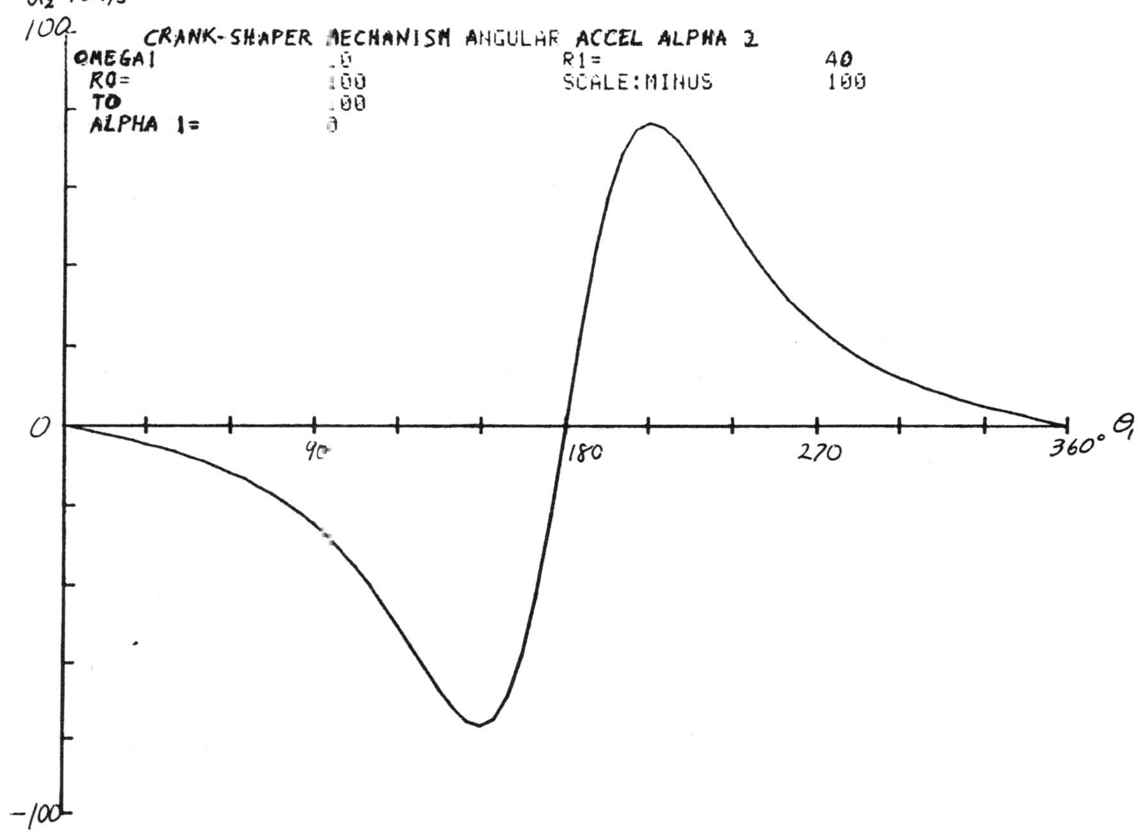
α2 rad/s²
100
CRANK-SHAPER MECHANISM ANGULAR ACCEL ALPHA 2
OMEGA1
R0=
TO
ALPHA 1=
R1= 40
SCALE:MINUS 100
0
90
180
270
360° θ1
−100

4.56

```
LIST
100 PAGE
110 D=32
120 W1=10
125 W3=W1↑2
130 WINDOW 0,2*PI,-W3,W3
140 VIEWPORT 0,130,0,100
150 AXIS @D:PI/6,W3/5
170 R1=40
180 R0=100
181 A1=0
182 HOME
184 PRINT @D:"       CRANK-SHAPER MECHANISM ANGULAR ACCEL ALPHA 2"
185 PRINT @D:" OMEGA1",W1,"R1=",R1,"  R0=",R0,"SCALE:MINUS",W3,"  TO",W3
186 PRINT @D:"  ALPHA 1=",A1
190 MOVE @D:0,0
200 FOR T1=0 TO 2*PI STEP PI/36
210 R2=(R0↑2+R1↑2+2*R0*R1*COS(T1))↑0.5
220 T2=ASN(R1*SIN(T1)/R2)
230 W2=W1*R1*COS(T1-T2)/R2
240 V1=W1*R1*SIN(T2-T1)
250 A2=(W1↑2*R1*SIN(T2-T1)+A1*R1*COS(T2-T1)-2*W2*V1)/R2
260 DRAW @D:T1,A2
270 NEXT T1
```

4.56 Let $\theta_1=45^\circ=\pi/4$

$R_o + R_1e^{j\theta_1} = R_2e^{j\theta_2}$

$100 + 40e^{j\pi/4} = R_2e^{j\theta_2}$ from which

$R_2 = 128.28 + j28.28 = 131.37\angle\theta_2 = 12.43^\circ$

$\omega_2 = \omega_1R_1\cos(\theta_1 - \theta_2)/R_2 = 10x40\cos(45-12.43)/131.37$

$\omega_2 = 2.566$ rad/s. $V_C = \omega_2O_2C = 2.566x150 = 384.9$ mm/s

$V_{B2} = \omega_2R_2 = 2.566x131.37 = 329$ mm/s

$V_{B1B2} = -\omega_1R_1\sin(\theta_1 - \theta_2) = -10x40\sin(45 - 12.43) = -215.31$ mm/s

$\alpha_2 = [\omega_1^2R_1\sin(\theta_2 - \theta_1)+ \alpha_1R_1\cos(\theta_2 - \theta_1)- 2\omega_2V_{B1B2}]/R_2$

$\alpha_2 = [10^2x40\sin(12.43 - 45)- 2x2.566x(-215.31)]/131.37$

$\alpha_2 = -7.98$ rad/s^2

$a_C = \omega_2^2O_2C + j'\alpha_2O_2C$ (axes referred to link 2)

$= 2.566^2x150 + j'7.98x150 = 1551.7$ mm/s^2

For other values of θ_1, see microcomputer program and plot of ω_2 and α_2 vs θ_1.

4.57 See figure.

Velocity

$\underset{\sim}{oc} = \underset{\sim}{ob} + \underset{\sim}{bc}$

$ob = \omega_1O_1B = 150x0.060 = 9$ m/s$\perp O_1B$ $bc\perp BC$

$bc = 4.8$ $oc = 6.1$ m/s (from polygon)

$\omega_2 = bc/BC = 4.8/0.150$

$\omega_2 = 32$ rad/s ccw

Acceleration

$\underset{\sim}{a}_C = \underset{\sim}{a}^n{}_B + \underset{\sim}{a}^t{}_B + \underset{\sim}{a}^n{}_{CB} + \underset{\sim}{a}^t{}_{CB}$

$a^n{}_B = \omega_1^2O_1B = 150^2x0.060 = 1350$ m/s^2 $||O_1B$

$a^t{}_B = 0$ $a^n{}_{CB} = bc^2/BC = 4.8^2/0.15 = 154$ m/s^2 $||BC$

$a^t{}_{CB} \perp BC = 1190$ m/s^2 (from polygon)

$a_C = o'c' = 940$ m/s^2 to the right (from polygon).

$\alpha_2 = a^t{}_{CB}/CB = 1190/0.150 = 7930$ rad/s^2 ccw.

(Compare solution 4.59).

4.58 See figure and above solution.

Velocity

$ob = \omega_1O_1B = 150x0.060 = 9$ m/s $\perp O_1B$ (vertical)

$bc \perp BC$ (vertical); oc horizontal

$bc = 9$ m/s and $oc = 0$ (from polygon)

Acceleration

$a^n{}_B = \omega_1O_1B = 150^2x0.060 = 1350$ m/s^2 $||O_1B$

4.57

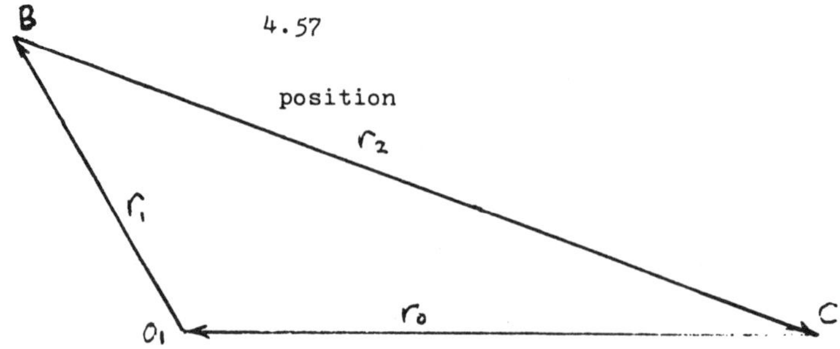

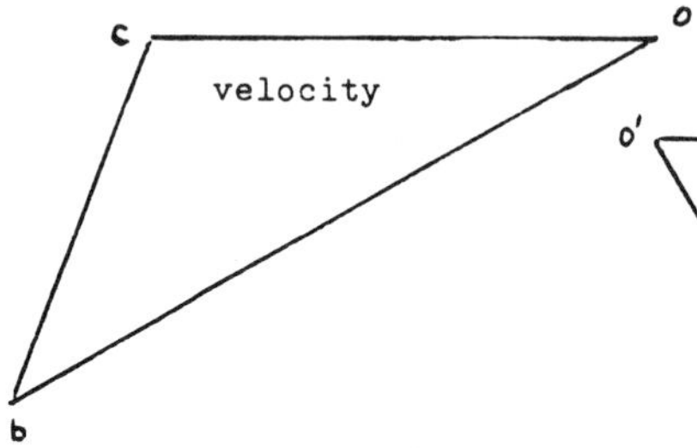

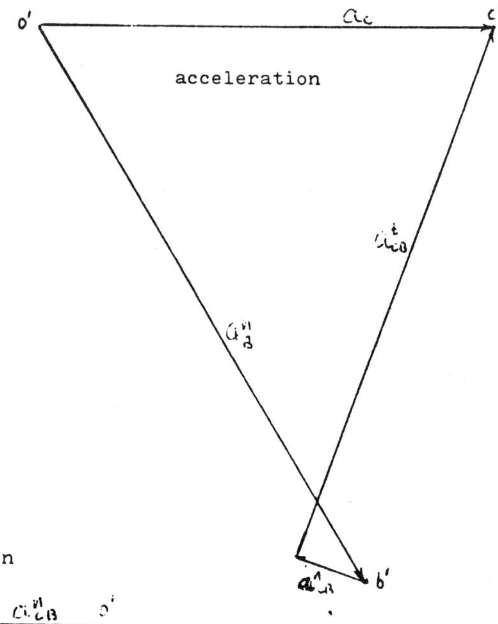

4.58

velocity

acceleration

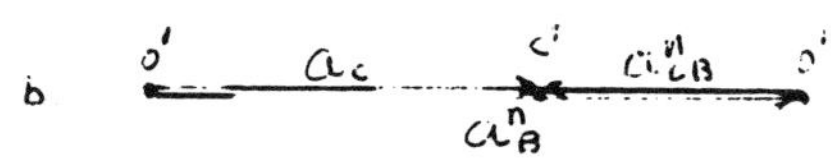

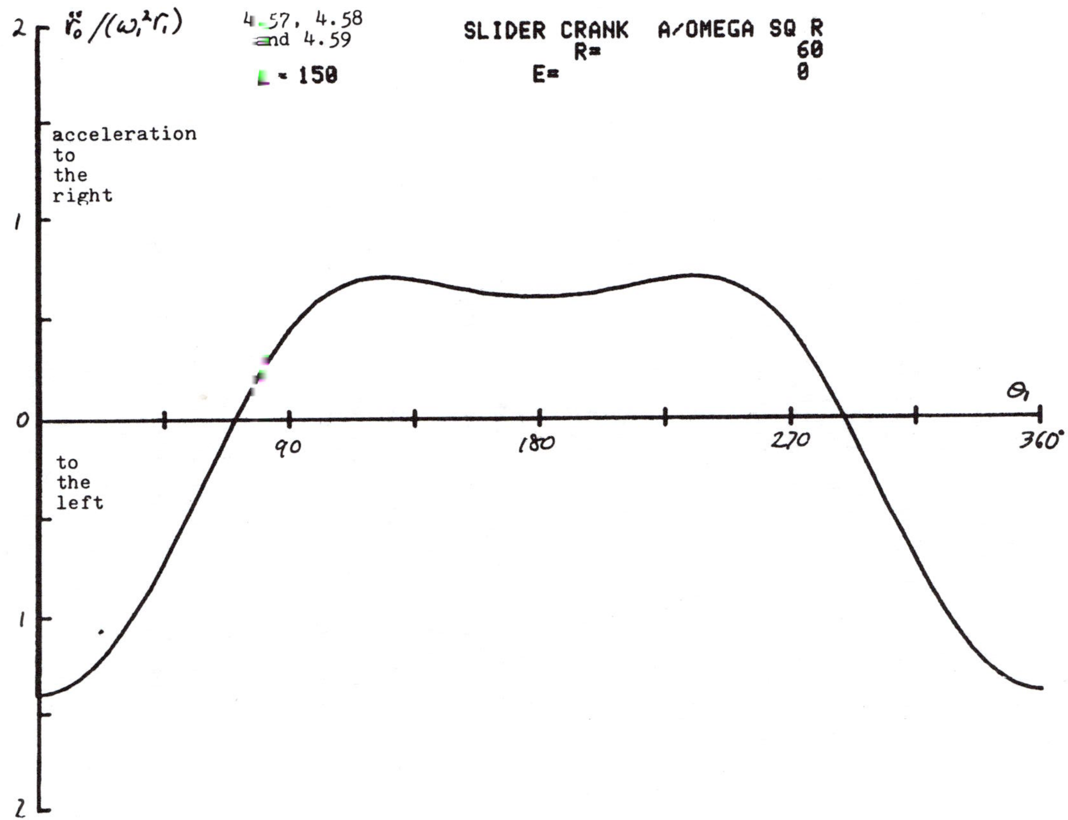

$\ddot{r}_0/(\omega_1^2 r_1)$
4.57, 4.58
and 4.59
L = 150
SLIDER CRANK A/OMEGA SQ R
R= 60
E= 0
acceleration
to
the
right
to
the
left
2
1
0
1
2
90
180
270
360°
θ_1

4.58 (Cont'd.)

$a^t_B = 0 \qquad a^n_{CB} = bc^2/BC = 9^2/0.150 = 540\ m/s^2 || BC$

$a_C = o'c' = 810\ m/s^2$ to the right $\qquad a^t_{CB} = 0$ (from polygon)

$\alpha_2 = a^t_{CB}/CB = 0$ See plot of $\ddot{r}_o/(\omega_1{}^2 r_1)$ vs θ_1 where $\ddot{r}_o = a_C$.

4.59 Position

Refer to Fig. 2.11 and Section 2.2.8. Using eq. 2.24 where $A = r_o \qquad B = r_2$ and $\qquad C = r_1$

$$\underset{\sim}{r}_o + \underset{\sim}{r}_1 + \underset{\sim}{r}_2 = 0 \qquad \text{(a)}$$

$$\underset{\sim}{r}_2 = -[\underset{\sim}{r}_1 \cdot (\underset{\sim}{r}_o{}^u x\underset{\sim}{k})](\underset{\sim}{r}_o{}^u x\underset{\sim}{k}) \pm \sqrt{r_2{}^2 - [\underset{\sim}{r}_1 \cdot (\underset{\sim}{r}_o{}^u x\underset{\sim}{k})]^2}\, \underset{\sim}{r}_o{}^u \qquad \text{(b)}$$

$\underset{\sim}{r}_o{}^u = -\underset{\sim}{i} \qquad \underset{\sim}{r}_o{}^u x\underset{\sim}{k} = -\underset{\sim}{i}x\underset{\sim}{k} = \underset{\sim}{j}$

$\underset{\sim}{r}_1 = 60(\underset{\sim}{i}\cos 120^\circ + \underset{\sim}{j}\sin 120^\circ) = -\underset{\sim}{i}30 + \underset{\sim}{j}51.96$

$r_2 = -60\sin 120^\circ \underset{\sim}{j} \pm \sqrt{150^2 - (60\sin 120^\circ)^2}\,(-\underset{\sim}{i})$

$\quad = \underset{\sim}{i}140.71 - \underset{\sim}{j}51.96$ (for the assembly position shown)

Velocity

$$\dot{\underset{\sim}{r}}_o + \underset{\sim}{\omega}_1 x\underset{\sim}{r}_1 + \underset{\sim}{\omega}_2 x\underset{\sim}{r}_2 = 0 \qquad \text{(c)}$$

Using solution 3.62,

$$\omega_2 = -\omega_1 r_{1x}/r_{2x} = -150x(-30)/140.71 = +31.98 \text{ rad/s ccw} \qquad \text{(d)}$$

$$\dot{r}_o = \omega_1 r_{1y} + \omega_2 r_{2y} = 150x51.96 + 31.98x(-51.96) = 6132 \text{ mm/s (to left)} \qquad \text{(e)}$$

Acceleration

$\ddot{\underset{\sim}{r}}_o + \underset{\sim}{\omega}_1 x(\underset{\sim}{\omega}_1 x\underset{\sim}{r}_1) + \underset{\sim}{\alpha}_1 x\underset{\sim}{r}_1 + \underset{\sim}{\omega}_2 x(\underset{\sim}{\omega}_2 x\underset{\sim}{r}_2) + \underset{\sim}{\alpha}_2 x\underset{\sim}{r}_2 = 0$ where

$\underset{\sim}{\omega}x(\underset{\sim}{\omega}x\underset{\sim}{r}) = -\omega^2(\underset{\sim}{i}r_x + \underset{\sim}{j}r_y) \qquad \underset{\sim}{\alpha}x\underset{\sim}{r} = \alpha(-\underset{\sim}{i}r_y + \underset{\sim}{j}r_x)$

Equating the $\underset{\sim}{j}$ components

$-\omega_1{}^2 r_{1y} + \alpha_1 r_{1x} - \omega_2{}^2 r_{2y} + \alpha_2 r_{2x} = 0$

$\alpha_2 = (\omega_1{}^2 r_{1y} - \alpha_1 r_{1x} + \omega_2{}^2 r_{2y})/r_{2x}$

$\quad = (150^2 x51.96 - 0 + 31.98^2 x(-51.96))/140.71 = 7931\ \text{rad/s}^2$ ccw

Equating the $\underset{\sim}{i}$ components

$\ddot{r}_o - \omega_1{}^2 r_{1x} - \alpha_1 r_{1y} - \omega_2{}^2 r_{2x} - \alpha_2 r_{2y}$

$\ddot{r}_c = 150^2 x(-30) + 0 + 31.98^2 x140.71 + 7931x(-51.96)$

$\quad = -943{,}197$ (943,197 mm/s^2 to right)

See plot of $\ddot{r}_o/(\omega_1{}^2 r_1)$ for other values of θ_1

4.60

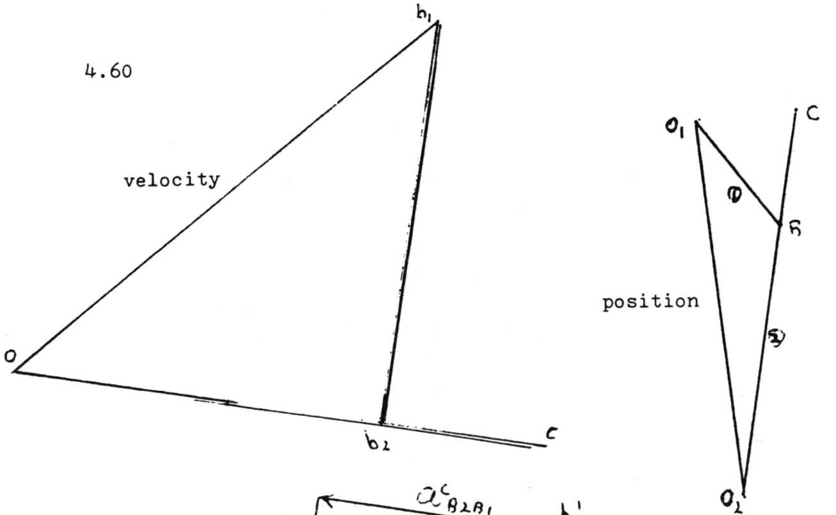

velocity
o
b_1
b_2
c
position
O_1
C
B
①
②
O_2

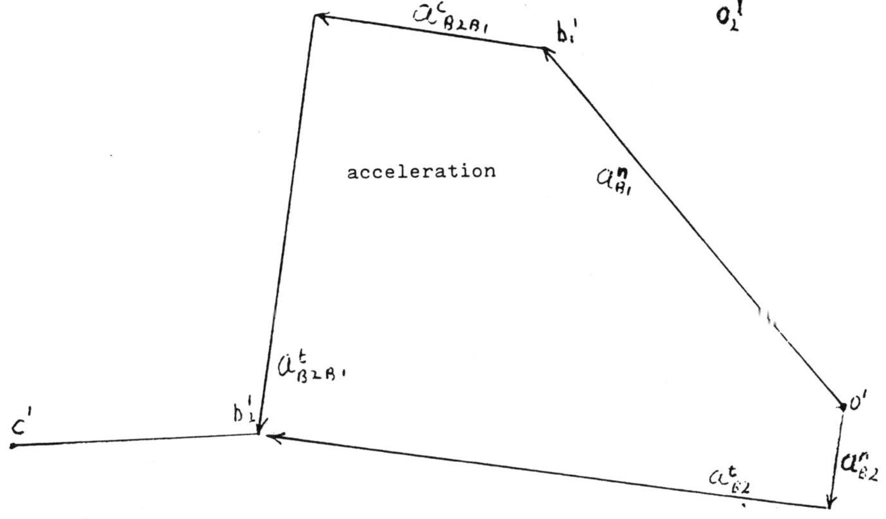

acceleration
a^c_{B2B1}
b_1'
a^n_{B1}
a^t_{B2B1}
o'
c'
b_2'
a^n_{B2}
a^t_{B2}

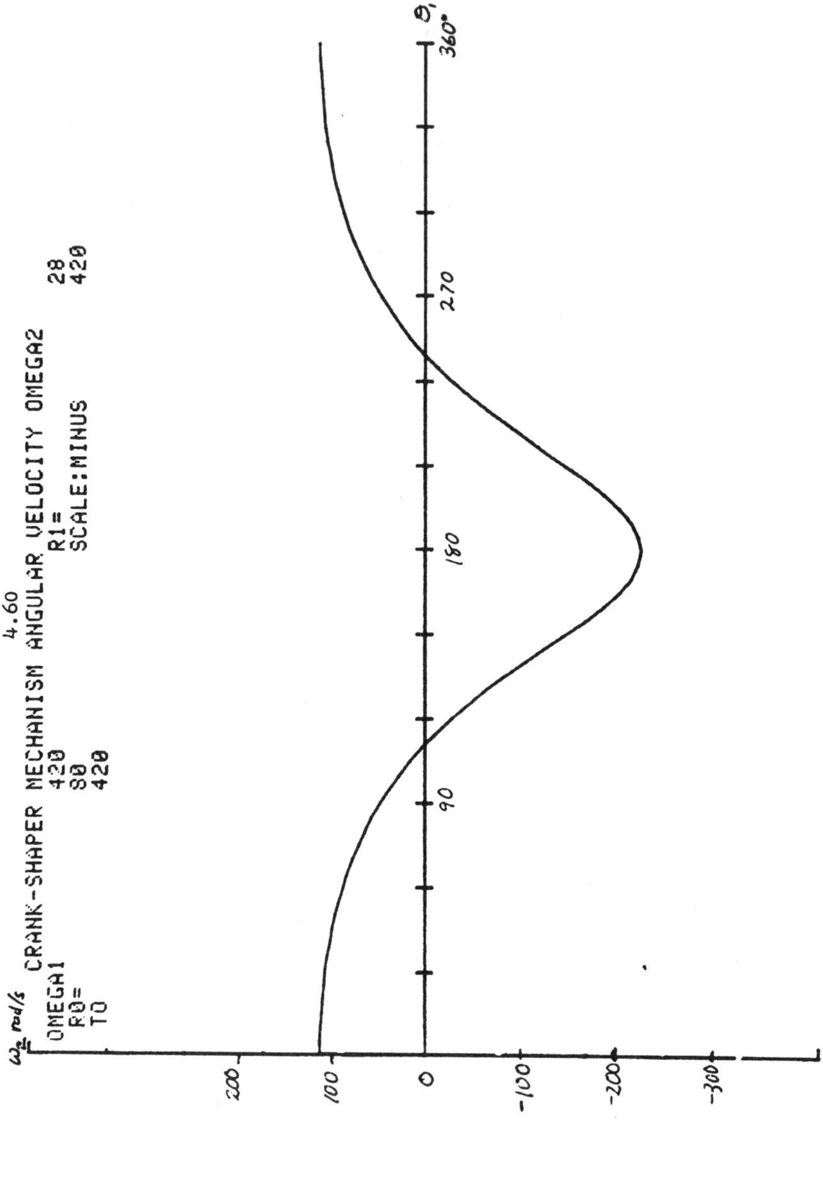

CRANK-SHAPER MECHANISM ANGULAR VELOCITY OMEGA2
4.60
OMEGA1 420
R0= 30
TO 420
R1= 28
SCALE:MINUS 420
ω_2 rad/s
θ_1
200
100
0
-100
-200
-300
90
180
270
360°

.60 See figure.

Velocity

$ob_2 = ob_1 + b_1b_2$

$ob_1 = \omega_1 O_1B_1 = 420 \times 0.028 = 11.76$ m/s $\perp O_1B_1$

$ob_2 \perp O_2B_2$ $\quad b_1b_2$ is parallel to link 2

$ob_2 = 8$ m/s and $b_1b_2 = 8.7$ (from polygon)

$oc = ob_2 \times O_2C/O_2B_2 = 8 \times 83/58 = 11.7$ m/s

$\omega_2 = ob_2/O_2B_2 = 8/0.058 = 140$ rad/s cw

Acceleration

$a^n_{B2} + a^t_{B2} = a^n_{B1} + a^t_{B1} + a^t_{B2B1} + a^c_{B2B1}$

$a^n_{B2} = ob_2^2/O_2B_2 = 8^2/0.058 = 1100$ m/s$^2 \| O_2B_2$

$a^t_{B2} \perp O_2B_2$

$a^n_{B1} = \omega_1^2\, O_1B_1 = 420^2 \times 0.028 = 4939$ m/s$^2 \| BO_1$

$a^t_{B1} = \alpha_1 O_1B_1 = 0$

$a^t_{B2B1} \|$ link 2

$a^c_{B2B1} = 2\omega_2 \times V_{B2B1} = 2 \times 140 \times 8.7 = 2400$ m/s$^2 \perp$ link 2

$a^t_{B2} = 6000$ m/s^2 and $a_{B2} = o'b_2' = 6050$ m/s^2 from polygon

$\alpha_2 = a^t_{B2}/O_2B_2 = 6000/0.058 = 103{,}000$ rad/s^2 ccw

$a_c = o'c' = o'b_2'O_2C/O_2B_2 = 6050 \times 83/58 = 8700$ m/s^2

See plots of ω_2 and α_2 for other values of θ_1.

.61 See solution 3.19 and section 4.3. Differentiating the velocity equation with respect to time:

$$-\omega_1^2R_1e^{j\theta_1} + j\alpha_1R_1e^{j\theta_1} = -\omega_2^2R_2e^{j\theta_2} + j2\omega_2V_{B1B2}e^{j\theta_2} + j\alpha_2R_2e^{j\theta_2} + a^t_{B1B2}\,e^{j\theta_2} \qquad (a)$$

Separating the real and imaginary parts and solving simultaneously the result is:

$$a^t_{B1B2} = -\omega_1^2R_1\cos(\theta_1-\theta_2) + \alpha_1R_1\sin(\theta_1-\theta_2) + \omega_2^2R_2 \qquad (b)$$

and

$$\alpha_2 = [\omega_1^2R_1\sin(\theta_2-\theta_1) + \alpha_1R_1\cos(\theta_2-\theta_1) - 2\omega_2V_{B1B2}]/R_2 \qquad (c)$$

where

$$V_{B1B2} = -\omega_1R_1\sin(\theta_1-\theta_2) \qquad (d)$$

From (b)

$a^t_{B1B2} = -10^2 \times 30\cos(15-5.615) + 0 + 3.730^2 \times 79.36 = -1856$ mm/s^2

From (c)

$\alpha_2 = [10^2 \times 30\sin(5.615-15) + 0 - 2 \times 3.730 \times (-10) \times 30\sin(15-5.615)]/79.36$
$= -1.566$ (1.566 rad/s^2 cw).

4.60

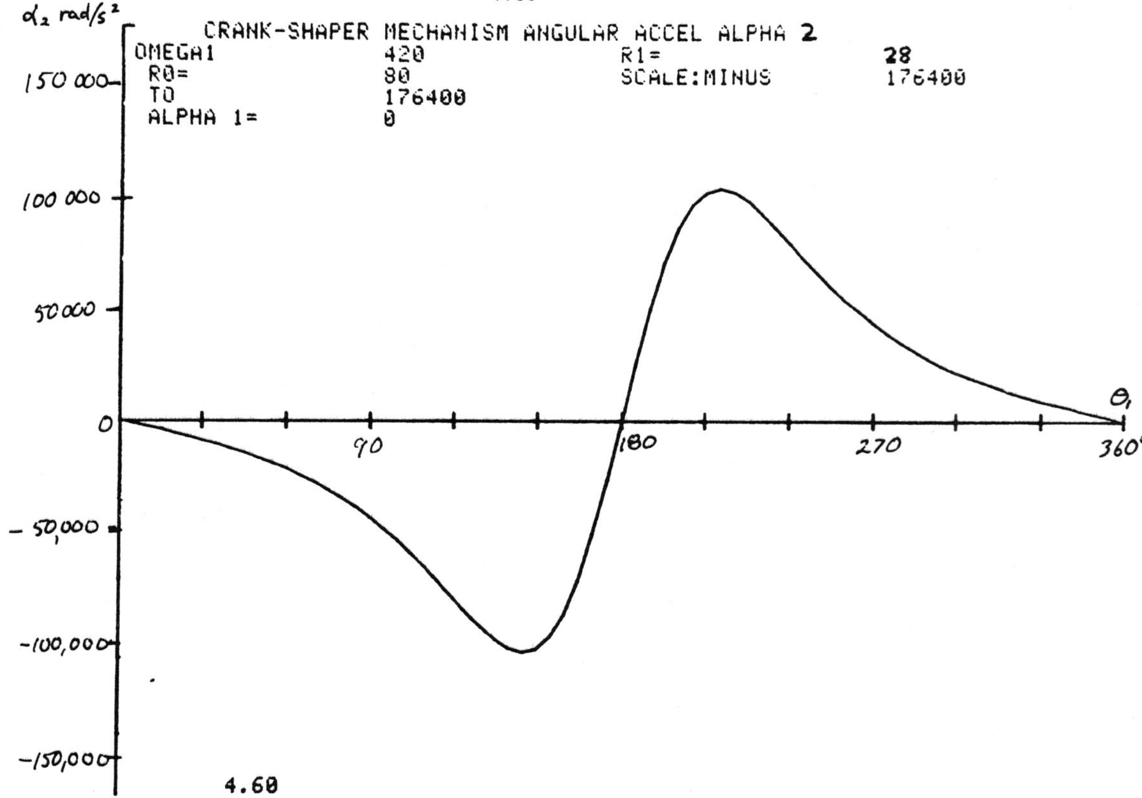

α_2 rad/s²
CRANK-SHAPER MECHANISM ANGULAR ACCEL ALPHA 2
OMEGA1 420
R0= 80
TO 176400
ALPHA 1= 0
R1= 28
SCALE:MINUS 176400
150 000
100 000
50 000
0
-50,000
-100,000
-150,000
90
180
270
360°
θ_1
4.60

4.62 See solutions 3.20 and 4.61

$a^t_{B1B2} = -10^2 x 30\cos(30-11.167) + 3.666^2 x 77.45 = -1798$ mm/s^2

$\alpha_2 = [10^2 x 30\sin(11.167-30) - 2x3.666x(-10)x30\sin(30-11.167)]/77.45$

$= -3.336$ (3.336 rad/s^2 cw)

4.63 See microcomputer results.

4.64 See solution 3.61 for position and velocities. Differentiating the velocity equation with respect to time:

$$(j\alpha_1 - \omega_1^2)r_1e^{j\theta_1} + (j\alpha_2 - \omega_2^2)r_2e^{j\theta_2} + (j\alpha_3 - \omega_3^2)r_3e^{j\theta_3} = 0 \quad \text{(a)}$$

Multiplying by $e^{-j\theta_3}$:

$$(j\alpha_1 - \omega_1^2)r_1e^{j(\theta_1-\theta_3)} + (j\alpha_2-\omega_2^2)r_2e^{j(\theta_2-\theta_3)} + j\alpha_3-\omega_3^2)r_3 = 0 \quad \text{(b)}$$

Substituting $e^{j\theta} = \cos\theta + j\sin\theta$:

$$(j\alpha_1 - \omega_1^2)r_1[\cos(\theta_1-\theta_3) + j\sin(\theta_1-\theta_3)]$$

$$+(j\alpha_2 - \omega_2^2)r_2[\cos(\theta_2-\theta_3)+j\sin(\theta_2-\theta_3)] + (j\alpha_3 - \omega_3^2)r_3 = 0 \quad \text{(c)}$$

Let α_1 be given

Equating the real parts eliminates α_3

$$-r_1[\alpha_1\sin(\theta_1-\theta_3) + \omega_1^2\cos(\theta_1-\theta_3)]$$

$$-r_2[\alpha_2\sin(\theta_2-\theta_3) + \omega_2^2\cos(\theta_2-\theta_3)] - r_3\omega_3^2 = 0 \quad \text{(d)}$$

from which

$$\alpha_2 = \frac{-r_1[\alpha_1\sin(\theta_1-\theta_3) + \omega_1^2\cos(\theta_1-\theta_3)] - r_2\omega_2^2\cos(\theta_2-\theta_3) - r_3\omega_3^2}{r_2\sin(\theta_2 - \theta_3)} \quad \text{(e)}$$

Multiplying equation (a) by $e^{-j\theta_2}$ and proceeding as above eliminates α_2 resulting in:

$$\alpha_3 = \frac{-r_1[\alpha_1\sin(\theta_1-\theta_2) + \omega_1^2\cos(\theta_1-\theta_2) - r_3\omega_3^2\cos(\theta_3-\theta_2) - r_2\omega_2^2}{r_3\sin(\theta_3 - \theta_2)} \quad \text{(f)}$$

4.65 See solutions 3.61 and 4.64 for position and velocity. Given and calculated values are as follows (with clockwise positive):

link	r(mm)	θ°	ω rad/s	α
0	30	180	0	0
1	10	45	30	0
2	35	65.997	-12.145	to be found
3	40	-77.450	-4.512	to be found

4.63

```
LIST
100 PAGE
110 D=32
115 HOME
120 PRINT @D:"CRANK SHAPER MECHANISM"
150 A1=0
160 W1=1
170 R1=1
180 R0=3
185 PRINT @D:" R0/R1  ",R0
195 PRINT @D:" THETA 1          THETA 2          W2/W1          A2/W1↑2"
200 FOR T1=0 TO PI STEP PI/12
210 R2=(R0↑2+R1↑2+2*R0*R1*COS(T1))↑0.5
220 T2=ASN(R1*SIN(T1)/R2)
230 W2=W1*R1*COS(T1-T2)/R2
240 H1=180*T1/PI
245 H2=180*T2/PI
260 V1=-W1*R1*SIN(T1-T2)
270 A2=(W1↑2*R1*SIN(T2-T1)+A1*R1*COS(T2-T1)-2*W2*V1)/R2
280 PRINT @D:H1,H2,W2,A2
290 NEXT T1
```

CRANK SHAPER MECHANISM

R0 R1 3

THETA 1	THETA 2	W2/W1	A2/W1↑2
0	0	0.25	0
15	3.73386716217	0.246764199757	-0.0248964654347
30	7.36925978757	0.236775475217	-0.0519653628355
45	10.7990804918	0.21915320425	-0.0836594890423
60	13.897886248	0.192307692308	-0.122985856159
75	16.5100151745	0.153767014426	-0.17368884154
90	18.4349488229	0.1	-0.24
105	19.4112241516	0.0264639038663	-0.324893644688
120	19.1066053509	-0.0714285714286	-0.424175707976
135	17.1392722378	-0.194762960347	-0.511975967329
150	13.186785432	-0.332665886176	-0.519999358501
165	7.25144491523	-0.451374071914	-0.351390577857
180	1.918465387E-12	-0.5	-8.928943355E-14

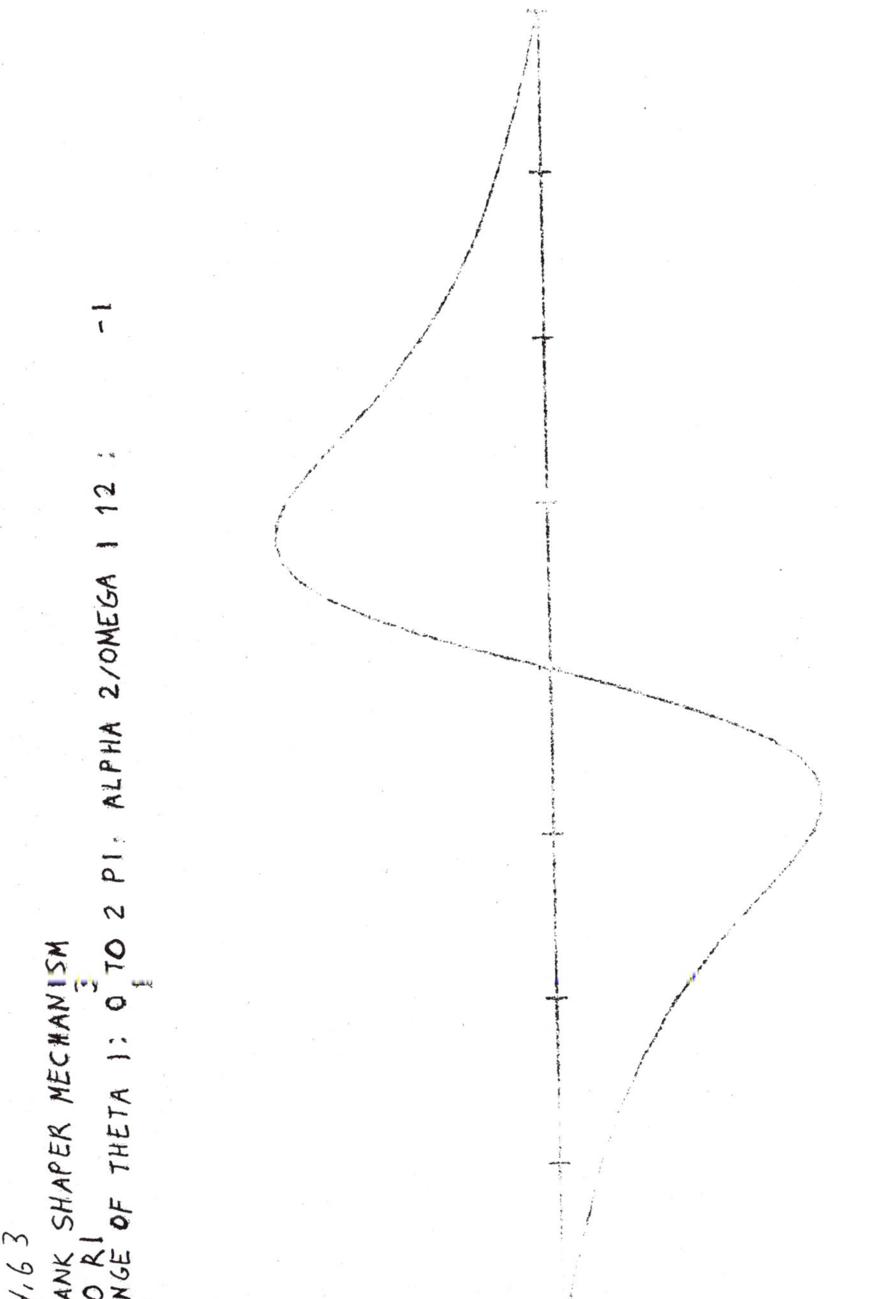
4.63
CRANK SHAPER MECHANISM
RO R1
RANGE OF THETA 1: 0 TO 2 PI. ALPHA 2/OMEGA 1 12 : -1
PO

4.63

```
LIST
100 PAGE
110 D=32
115 HOME
120 PRINT @D:"CRANK SHAPER MECHANISM"
150 A1=0
160 W1=1
170 R1=1
180 R0=3
185 PRINT @D:" R0/R1  ",R0
195 PRINT @D:" THETA 1            THETA 2                 W2/W1          A2/W1↑2"
200 FOR T1=0 TO PI STEP PI/12
210 R2=(R0↑2+R1↑2+2*R0*R1*COS(T1))↑0.5
220 T2=ASN(R1*SIN(T1)/R2)
230 W2=W1*R1*COS(T1-T2)/R2
240 H1=180*T1/PI
245 H2=180*T2/PI
260 V1=-W1*R1*SIN(T1-T2)
270 A2=(W1↑2*R1*SIN(T2-T1)+A1*R1*COS(T2-T1)-2*W2*V1)/R2
280 PRINT @D:H1,H2,W2,A2
290 NEXT T1
```

4.63

CRANK SHAPER MECHANISM

RO/R1

THETA 1	THETA 2	W2/W1	A2/W1↑2
0	[illegible]	0.25	0
15	[illegible].73306716217	0.246764199757	-0.0248964654347
30	[illegible].35925978757	0.236775475217	-0.0519653628355
45	0.7990804918	0.21915320425	-0.0836594896423
60	3.397086249	0.192307692308	-0.122985856159
75	6.5100151745	0.153767014426	-0.17368804154
90	8.4349488229	0.1	-0.24
105	9.4112241516	0.0264630039663	-0.324893644688
120	9.1066053509	-0.0714285714286	-0.424175707976
135	7.1392722378	-0.194762960347	-0.511975967329
150	3.186785432	-0.332663886176	-0.519999358501
165	7.25144491523	-0.451374071914	-0.351390577857
180	[illegible].91846538 7E-12	-0.5	-8.92894335 5E-14

4.65 (Cont'd.)

Using solution 4.64

$$\alpha_2 = \frac{-10\times30^2\cos(45+77.45)-35\times12.145^2\cos(65.997+77.45)-40\times4.512^2}{35\sin(65.997+77.45)}$$

$\alpha_2 = 391.55$ rad/s^2

$$\alpha_3 = \frac{-10\times30^2\cos(45-65.997)-40\times4.512^2\cos(-77.45-65.997)-35\times12.145^2}{40\sin(-77.45-65.997)}$$

$\alpha_3 = 541.95$ rad/s^2

4.66 The procedure follows the above problem. Results are as follows:

Link	r(mm)	θ°	ω rad/s	α rad/s^2
0	30	180	0	0
1	15	45	30	0
2	35	57.114	-16.656	872.03
3	40	-89.442	-4.284	1011.43

The acceleration of the pin joining links 2 and 3 is

$a = \underset{\sim}{\alpha}_3\times(-\underset{\sim}{r}_3)+\underset{\sim}{\omega}_3\times(\underset{\sim}{\omega}_3\times(-\underset{\sim}{r}_3)) = -40{,}448 - j1128$

$= 40{,}464$ mm/s$^2\underline{/-178.4^\circ}$

4.67

For $\gamma = 45^\circ$ and $\theta = 60^\circ$, $r_3 = 0.457$,

$V_{S2} = -130.7$ and $a_{S2} = -8117$.

(Negative values of V_{S2} indicate decreasing r_3)

4.68 See results obtained using a programmable calculator. The procedure is similar to example problem where each value of θ is followed by $R3 = r_3$, $V3 = \dot{r}_3/(\omega_1 r_1)$ and

$A3 = \ddot{r}_3/(\omega_1^2 r_1)$.

Note that R3 and A3 are symmetric about $\theta = 180^\circ$ and V3 is antisymmetric about $\theta = 180^\circ$.

4.69 See calculated results and note in above problem.

4.68

```
THETA      =
           0.
  R3=
 1.203186113
  V3=
           0.
  A3=
-1.480127464
THETA      =
          30.
  R3=
 1.004481292
  V3=
-.7431620651
  A3=
-1.299600223
THETA      =
          60.
  R3=
 0.456615131
  V3=
-1.306737511
  A3=
-.8117155196
THETA      =
          90.
  R3=
-.3100289793
  V3=
-1.564599707
  A3=
-.1614104331
THETA      =
         120.
  R3=
-1.121320344
  V3=
-1.478397839
  A3=
 .4785533906
THETA      =
         150.
  R3=
 -1.80140241
  V3=
-1.059389831
  A3=
 1.190055712
THETA      =
         180.
  R3=
-2.121320344
  V3=
           0.
```

4.69

```
THETA      =
           0.
  R3=
 7.483314774
  V3=
           0.
  A3=
-.8017837257
THETA      =
          30.
  R3=
 7.375113807
  V3=
 .4067723705
  A3=
[illegible]
THETA      =
          60
  R3=
 7.071067812
  V3=
-.7348469228
  A3=
-.5006316011
THETA      =
          90.
  R3=
 6.633249581
  V3=
-.9045340337
  A3=
-.1233455501
THETA      =
         120.
  R3=
 6.104414007
  V3=
-.8429272304
  A3=
 .3714016747
THETA      =
         150.
  R3=
 5.797214431
  V3=
-.5174899145
  A3=
 .8501249471
THETA      =
         180.
  R3=
```

4.70 The solution utilizes spreadsheet formulas similar to those given in section 4.3. Results for $\theta_1 = 0°$ to $35°$ are given in spreadsheet table S4.3.1. Figure S4.3.1 shows a plot of normalized angular velocity and angular acceleration for links 2 and 3 for $\theta_1 = 0°$ to $360°$.

4.71 The solution utilizes spreadsheet formulas similar to those given in section 4.3. Results for $\theta_1 = 0°$ to $35°$ are given in spreadsheet table S4.3.2. Figure S4.3.2 shows a plot of normalized angular velocity and angular acceleration for links 2 and 3 for $\theta_1 = 0°$ to $360°$.

4.72 The solution utilizes spreadsheet formulas similar to those given in section 4.3. Results for $\theta_1 = 0°$ to $35°$ are given in spreadsheet table S4.3.3. Figure S4.3.3 shows a plot of normalized angular velocity and angular acceleration for links 2 and 3 for $\theta_1 = 0°$ to $360°$.

4.73 The solution utilizes spreadsheet formulas similar to those given in section 4.3. Results for $\theta_1 = 0°$ to $35°$ are given in spreadsheet table S4.3.4. Figure S4.3.4 shows a plot of normalized angular velocity and angular acceleration for links 2 and 3 for $\theta_1 = 0°$ to $360°$.

4.74 The solution is based on the vector equations of sections 3.6 and 4.4, utilizing mathematics software. Vector equations, with tabulated and plotted results are shown in figure S4.4.1.

4.75 The solution is based on the vector equations of sections 3.6 and 4.4, utilizing mathematics software. Vector equations, with tabulated and plotted results are shown in figure S4.4.2.

4.76 The solution is based on the vector equations of sections 3.6 and 4.4, utilizing mathematics software. Vector equations, with tabulated and plotted results are shown in figure S4.4.3.

54,70

	A	B	C	D	E	F	G
1	Four-bar linkage. Acceleration analysis.						
2	Position analysis based on the vector cross product method.						
3	If Rd-R2-R3 loop clockwise, config = 1; -1 if ccw.						
4	R0	deg	R1	R2	R3	config	
5	3.5	180	1.2	2.2	3.1	1	
6							
7	Theta1	Rdx	Rdy	Rd	Rdux	Rduy	a
8	deg						
9	0	-2.3	0	2.3	-1	0	2.186957
10	5	-2.30457	.1045869	2.306938	-.998972	.0453358	2.187307
11	10	-2.31823	.2083778	2.327577	-.995985	.0895256	2.188459
12	15	-2.34089	.3105829	2.361403	-.991313	.1315247	2.190694
13	20	-2.37237	.4104242	2.407609	-.985363	.1704696	2.194414
14	25	-2.41243	.5071419	2.46516	-.97861	.2057237	2.200063
15	30	-2.46077	.6	2.532861	-.971537	.2368862	2.208053
16	35	-2.51702	.6882917	2.60943	-.964585	.2637709	2.218708

	H	I	J	K	L	M	N
1							
2							
3							
4	Omega1	Alpha1					
5	400	0					
6							
7	r2x	r2y	theta2	r3x	r3y	theta3	theta3
8			deg			deg	deg
9	.1130435	2.197094	87.05465	2.186957	-2.19709	-45.1325	314.8675
10	.2190996	2.189063	84.2844	2.085467	-2.29365	-47.7218	312.2782
11	.3351214	2.174326	81.23814	1.983109	-2.3827	-50.2296	309.7704
12	.4577075	2.151861	77.99196	1.883181	-2.46244	-52.5926	307.4074
13	.5833428	2.121252	74.62379	1.789026	-2.53168	-54.7528	305.2472
14	.7087213	2.082718	71.20724	1.703709	-2.58986	-56.6615	303.3385
15	.8310015	2.037017	67.80698	1.629768	-2.63702	-58.2825	301.7175
16	.9479556	1.985291	64.47602	1.569062	-2.67358	-59.5923	300.4077

$\dot{5}$4.20 cont'd

	O	P	Q	R	S	T	U
1							
2							
3							
4							
5							
6							
7	Trans	r1x	r1y	Omega2	Omega3	a1	b1
8	angle						
9	47.81286	1.2	0	-208.696	-208.696	-292174	0
10	47.99382	1.195434	.1045869	-233.635	-204.743	-290651	40076.27
11	48.53225	1.181769	.2083778	-252.742	-195.657	-286407	81019.37
12	49.4154	1.159111	.3105829	-265.599	-181.649	-279884	120240.6
13	50.6234	1.127631	.4104242	-272.321	-163.327	-271404	155442.5
14	52.13121	1.087569	.5071419	-273.461	-141.585	-261163	184973.2
15	53.91053	1.03923	.6	-269.873	-117.457	-249284	207978.2
16	55.93164	.9829825	.6882917	-262.541	-91.9756	-235891	224351.6

	V	W	X	Y	Z	AA	AB
1							
2				Omega* = Omega/Omega1			
3				Alpha* = Alpha/Omega1^2			
4							
5							
6							
7	D	Alpha2	Alpha3	Omega2*	Omega3*	Alpha2*	Alpha3*
8		rad/s^2	rad/s^2				
9	5.053316	-126446	6535.977	-.521739	-.521739	-.790288	.0408499
10	[illegible]	[illegible]	[illegible]	[illegible]	[illegible]	[illegible]	[illegible]
11	5.110421	-73365.8	53252.66	-.631854	-.489142	-.458537	.3328291
12	5.179423	-44596.9	74689.01	-.663999	-.454122	-.278731	.4668063
13	5.271811	-17455	92578.17	-.680802	-.408317	-.109093	.5786136
14	5.383835	6335.366	105935.4	-.683653	-.353963	.039596	.6620965
15	5.511229	25795.75	114459.2	-.674682	-.293642	.1612235	.7153698
16	5.649482	40657.59	118421.1	-.656353	-.229939	.2541099	.7401316

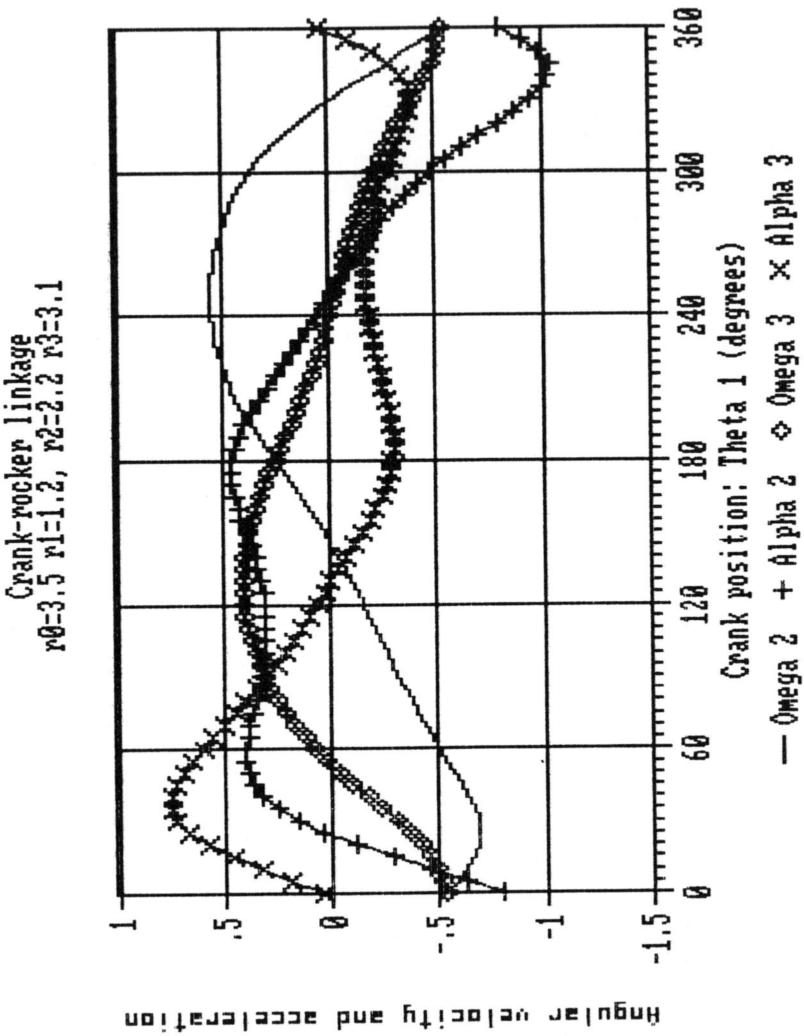
Crank-rocker linkage
r0=3.5 r1=1.2, r2=2.2 r3=3.1
Crank position: Theta 1 (degrees)
Angular velocity and acceleration
— Omega 2 + Alpha 2 ◇ Omega 3 × Alpha 3
0
60
120
180
240
300
360
1
.5
0
-.5
-1
-1.5

54.71	A	B	C	D	E	F	G
1	Four-bar linkage.			Acceleration analysis.			
2	Position analysis based on the vector cross product method.						
3	If Rd-R2-R3 loop clockwise, config = 1; -1 if ccw.						
4	R0	deg	R1	R2	R3	config	
5	38	180	13	27	41	1	
6							
7	Theta1	Rdx	Rdy	Rd	Rdux	Rduy	a
8	deg						
9	0	-25	0	25	-1	0	31.54
10	5	-25.0495	1.133025	25.07508	-.998979	.0451853	31.52053
11	10	-25.1975	2.257426	25.29842	-.996011	.0892319	31.46461
12	15	-25.443	3.364648	25.66448	-.991369	.1311014	31.37928
13	20	-25.784	4.446262	26.16455	-.985455	.1699346	31.27483
14	25	-26.218	5.494037	26.78746	-.978742	.2050974	31.16324
15	30	-26.7417	6.5	27.5203	-.971707	.2361893	31.05647
16	35	-27.351	7.456494	28.34921	-.96479	.263023	30.9652

	H	I	J	K	L	M	N
1							
2							
3							
4	Omega1	Alpha1					
5	50	0					
6							
7	r2x	r2y	theta2	r3x	r3y	theta3	theta3
8			deg			deg	deg
9	-6.54	26.19596	104.0177	31.54	-26.196	-39.7118	320.2882
10	-5.25414	26.48385	101.2212	30.30361	-27.6169	-42.3441	317.6559
11	-3.79601	26.73182	98.08216	28.99351	-28.9892	-44.9958	315.0042
12	-2.20594	26.90974	94.68637	27.6489	-30.2744	-47.5953	312.4047
13	-.53065	26.99478	91.12615	26.31465	-31.441	-50.0723	309.9277
14	1.181663	26.97413	87.49163	25.03634	-32.4682	-52.364	307.636
15	2.886056	26.84531	83.86388	23.85561	-33.3453	-54.4196	305.5804
16	4.544334	26.61483	80.3105	22.80669	-34.0713	-56.2024	303.7976

	O	P	Q	R	S	T	U
1							
2							
3							
4							
5							
6							
7	Trans	r1x	r1y	Omega2	Omega3	a1	b1
8	angle						
9	36.27049	13	0	-26	-26	-49400	0
10	36.43463	12.95053	1.133025	-29.8111	-26.5367	-49046.7	6921.056
11	36.92205	12.8025	2.257426	-32.8257	-26.376	-48086.5	14280.24
12	37.71837	12.55704	3.364648	-34.9348	-25.4953	-46672.4	21574.7
13	38.80157	12.216	4.446262	-36.1182	-23.9398	-44929	28311.85
14	40.14433	11.782	5.494037	-36.4362	-21.8101	-42933.1	34101.45
15	41.71649	11.25833	6.5	-36.0059	-19.2408	-40718.9	38708.29
16	43.48709	10.64898	7.456494	-34.974	-16.3775	-38298.2	42057.27

Table 5.4.7/cont'd.

	V	W	X	Y	Z	AA	AB
1							
2				Omega* = Omega/Omega1			
3				Alpha* = Alpha/Omega1^2			
4							
5							
6							
7	D	Alpha2	Alpha3	Omega2*	Omega3*	Alpha2*	Alpha3*
8		rad/s^2	rad/s^2				
9	654.899	-2379.11	-493.322	-.52	-.52	-.951643	-.197329
10	657.4532	-1969.96	-113.167	-.596222	-.530734	-.787982	-.045267
11	665.0059	-1474.01	299.5463	-.656514	-.527519	-.589602	.1198185
12	677.2412	-940.993	705.2333	-.698695	-.509905	-.376397	.2820933
13	693.674	-421.144	1067.404	-.722365	-.478795	-.168458	.4269618
14	713.6998	45.29231	1359.941	-.728725	-.436202	.0181169	.5439763
15	736.6478	433.5375	1570.158	-.720119	-.384816	.173415	.6280631
16	761.8276	734.4072	1697.742	-.699479	-.327549	.2937629	.6790969

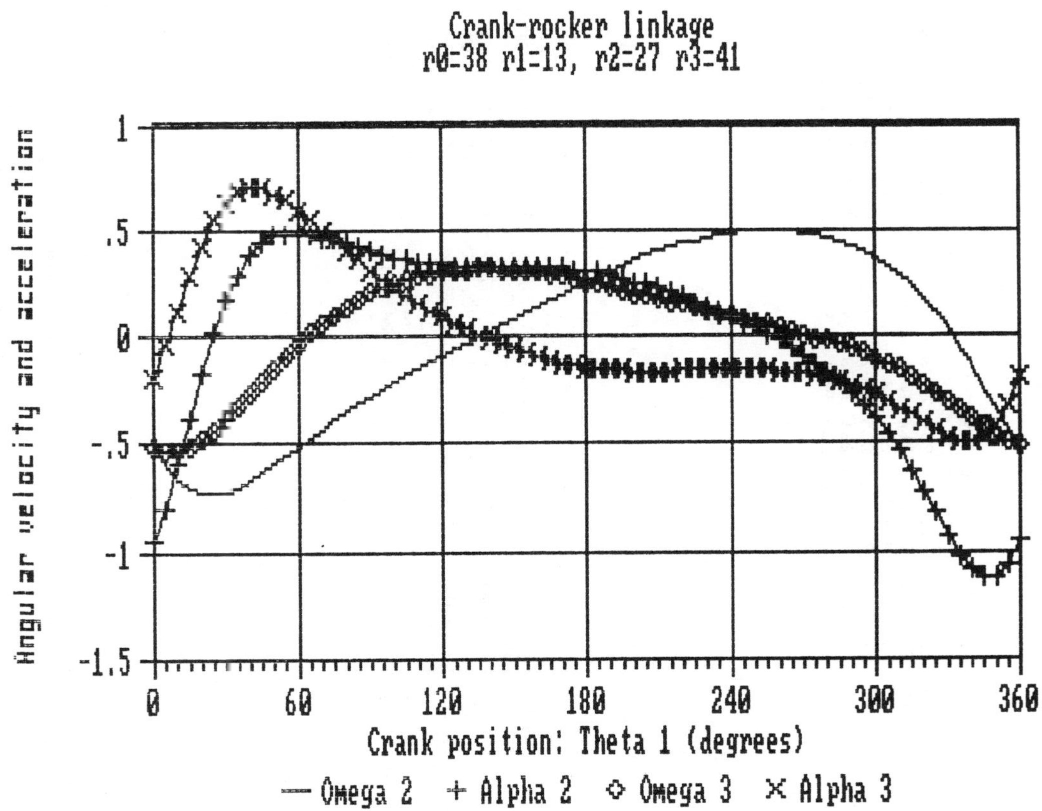
Crank-rocker linkage
r0=38 r1=13, r2=27 r3=41
Angular velocity and acceleration
1
.5
0
-.5
-1
-1.5
0
60
120
180
240
300
360
Crank position: Theta 1 (degrees)
— Omega 2 + Alpha 2 ◇ Omega 3 × Alpha 3

Fig 54.71

54,72

	A	B	C	D	E	F	G
1	Four-bar linkage.		Acceleration analysis.				
2	Position analysis based on the vector cross product method.						
3	If Rd-R2-R3 loop clockwise, config = 1; -1 if ccw.						
4	R0	deg	R1	R2	R3	config	
5	312.48	180	100	200	300	1	
6							
7	Theta1	Rdx	Rdy	Rd	Rdux	Rduy	a
8	deg						
9	0	-212.48	0	212.48	-1	0	223.8981
10	5	-212.861	8.715574	213.0389	-.999163	.0409107	223.8689
11	10	-213.999	17.36482	214.7026	-.996724	.0808785	223.7914
12	15	-215.887	25.8819	217.4333	-.99289	.1190338	223.6944
13	20	-218.511	34.20201	221.1712	-.987971	.1546404	223.6202
14	25	-221.849	42.26183	225.8387	-.982335	.1871328	223.6178
15	30	-225.877	50	231.3453	-.976365	.2161272	223.7362
16	35	-230.565	57.35764	237.5921	-.970423	.2414122	224.0184

	H	I	J	K	L	M	N
1							
2							
3							
4	Omega1	Alpha1					
5	250	0					
6							
7	r2x	r2y	theta2	r3x	r3y	theta3	theta3
8			deg			deg	deg
9	-11.4181	199.6738	93.27283	223.8981	-199.674	-41.7268	318.2732
10	-2.65082	199.9824	90.75943	215.5114	-208.698	-44.0798	315.9202
11	7.099918	199.8739	87.9656	206.8993	-217.239	-46.3965	313.6035
12	17.57848	199.226	84.95763	198.3089	-225.108	-48.6215	311.3785
13	28.50626	197.9581	81.80565	190.0045	-232.16	-50.7024	309.2976
14	39.60593	196.0392	78.57824	182.2433	-238.301	-52.5927	307.4073
15	50.62334	193.4872	75.33798	175.2541	-243.487	-54.2549	305.7451
16	61.34337	190.3602	72.13852	169.2214	-247.718	-55.6622	304.3378

	O	P	Q	R	S	T	U
1							
2							
3							
4							
5							
6							
7	Trans	r1x	r1y	Omega2	Omega3	a1	b1
8	angle						
9	45.00038	100	0	-117.658	-117.658	-9191453	0
10	45.16074	99.61947	8.715574	-133.203	-117.2	-9139421	1226389
11	45.63795	98.48078	17.36482	-145.623	-113.999	-8994419	2500641
12	46.42083	96.59258	25.8819	-154.585	-108.068	-8773074	3749480
13	47.49193	93.96926	34.20201	-160.038	-99.6305	-8489212	4903257
14	48.8291	90.63078	42.26183	-162.18	-89.081	-8152327	5906609
15	50.40713	86.60254	50	-161.398	-76.9176	-7768222	6724663
16	52.19932	81.9152	57.35764	-158.188	-63.6742	-7340808	7343955

54.72 cont'd

	V	W	X	Y	Z	AA	AB
1							
2				Omega* =	Omega/Omega1		
3				Alpha* =	Alpha/Omega1^2		
4							
5							
6							
7	D	Alpha2	Alpha3	Omega2*	Omega3*	Alpha2*	Alpha3*
8		ra[illegible]/s^2	rad/s^2				
9	42426.69	-[illegible]8506	-2473.66	-.470633	-.470633	-.776096	-.039579
10	42545.26	-40[illegible]79.5	5195.156	-.532813	-.468801	-.644473	.0831225
11	42896.16	-30[illegible]18.4	13140.4	-.582491	-.455995	-.491495	.2102464
12	43465.35	-20[illegible]08.2	20734.02	-.61834	-.432271	-.329731	.3317443
13	44230.93	-10[illegible]31.1	27415.99	-.64015	-.398522	-.171698	.4386559
14	45164.96	-17[illegible]0.45	32786.64	-.648718	-.356324	-.027687	.5245863
15	46235.55	596[illegible].485	36646.89	-.645592	-.30767	.0954958	.5863502
16	47408.86	121[illegible]0.85	38986.51	-.632751	-.254697	.1947336	.6237842

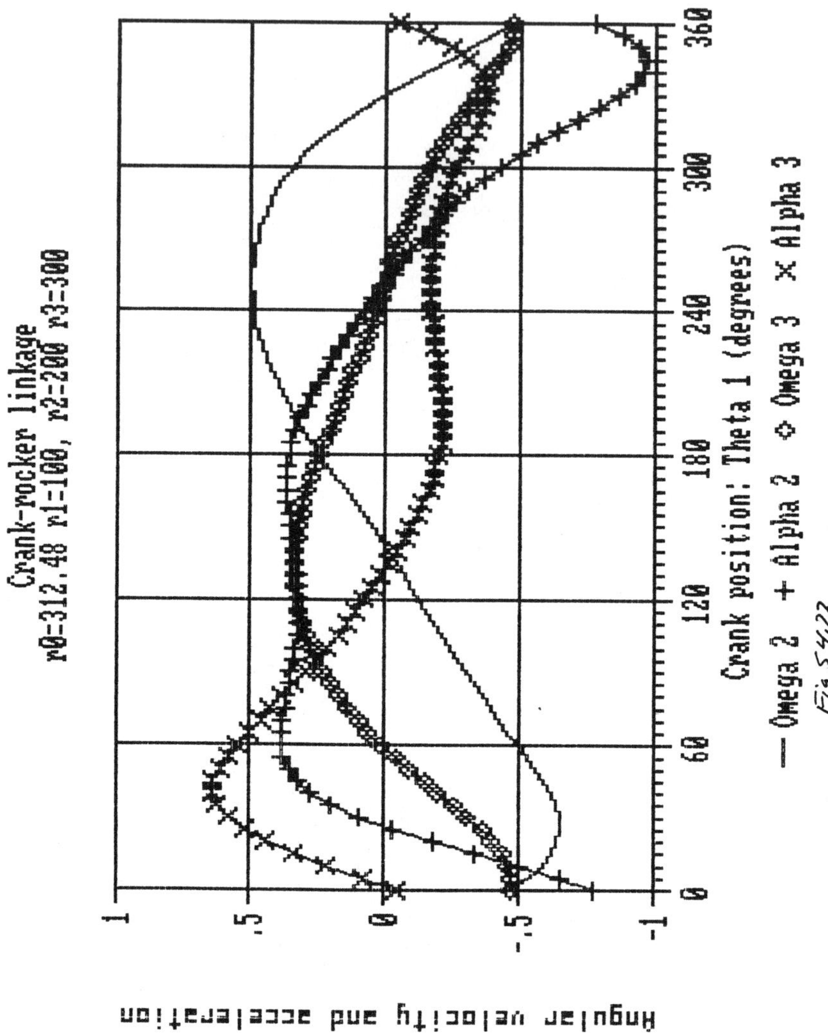
Crank-rocker linkage
r0=312.48 r1=100, r2=200 r3=300
Angular velocity and acceleration
1
.5
0
-.5
-1
0
60
120
180
240
300
360
Crank position: Theta 1 (degrees)
— Omega 2 + Alpha 2 ◇ Omega 3 × Alpha 3
Fig S4.12

Table S4.23

	A	B	C	D	E	F	G
1	Four-bar linkage.			Acceleration analysis.			
2	Position analysis based on the vector cross product method.						
3	If Rd-R2-R3 loop clockwise, config = 1; -1 if ccw.						
4	R0	deg	R1	R2	R3	config	
5	37	180	10	25	40	1	
6							
7	Theta1	Rdx	Rdy	Rd	Rdux	Rduy	a
8	deg						
9	0	-27	0	27	-1	0	31.55556
10	5	-27.0381	.8715574	27.0521	-.999481	.0322177	31.54683
11	10	-27.1519	1.736482	27.20739	-.997961	.0638239	31.52162
12	15	-27.3407	2.58819	27.46297	-.995549	.0942429	31.48266
13	20	-27.6031	3.420201	27.81416	-.992411	.1229662	31.43412
14	25	-27.9369	4.226183	28.25477	-.988751	.1495741	31.38111
15	30	-28.3397	5	28.77744	-.98479	.1737472	31.32907
16	35	-28.8085	5.735764	29.37393	-.98075	.1952672	31.28331

	H	I	J	K	L	M	N
1							
2							
3							
4	Omega1	Alpha1					
5	100	0					
6							
7	r2x	r2y	theta2	r3x	r3y	theta3	theta3
8			deg			deg	deg
9	-4.55556	24.58143	100.4992	31.55556	-24.5814	-37.9182	322.0818
10	-3.70008	24.72467	98.51124	30.73814	-25.5962	-39.7848	320.2152
11	-2.73377	24.85008	96.2779	29.88569	-26.5866	-41.6566	318.3434
12	-1.67638	24.94373	93.84486	29.01712	-27.5319	-43.4955	316.5045
13	-.550735	24.99393	91.26229	28.15381	-28.4141	-45.2637	314.7363
14	.6188338	24.99234	88.58159	27.31809	-29.2185	-46.9252	313.0748
15	1.808175	24.93452	85.85235	26.53157	-29.9345	-48.4488	311.5512
16	2.994788	24.81998	83.11993	25.81369	-30.5557	-49.8087	310.1913

Table 54.73 cont'd

	O	P	Q	R	S	T	U
1							
2							
3							
4							
5							
6							
7	Trans	r1x	r1y	Omega2	Omega3	a1	b1
8	angle						
9	41.58257	10	0	-37.037	-37.037	-137037	0
10	41.70397	9.961947	.8715574	-42.3547	-37.5075	-136225	17060.6
11	42.06555	9.848078	1.736482	-46.8256	-37.2358	-133923	34989.65
12	42.6596	9.659258	2.58819	-50.3275	-36.1957	-130363	52990.48
13	43.47403	9.396926	3.420201	-52.8029	-34.41	-125769	70244.9
14	44.49318	9.063078	4.226183	-54.2591	-31.947	-120334	86019.81
15	45.69887	8.660254	5	-54.759	-28.9094	-114198	99749.43
16	47.07141	8.19152	5.735764	-54.4056	-25.4214	-107462	111077.4

	V	W	X	Y	Z	AA	AB
1							
2				Omega* =	Omega/Omega1		
3				Alpha* =	Alpha/Omega1^2		
4							
5							
6							
7	D	Alpha2	Alpha3	Omega2*	Omega3*	Alpha2*	Alpha3*
8		rad/s^2	rad/s^2				
9	663.6987	-6515.43	-940.607	-.37037	-.37037	-.651543	-.094061
10	665.2821	-5637.61	-123.594	-.423547	-.375075	-.563761	-.012359
11	669.9803	-4585.41	751.3353	-.468256	-.372358	-.458541	.0751335
12	677.6413	-3429.27	1628.064	-.503275	-.361957	-.342927	.1628064
13	688.0258	-2245.46	2451.116	-.528029	-.3441	-.224546	.2451116
14	700.8244	-1104.29	3173.838	-.542591	-.31947	-.110429	.3173838
15	715.679	-61.3515	3763.831	-.54759	-.289094	-.006135	.3763831
16	732.2032	846.8559	4204.794	-.544056	-.254214	.0846856	.4204794

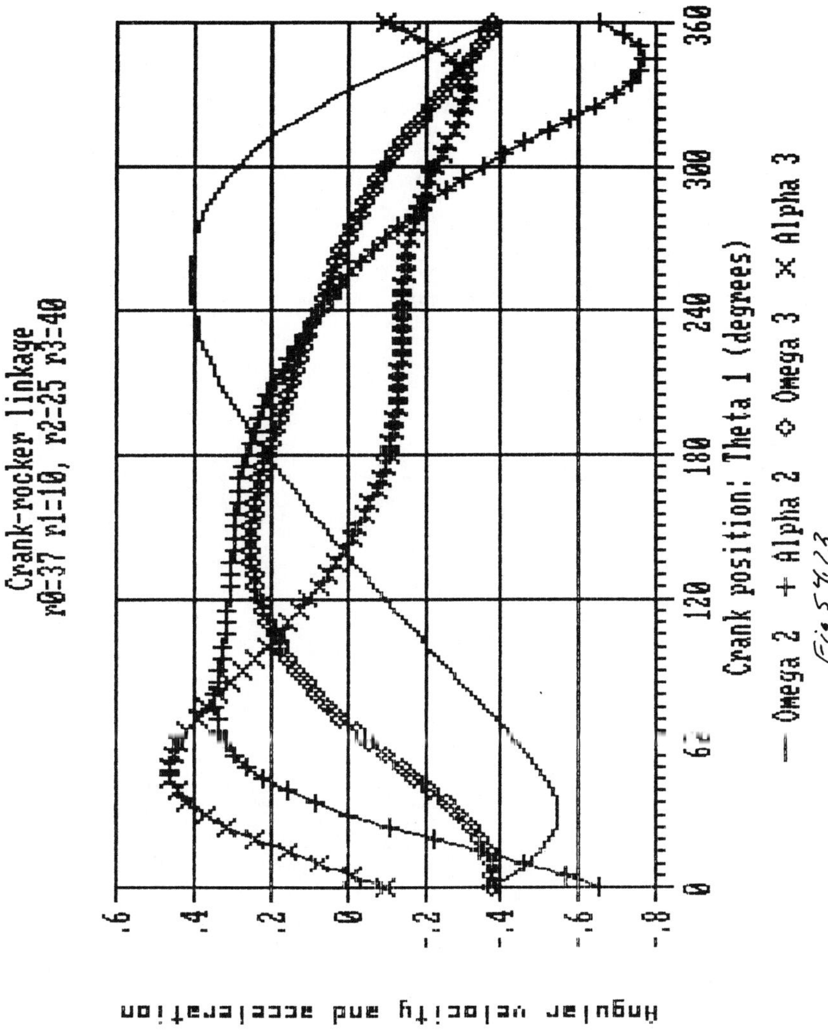
Crank-rocker linkage
r0=37 r1=10, r2=25 r3=40
Angular velocity and acceleration
Crank position: Theta 1 (degrees)
— Omega 2 + Alpha 2 ◇ Omega 3 × Alpha 3

54.74

Slider crank linkage $\quad c_o := 1 \quad$ (clockwise configuration positive)

$$\theta_1 := 0, \frac{\pi}{9} .. 2\cdot\pi \qquad R_1 := 1 \qquad R_2 := 2 \qquad e := \begin{bmatrix} 0 \\ -0.5 \\ 0 \end{bmatrix} \qquad r_{0u} := \begin{bmatrix} -1 \\ 0 \\ 0 \end{bmatrix}$$

$$r_1(\theta_1) := \begin{bmatrix} R_1 \cdot \cos(\theta_1) \\ R_1 \cdot \sin(\theta_1) \\ 0 \end{bmatrix} \qquad i := \begin{bmatrix} 1 \\ 0 \\ 0 \end{bmatrix} \qquad j := \begin{bmatrix} 0 \\ 1 \\ 0 \end{bmatrix} \qquad k := \begin{bmatrix} 0 \\ 0 \\ 1 \end{bmatrix} \qquad \omega_1 := 1$$

$$r_3(\theta_1) := r_1(\theta_1) + e \qquad \alpha_1 := 0$$

$$r_0(\theta_1) := \left[-r_3(\theta_1)\cdot r_{0u} + c_o \cdot \left[R_2^{\,2} - \left[r_3(\theta_1)\cdot\left(r_{0u} \times k \right) \right]^2 \right]^{.5} \right] \cdot r_{0u}$$

$$a(\theta_1) := -\left[r_3(\theta_1)\cdot\left(r_{0u} \times k \right) \right] \cdot \left(r_{0u} \times k \right)$$

$$r_2(\theta_1) := a(\theta_1) - c_o \cdot \left[R_2^{\,2} - \left[r_3(\theta_1)\cdot\left(r_{0u} \times k \right) \right]^2 \right]^{.5} \cdot r_{0u}$$

$$\omega_2(\theta_1) := -\omega_1 \cdot \frac{r_1(\theta_1)\cdot i}{r_2(\theta_1)\cdot i}$$

$$v_c(\theta_1) := \omega_1 \cdot k \times r_1(\theta_1) + \omega_2(\theta_1)\cdot k \times r_2(\theta_1) \qquad V_c(\theta_1) := \left| v_c(\theta_1) \right|$$

$$\alpha_2(\theta_1) := \frac{-\omega_1^{\,2} \cdot \left[k \times \left(k \times r_1(\theta_1) \right) \right] \cdot j - \omega_2(\theta_1)^2 \cdot \left[k \times \left(k \times r_2(\theta_1) \right) \right] \cdot j}{\left[k \times r_2(\theta_1) \right] \cdot j}$$

$$b(\theta_1) := \omega_1^{\,2} \cdot \left[k \times \left(k \times r_1(\theta_1) \right) \right] + \omega_2(\theta_1)^2 \cdot \left[k \times \left(k \times r_2(\theta_1) \right) \right]$$

$$A_c(\theta_1) := \left[b(\theta_1) + \alpha_2(\theta_1) \cdot \left(k \times r_2(\theta_1) \right) \right] \cdot i$$

$\theta_1 \cdot \frac{180}{\pi}$	$\lvert r_0(\theta_1) \rvert$	$\omega_2(\theta_1)$	$V_c(\theta_1)$	$\alpha_2(\theta_1)$	$A_c(\theta_1)$
0	2.936	-0.516	0.258	0.069	-1.551
20	2.933	-0.471	-0.268	0.189	-1.412
40	2.761	-0.384	-0.698	0.312	-1.016
60	2.466	-0.254	-0.959	0.428	-0.47
80	2.114	-0.089	-1.028	0.506	0.056
100	1.767	0.089	-0.941	0.506	0.403
120	1.466	0.254	-0.773	0.428	0.53
140	1.229	0.384	-0.588	0.312	0.516
160	1.054	0.471	-0.416	0.189	0.467
180	0.936	0.516	-0.258	0.069	0.449
200	0.874	0.518	-0.094	-0.064	0.507

220	0.875	0.467	0.109	-0.24	0.683
240	0.961	0.342	0.398	-0.483	0.989
260	1.166	0.13	0.792	-0.716	1.215
280	1.514	-0.13	1.177	-0.716	0.868
300	1.961	-0.342	1.334	-0.483	-0.011
320	2.407	-0.467	1.176	-0.24	-0.849
340	2.754	-0.518	0.778	-0.064	-1.373
360	2.936	-0.516	0.258	0.069	-1.551

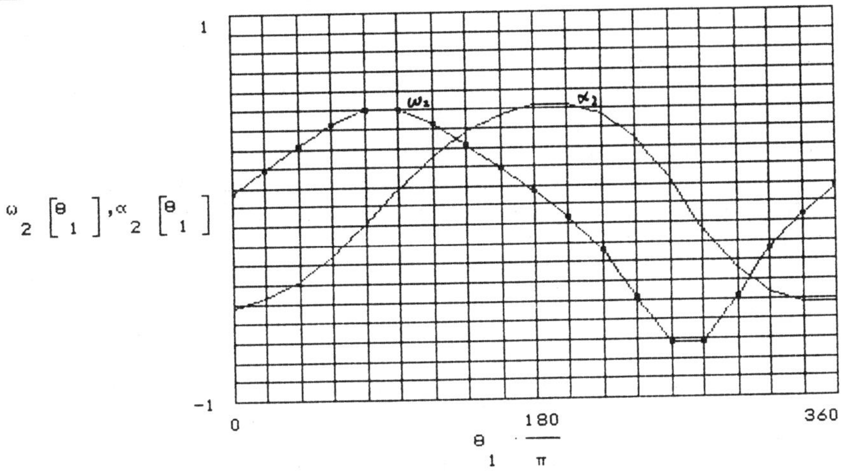

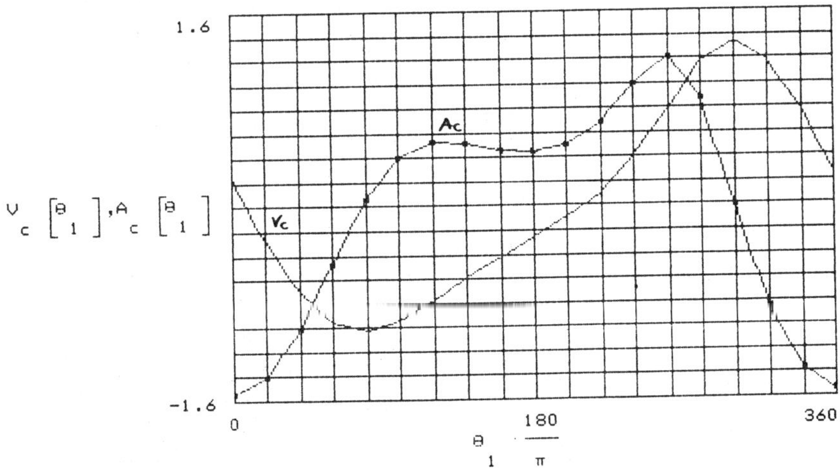

5.74 cont'd

Slider crank linkage $\qquad c_o := 1 \qquad$ (clockwise configuration positive)

$$\theta_1 := 0, \frac{\pi}{9} .. 2\cdot\pi \qquad R_1 := 1 \qquad R_2 := 3 \qquad e := \begin{bmatrix} 0 \\ -0.5 \\ 0 \end{bmatrix} \qquad r_{0u} := \begin{bmatrix} -1 \\ 0 \\ 0 \end{bmatrix}$$

$$r_1\left[\theta_1\right] := \begin{bmatrix} R_1\cdot\cos\left[\theta_1\right] \\ R_1\cdot\sin\left[\theta_1\right] \\ 0 \end{bmatrix} \qquad i := \begin{bmatrix} 1 \\ 0 \\ 0 \end{bmatrix} \quad j := \begin{bmatrix} 0 \\ 1 \\ 0 \end{bmatrix} \quad k := \begin{bmatrix} 0 \\ 0 \\ 1 \end{bmatrix} \qquad \omega_1 := 1$$

$$r_3\left[\theta_1\right] := r_1\left[\theta_1\right] + e \qquad \alpha_1 := 0$$

$$r_0\left[\theta_1\right] := \left[-r_3\left[\theta_1\right]\cdot r_{0u} + c_o\cdot\left[R_2{}^2 - \left[r_3\left[\theta_1\right]\cdot\left[r_{0u}\times k\right]\right]^2\right]^{.5}\right]\cdot r_{0u}$$

$$a\left[\theta_1\right] := -\left[r_3\left[\theta_1\right]\cdot\left[r_{0u}\times k\right]\right]\cdot\left[r_{0u}\times k\right]$$

$$r_2\left[\theta_1\right] := a\left[\theta_1\right] - c_o\cdot\left[R_2{}^2 - \left[r_3\left[\theta_1\right]\cdot\left[r_{0u}\times k\right]\right]^2\right]^{.5}\cdot r_{0u}$$

$$\omega_2\left[\theta_1\right] := -\omega_1\cdot\frac{r_1\left[\theta_1\right]\cdot i}{r_2\left[\theta_1\right]\cdot i}$$

$$v_c\left[\theta_1\right] := \omega_1\cdot k\times r_1\left[\theta_1\right] + \omega_2\left[\theta_1\right]\cdot k\times r_2\left[\theta_1\right] \qquad V_c\left[\theta_1\right] := v_c\left[\theta_1\right]$$

$$\alpha_2\left[\theta_1\right] := \frac{-\omega_1{}^2\cdot\left[k\times\left[k\times r_1\left[\theta_1\right]\right]\right]\cdot j - \omega_2\left[\theta_1\right]^2\cdot\left[k\times\left[k\times r_2\left[\theta_1\right]\right]\right]\cdot j}{\left[k\times r_2\left[\theta_1\right]\right]\cdot j}$$

$$b\left[\theta_1\right] := \omega_1{}^2\cdot\left[k\times\left[k\times r_1\left[\theta_1\right]\right]\right] + \omega_2\left[\theta_1\right]^2\cdot\left[k\times\left[k\times r_2\left[\theta_1\right]\right]\right]$$

$$A_c\left[\theta_1\right] := \left[b\left[\theta_1\right] + \alpha_2\left[\theta_1\right]\cdot\left[k\times r_2\left[\theta_1\right]\right]\right]\cdot i$$

$\theta_1\cdot\frac{180}{\pi}$	$\lvert r_0\left[\theta_1\right]\rvert$	$\omega_2\left[\theta_1\right]$	$V_c\left[\theta_1\right]$	$\alpha_2\left[\theta_1\right]$	$A_c\left[\theta_1\right]$
0	3.958	-0.338	0.169	0.019	-1.348
20	3.936	-0.314	-0.292	0.119	-1.253
40	3.763	-0.256	-0.679	0.211	-0.932
60	3.478	-0.168	-0.927	0.287	-0.479
80	3.134	-0.059	-1.013	0.332	-0.023
100	2.787	0.059	-0.956	0.332	0.324
120	2.478	0.168	-0.805	0.287	0.521
140	2.231	0.256	-0.606	0.211	0.6
160	2.056	0.314	-0.392	0.119	0.626
180	1.958	0.338	-0.169	0.019	0.652
200	1.94	0.326	0.067	-0.088	0.707

54.75

220	2.008	0.276	0.327	-0.2	0.783
240	2.171	0.187	0.61	-0.306	0.825
260	2.433	0.067	0.886	-0.375	0.719
280	2.78	-0.067	1.084	-0.375	0.372
300	3.171	-0.187	1.122	-0.306	-0.175
320	3.54	-0.276	0.958	-0.2	-0.749
340	3.819	-0.326	0.617	-0.088	-1.173
360	3.958	-0.338	0.169	0.019	-1.348

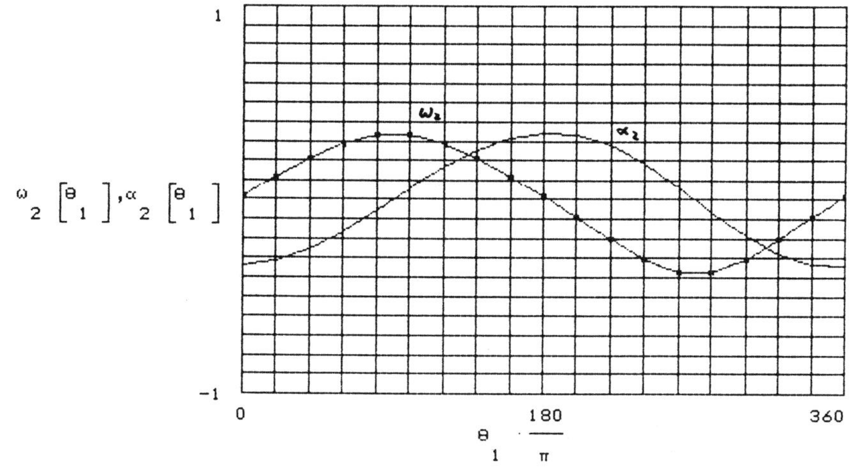

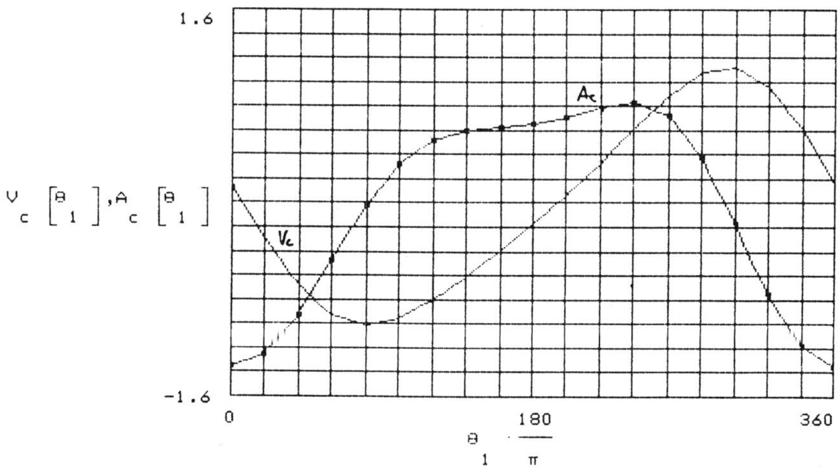

54.75 cont'd

Slider crank linkage $\quad c_o := 1 \quad$ (clockwise configuration positive)

$$\theta_1 := 0, \frac{\pi}{9} .. 2\cdot\pi \qquad R_1 := 1 \qquad R_2 := 2.5 \qquad e := \begin{bmatrix} 0 \\ -0.6 \\ 0 \end{bmatrix} \qquad r_{0u} := \begin{bmatrix} -1 \\ 0 \\ 0 \end{bmatrix}$$

$$r_1\left[\theta_1\right] := \begin{bmatrix} R_1\cdot\cos\left[\theta_1\right] \\ R_1\cdot\sin\left[\theta_1\right] \\ 0 \end{bmatrix} \qquad i := \begin{bmatrix} 1 \\ 0 \\ 0 \end{bmatrix} \quad j := \begin{bmatrix} 0 \\ 1 \\ 0 \end{bmatrix} \quad k := \begin{bmatrix} 0 \\ 0 \\ 1 \end{bmatrix} \qquad \omega_1 := 1$$

$$r_3\left[\theta_1\right] := r_1\left[\theta_1\right] + e \qquad \alpha_1 := 0$$

$$r_0\left[\theta_1\right] := \left[-r_3\left[\theta_1\right]\cdot r_{0u} + c_o\cdot\left[R_2^{\;2} - \left[r_3\left[\theta_1\right]\cdot\left[r_{0u}\times k\right]\right]^2\right]^{.5}\right]\cdot r_{0u}$$

$$a\left[\theta_1\right] := -\left[r_3\left[\theta_1\right]\cdot\left[r_{0u}\times k\right]\right]\cdot\left[r_{0u}\times k\right]$$

$$r_2\left[\theta_1\right] := a\left[\theta_1\right] - c_o\cdot\left[R_2^{\;2} - \left[r_3\left[\theta_1\right]\cdot\left[r_{0u}\times k\right]\right]^2\right]^{.5}\cdot r_{0u}$$

$$\omega_2\left[\theta_1\right] := -\omega_1\cdot\frac{r_1\left[\theta_1\right]\cdot i}{r_2\left[\theta_1\right]\cdot i}$$

$$v_c\left[\theta_1\right] := \omega_1\cdot k\times r_1\left[\theta_1\right] + \omega_2\left[\theta_1\right]\cdot k\times r_2\left[\theta_1\right] \qquad V_c\left[\theta_1\right] := v_c\left[\theta_1\right]$$

$$\alpha_2\left[\theta_1\right] := \frac{-\omega_1^{\;2}\cdot\left[k\times\left[k\times r_1\left[\theta_1\right]\right]\right]\cdot j - \omega_2\left[\theta_1\right]^2\cdot\left[k\times\left[k\times r_2\left[\theta_1\right]\right]\right]\cdot j}{\left[k\times r_2\left[\theta_1\right]\right]\cdot j}$$

$$b\left[\theta_1\right] := \omega_1^{\;2}\cdot\left[k\times\left[k\times r_1\left[\theta_1\right]\right]\right] + \omega_2\left[\theta_1\right]^2\cdot\left[k\times\left[k\times r_2\left[\theta_1\right]\right]\right]$$

$$A_c\left[\theta_1\right] := \left[b\left[\theta_1\right] + \alpha_2\left[\theta_1\right]\cdot\left[k\times r_2\left[\theta_1\right]\right]\right]\cdot i$$

$\theta_1\cdot\frac{180}{\pi}$	$r_0\left[\theta_1\right]$	$\omega_2\left[\theta_1\right]$	$V_c\left[\theta_1\right]$	$\alpha_2\left[\theta_1\right]$	$A_c\left[\theta_1\right]$
0	3.427	-0.412	0.247	0.042	-1.437
20	3.426	-0.378	-0.245	0.152	-1.334
40	3.266	-0.306	-0.656	0.256	-0.99
60	2.986	-0.201	-0.92	0.344	-0.509
80	2.644	-0.07	-1.012	0.398	-0.033
100	2.297	0.07	-0.958	0.398	0.315
120	1.986	0.201	-0.813	0.344	0.491
140	1.734	0.306	-0.63	0.256	0.542
160	1.547	0.378	-0.44	0.152	0.545
180	1.427	0.412	-0.247	0.042	0.563
200	1.376	0.406	-0.04	-0.081	0.634

54.76

220	1.403	0.353	0.204	-0.225	0.775
240	1.525	0.247	0.504	-0.384	0.939
260	1.76	0.09	0.842	-0.503	0.955
280	2.107	-0.09	1.127	-0.503	0.607
300	2.525	-0.247	1.228	-0.384	-0.061
320	2.935	-0.353	1.082	-0.225	-0.757
340	3.255	-0.406	0.724	-0.081	-1.245
360	3.427	-0.412	0.247	0.042	-1.437

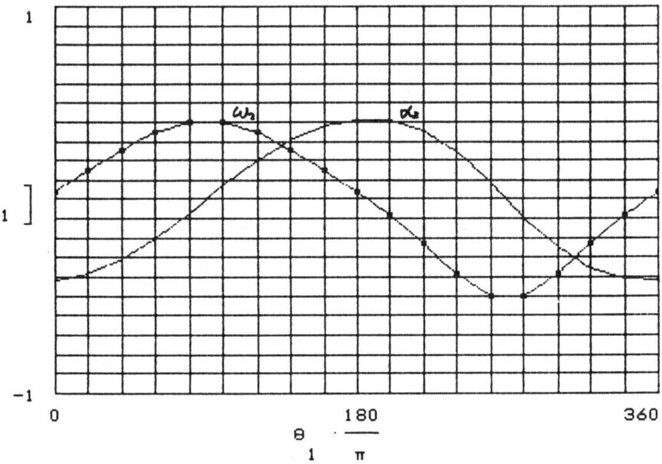

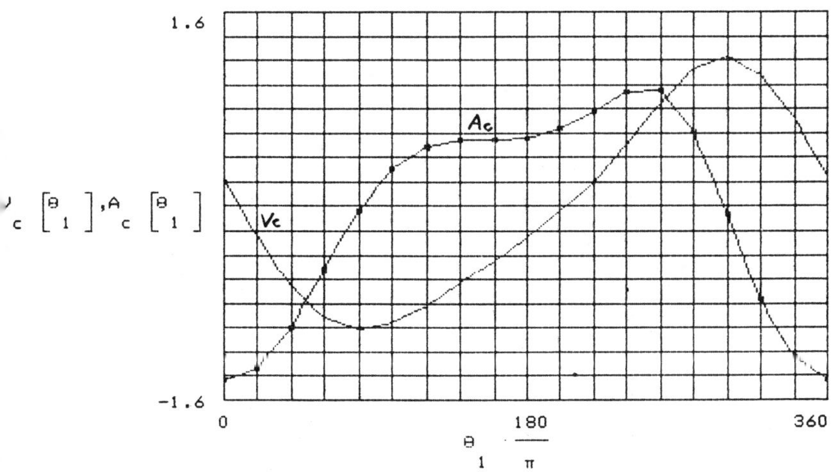

S4.76 con t'd

5.1

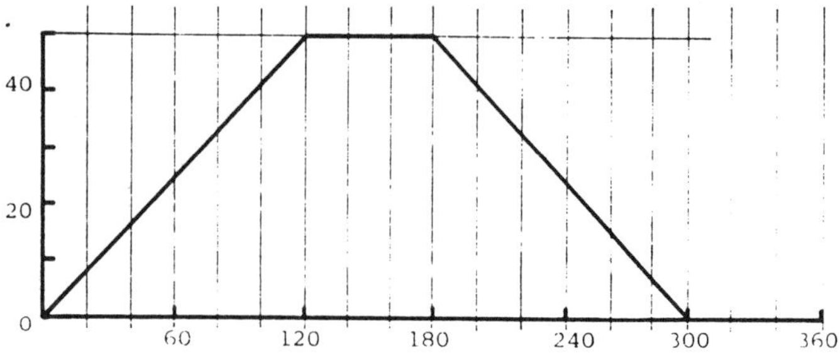
40
20
0
60
120
180
240
300
360

5.2

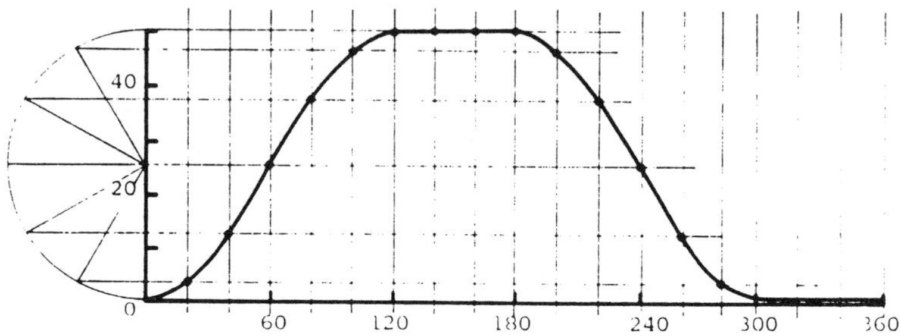
40
20
0
60
120
180
240
300
360

5.3

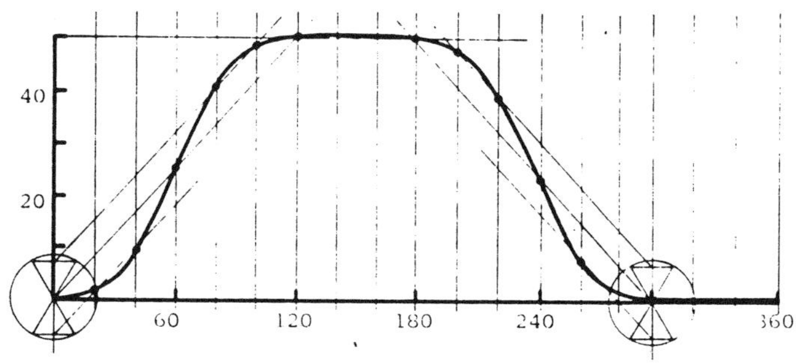
40
20
60
120
180
240
360

5.4

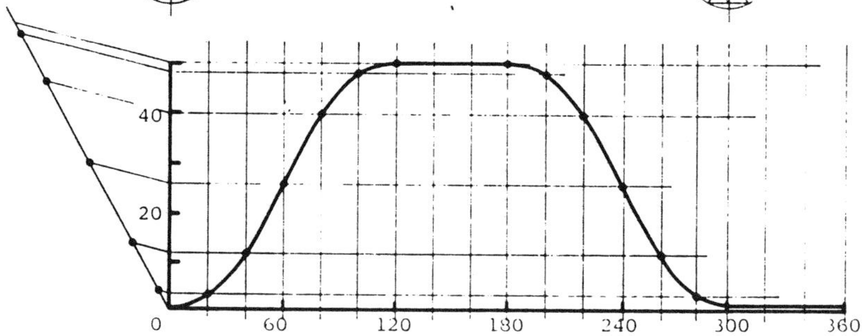
40
20
0
60
120
180
240
300
360

5.5

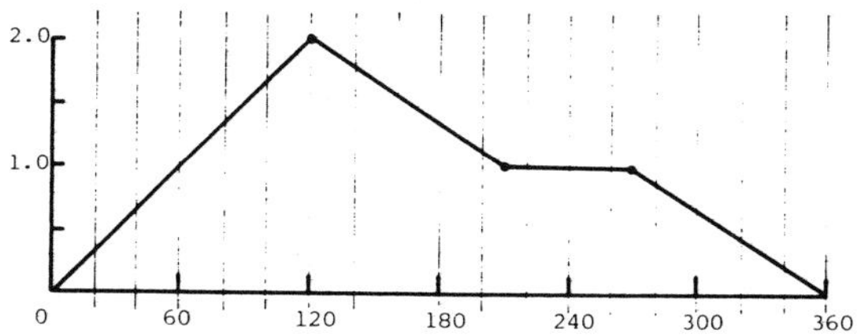
2.0
1.0
0
60
120
180
240
300
360

5.6

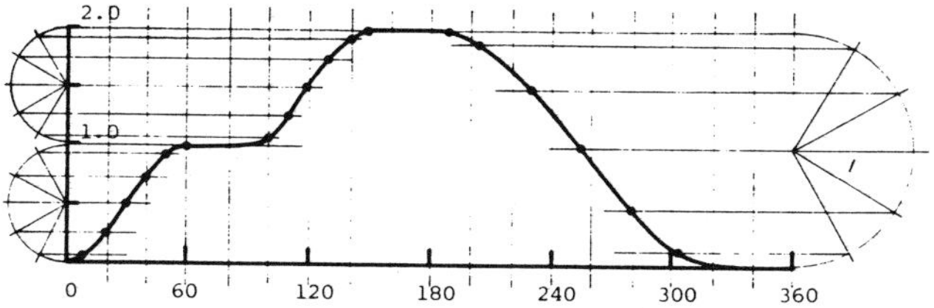
2.0
1.0
0
60
120
180
240
300
360

5.7

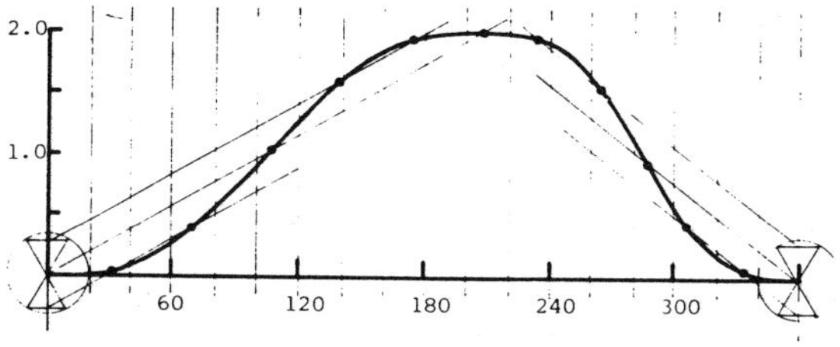
2.0
1.0
60
120
180
240
300

5.8

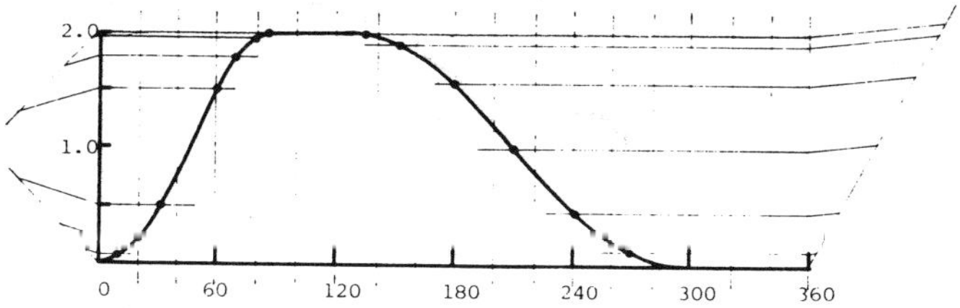
2.0
1.0
0
60
120
180
240
300
360

5.9

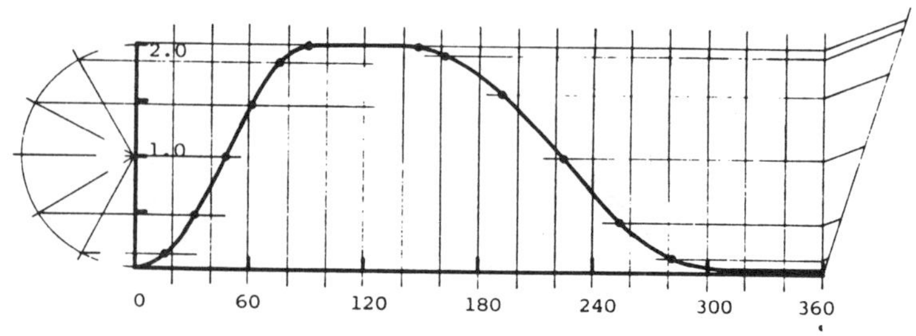

5.10

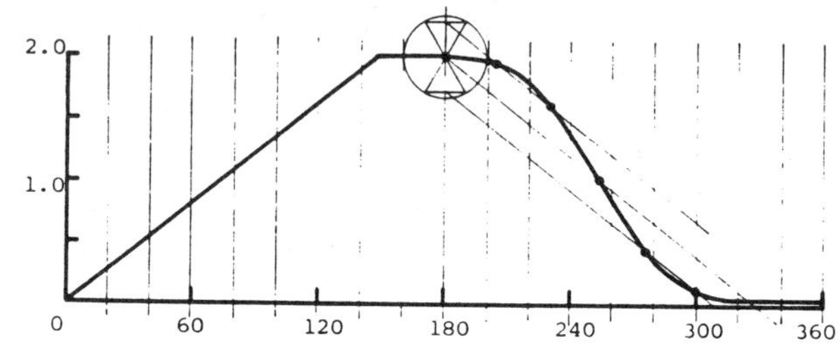

5.11

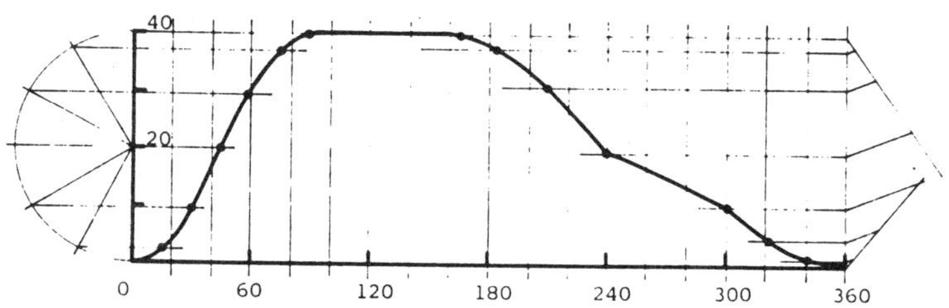

5.12

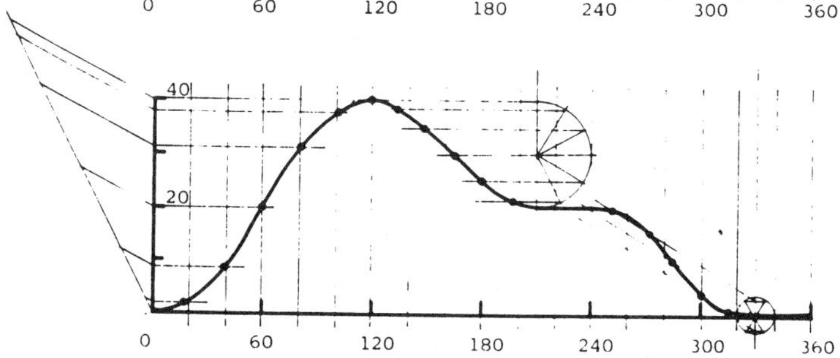

5.13

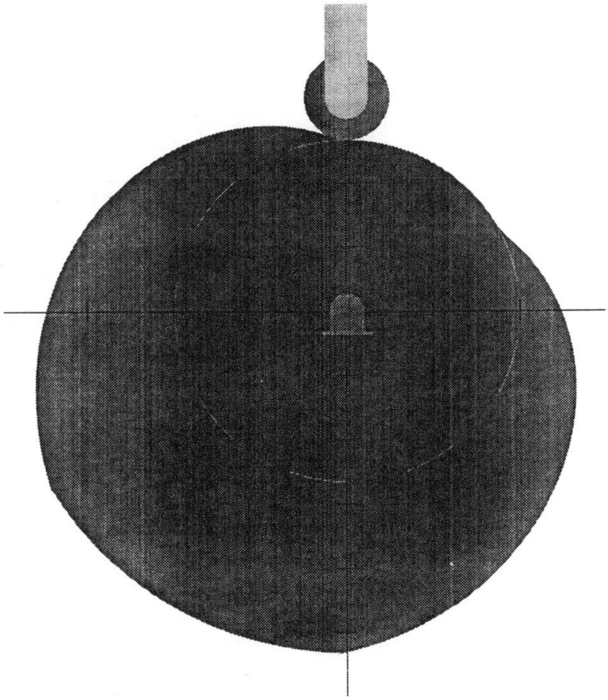

5.14

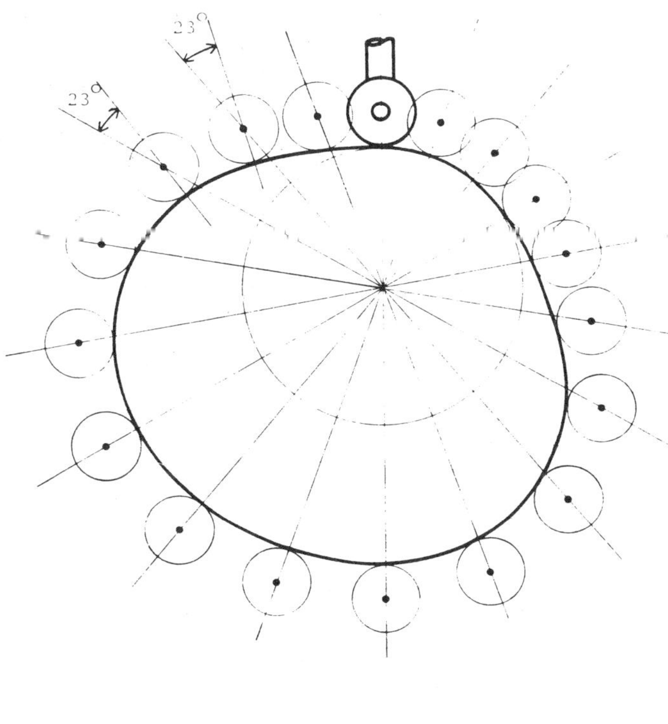

5.15

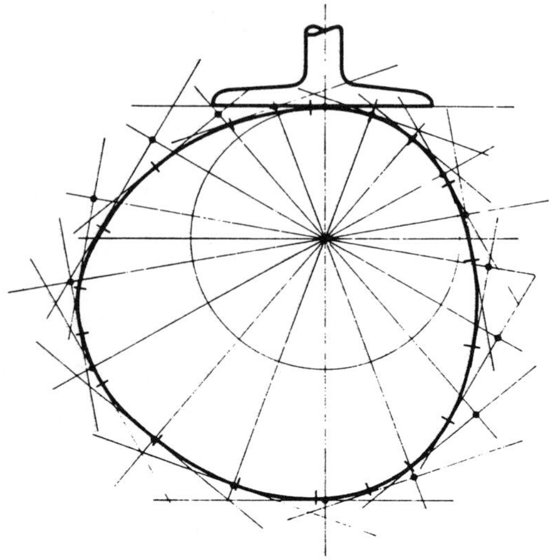

5.16

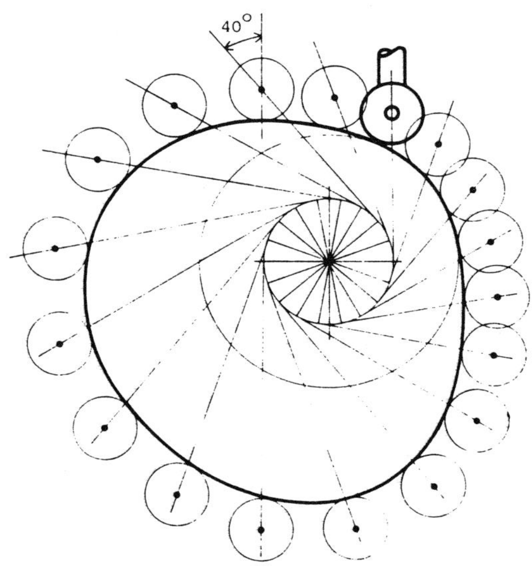

5.17

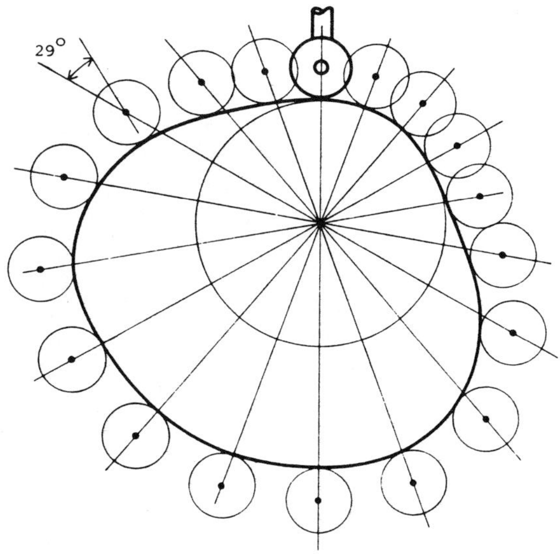

5.18

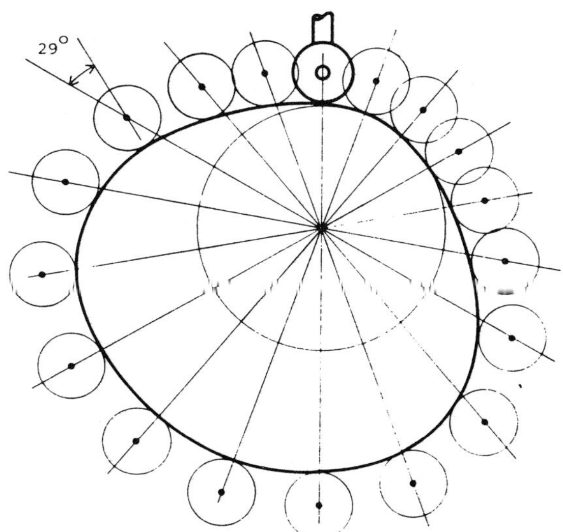

5.19

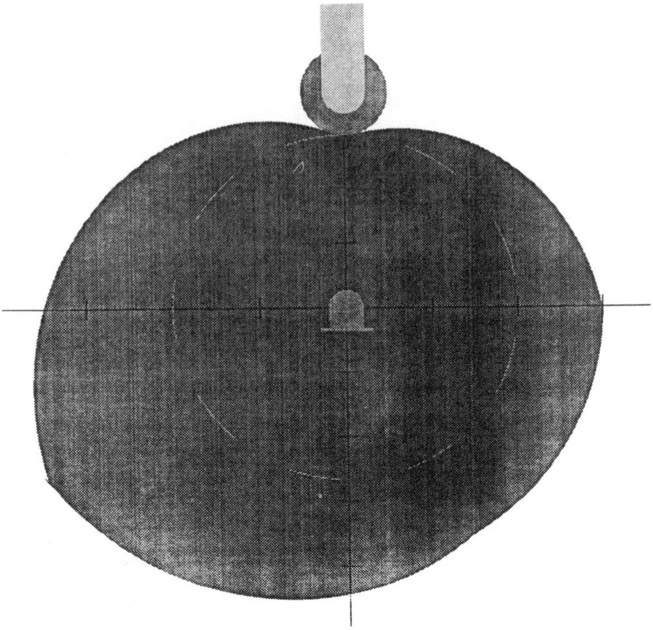

5.20

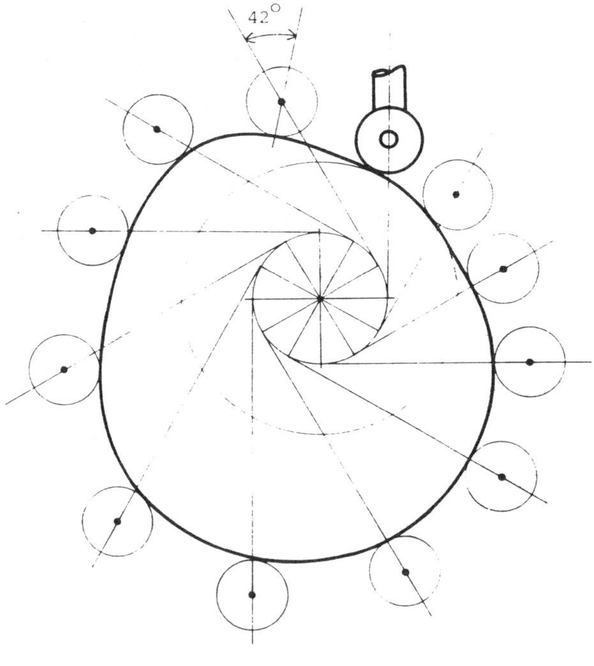

5.21

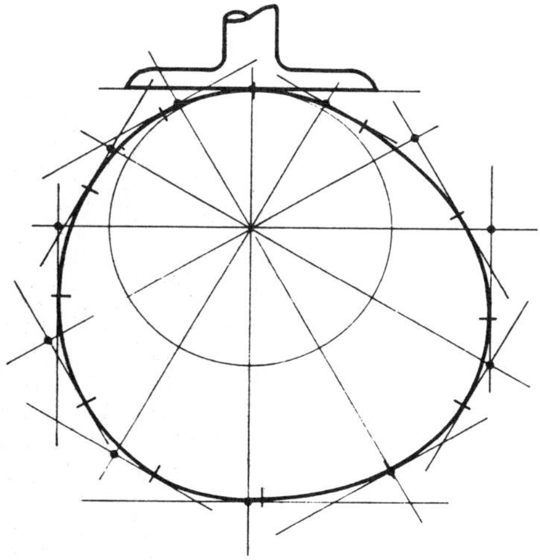

5.22

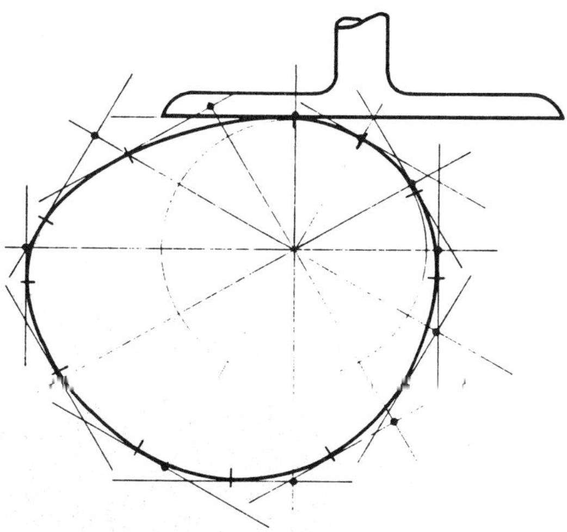

5.23

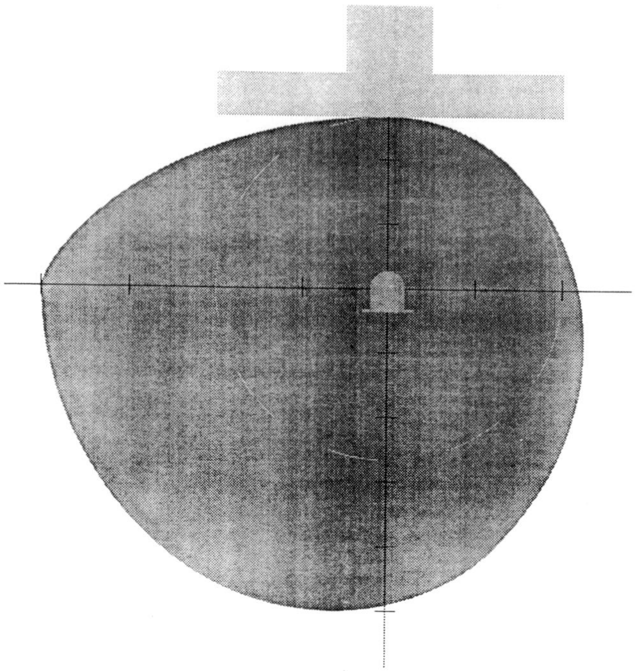

5.24

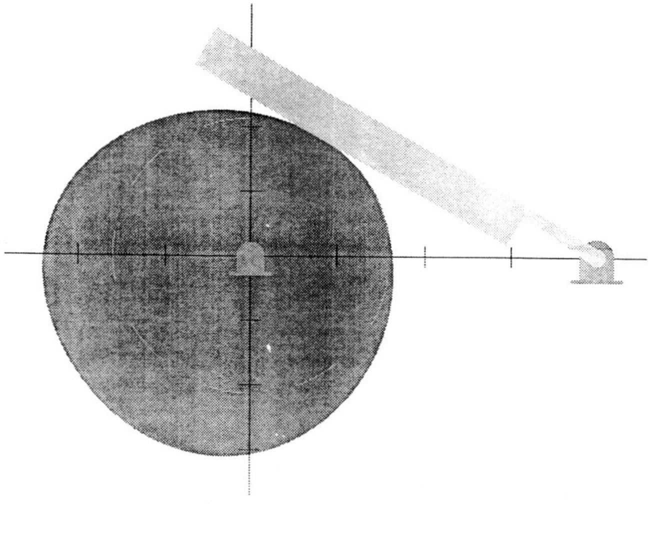

5.25

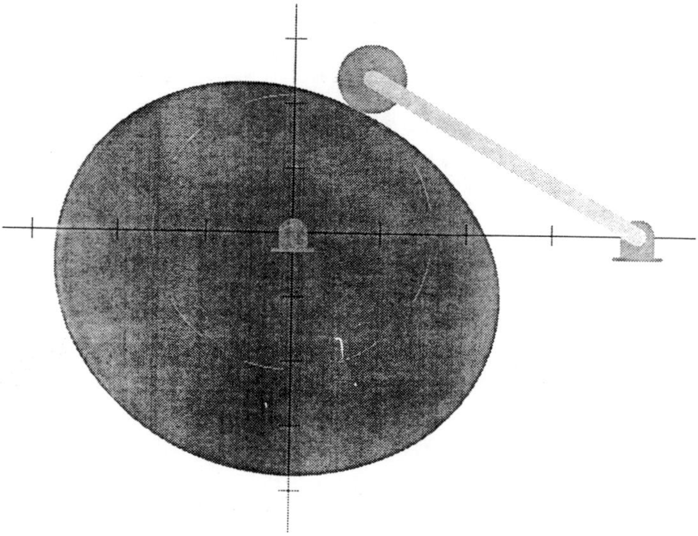

5.26 A) $s = C_o + C_1(\Theta - \Theta_i)$, For $0 \leq \Theta \leq \pi$, $\Theta_i = 0$, and $s = 0$ when $\Theta = 0$. $C_o = 0$,

Also, $s = 4$ in. when $\Theta = \pi$. $C_1 = \frac{4}{\pi}$ and $s = \frac{4}{\pi}\Theta$,

$v = \frac{4\omega}{\pi} = \frac{4}{\pi}(\frac{60}{60} \times 2\pi) = 8$ in/s, $a = 0$, $j = 0$ (a and j are infinite at $\Theta = 0$

and π), For $\pi \leq \Theta \leq 2\pi$, $\Theta_i = \pi$, and $s = 4$ when $\Theta = \pi$. $C_o = 4$ in,

Also, $s = 0$ when $\Theta = 2\pi$. $C_1 = -\frac{4}{\pi}$ and $s = 4 - \frac{4}{\pi}(\Theta - \pi) = 8 - \frac{4}{\pi}\Theta$,

$v = -\frac{4\omega}{\pi} = -8$ in/s, $a = 0$, $j = 0$, (a and j are infinite at $\Theta = \pi$ and 2π).

B) $v_{max} = 8$ in/s for $0 \leq \Theta \leq \pi$ and -8 in/s for $\pi \leq \Theta \leq 2\pi$ $a \to \infty$ for $\Theta = 0, \pi$, and 2π,

C) For $\Theta = 120^o = 2\pi/3$, $s = \frac{4}{\pi}(\frac{2\pi}{3}) = \frac{8}{3}$ in, $v = 8$ in/s, $a = j = 0$,

D) When $s = 3$ in, $v = 8$ in/s, and $a = j = 0$

5.27 A) For $0 \leq \Theta \leq \frac{\pi}{4}$, assuming constant positive acceleration, the follower

displacement will be $\frac{4}{6} = \frac{2}{3}$ in. Then, $s = C\Theta^2$ where $s = \frac{2}{3}$ in. when $\Theta = \frac{\pi}{4}$,

$C = s/\Theta^2 = \frac{2}{3}(\frac{4}{\pi})^2 = 32/3\pi^2$, $s = 32\Theta^2/3\pi^2$ in, $v = 64\omega\Theta/3\pi^2 = 128\Theta/3\pi$ in/s,

$a = 128\omega/3\pi = 256/3$ in/s^2, $j = 0$, For $\frac{\pi}{4} \leq \Theta \leq \frac{3\pi}{4}$, the velocity is

$\frac{128}{12} = \frac{32}{3}$ in/s and the displacement is $\frac{32}{3}(\frac{1}{4}) = \frac{8}{3}$ in. Then,

$s = C_o + C_1(\Theta - \frac{\pi}{4}) = \frac{2}{3} + C_1(\Theta - \frac{\pi}{4})$, $s = \frac{10}{3}$ when $\Theta = \frac{3\pi}{4}$,

$C_1 = \frac{2}{\pi}(\frac{10}{3} - \frac{2}{3}) = \frac{16}{3\pi}$, $s = \frac{2}{3} + \frac{16}{3\pi}(\Theta - \frac{\pi}{4}) = -\frac{2}{3} + \frac{16}{3\pi}\theta$ in, $v = \frac{16\omega}{3\pi} = \frac{32}{3}$ in/s,

$a = j = 0$, (j is infinite at $\Theta = \frac{\pi}{4}$ and $\frac{3\pi}{4}$), For $\frac{3\pi}{4} \leq \Theta \leq \pi$, from symmetry,

$s = 4 - C(\pi - \Theta)^2 = 4 - 32(\pi - \Theta)^2/3\pi^2$ in, $v = 64\omega(\pi - \Theta)/3\pi^2 = 128(\pi - \Theta)/3\pi$ in/s

$a = -128\omega/3\pi = -256/3$ in/s^2, $j = 0$, For $\pi \leq \Theta \leq 2\pi$, from symmetry,

$s = 4 - s(\Theta - \pi)$, $v = -v(\Theta - \pi)$, $a = -a(\Theta - \pi)$, and $j = 0$

B) $v_{max} = 10.7$ in/s for $\frac{\pi}{4} \leq \Theta \leq \frac{3\pi}{4}$ and -10.7 in/s for $\frac{5\pi}{4} \leq \Theta \leq \frac{7\pi}{4}$,

$a_{max} = 85.3$ in/s^2 for $0 \leq \Theta \leq \frac{\pi}{4}$ and $\frac{7\pi}{4} \leq \Theta \leq 2\pi$ and -85.3 in/s^2 for $\frac{3\pi}{4} \leq \Theta \leq \frac{5\pi}{4}$,

C) For $\Theta = 120^o = 2\pi/3$, $s = -\frac{2}{3} + \frac{16}{3\pi}(\frac{2\pi}{3}) = 2.89$ in, $v = 10.7$ in/s,

$a = j = 0$, D) When $s = 3$ in, $v = 10.7$ in/s, $a = j = 0$

5.28 A) For $0 \le \theta \le 2\pi$, $s = \frac{h}{2} - \frac{h}{2}\cos\frac{\pi}{\alpha}\theta$ where $h = 4$ in and $\alpha = \pi$, $s = 2 - 2\cos\theta$ in,

$v = 2\omega\sin\theta = 4\pi\sin\theta$ in/s, $a = 4\pi\omega\cos\theta = 8\pi^2\cos\theta$ in/s^2, $j = -8\pi^2\omega\sin\theta =$

$-16\pi^3\sin\theta$ in/s^3, B) $v_{max} = 12.6$ in/s at $\theta = \frac{\pi}{2}$ and -12.6 in/s at $\theta = \frac{3\pi}{2}$,

$a_{max} = 78.9$ in/s^2 at $\theta = 0$ and -78.9 in/s^2 at $\theta = \pi$, C) For $\theta = 120^\circ = \frac{2\pi}{3}$,

$s = 3$ in, $v = 10.9$ in/s, $a = -39.4$ in/s^2, $j = -429$ in/s^3,

D) When $s = 3$ in, $v = 10.9$ in/s, $a = -39.4$ in/s^2, $j = -429$ in/s^3

5.29 A) For $0 \le \theta \le \pi$, $s = \frac{h}{\alpha}\theta - \frac{h}{2\pi}\sin 2\pi\frac{\theta}{\alpha} = \frac{4}{\pi}\theta - \frac{2}{\pi}\sin 2\theta$, where $h = 4$ in. and $\alpha = \pi$,

Then, $v = \frac{4\omega}{\pi} - \frac{4\omega}{\pi}\cos 2\theta = 8 - 8\cos 2\theta$ in/s, $a = \frac{8\omega^2}{\pi}\sin 2\theta = 32\pi\sin 2\theta$ in/s^2,

$j = \frac{16\omega^3}{\pi}\cos 2\theta = 128\pi^2\cos 2\theta$ in/s^3, For $\pi \le \theta \le 2\pi$, from symmetry,

$s = h - \frac{h}{\alpha}(\theta - \pi) + \frac{h}{2\pi}\sin\frac{2\pi}{\alpha}(\theta - \pi)$, $s = 4 - \frac{4}{\pi}(\theta - \pi) + \frac{2}{\pi}\sin 2(\theta - \pi)$

$= 8 - \frac{4}{\pi}\theta + \frac{2}{\pi}\sin 2\theta$, Then, $v = -\frac{4\omega}{\pi} + \frac{4\omega}{\pi}\cos 2\theta = -8 + 8\cos 2\theta$ in/s,

$a = -\frac{8\omega^2}{\pi}\sin 2\theta = -32\pi\sin 2\theta$ in/s^2, $j = -\frac{16\omega^3}{\pi}\cos 2\theta = -128\pi^2\cos 2\theta$ in/s^3

B) $v_{max} = 16$ in/s when $\theta = \frac{\pi}{2}$, and -16 in/s when $\theta = \frac{3\pi}{2}$,

$a_{max} = 100.5$ in/s^2 when $\theta = \frac{\pi}{4}$ and $\frac{7\pi}{4}$, and -100.5 in/s^2 when $\theta = \frac{3\pi}{4}$ and $\frac{5\pi}{4}$,

C) For $\theta = 120^\circ = 2\pi/3$, $s = \frac{8}{3} - \frac{2}{\pi}\sin\frac{4\pi}{3} = 3.22$ in, $v = 8 - 8\cos\frac{4\pi}{3} = 12$ in/s,

$a = 32\pi\sin\frac{4\pi}{3} = -87.1$ in/s^2, $j = 128\pi^2\cos\frac{4\pi}{3} = -632$ in/s^3, D) By trial and

error, when $s = \frac{4}{\pi}\theta - \frac{2}{\pi}\sin 2\theta = 3$, $\theta = 1.99$ rad $= 114^\circ$, Then,

$v = 8 - 8\cos 228^\circ = 13.4$ in/s, $a = 32\pi\sin 228^\circ = -74.7$ in/s^2,

$j = 128\pi^2\cos 228^\circ = -845$ in/s^3

5.30 A) For $0 \le \theta \le \frac{\pi}{2}$, $s = \frac{2h}{\alpha^2}\theta^2 = \frac{8}{\pi^2}\theta^2$ where $h = 4$ in. and $\alpha = \pi$, Then,

$v = \frac{16\omega}{\pi^2}\theta = 10.20$ in/s, $a = \frac{16\omega^2}{\pi^2} = 64$ in/s^2,

$j = 0$ (j is infinite at $\theta = \frac{\pi}{2}$), For $\frac{\pi}{2} \le \theta \le \pi$, $s = -4 + \frac{16}{\pi}\theta - \frac{8}{\pi^2}\theta^2$,

$v = \frac{16\omega}{\pi} - \frac{16\omega}{\pi^2}\theta = 32 - 10.2\,\theta$ in/s, $a = -\frac{16\omega^2}{\pi^2} = -64$ in/s^2, $j = 0$,

For $\pi \le \theta \le \frac{3\pi}{2}$, $s = 4 - \frac{8}{\pi^2}(\theta - \pi)^2$, $v = \frac{-16\omega}{\pi^2}(\theta - \pi) = -10.2(\theta - \pi)$ in/s,

$a = -\frac{16\omega^2}{\pi^2} = -64$ in/s^2, $j = 0$ (j is infinite at $\theta = \frac{3\pi}{2}$), For $\frac{3\pi}{2} \le \theta \le 2\pi$,

$s = 4 + 4 - \frac{16}{\pi}(\theta - \pi) + \frac{8}{\pi^2}(\theta - \pi)^2$, $v = -\frac{16\omega}{\pi} + \frac{16\omega}{\pi^2}(\theta - \pi) = -32 + 10.2\,(\theta - \pi)$,

$a = \frac{16\omega^2}{\pi^2} = 64$ in/s^2, $j = 0$, B) $v_{max} = 16$ in/s at $\theta = \frac{\pi}{2}$,

and -16 in/s at $\theta = \frac{3\pi}{2}$, $a_{max} = 64$ in/s^2 for $-\frac{\pi}{2} \le \theta \le \frac{\pi}{2}$,

and -64 in/s^2 for $\frac{\pi}{2} \le \theta \le \frac{3\pi}{2}$, C) For $\theta = 120^\circ = \frac{2\pi}{3}$ rad,

$s = -4 + \frac{16}{\pi}\left(\frac{2\pi}{3}\right) - \frac{8}{\pi^2}\left(\frac{2\pi}{3}\right)^2 = 3.11$ in., $v = 32 - 21.3 = 10.7$ in/s,

$a = -64$ in/s^2, $j = 0$, D) When $s = -4 + \frac{16}{\pi}\theta - \frac{8}{\pi^2}\theta^2 = 3$, $8\theta^2 - 16\pi\theta + 7\pi^2 = 0$,

and $\theta = \frac{16\pi \pm \sqrt{(16\pi)^2 - 4(8)(7\pi^2)}}{16} = \pi \pm \pi/\sqrt{8} = 0.646\pi$, $v = 32 - 20.7 = 11.3$ in/s,

$a = -64$ in/s^2, $j = 0$

5.31 A) For $0 \le \theta \le \pi$, $s = 10h\left(\frac{\theta - \theta_i}{\alpha}\right)^3 - 15h\left(\frac{\theta - \theta_i}{\alpha}\right)^4 + 6h\left(\frac{\theta - \theta_i}{\alpha}\right)^5$,

where $h = 4$ in., $\alpha = \pi$, and $\theta_i = 0$, Then, $s = 1.29\theta^3 - 0.616\theta^4$

$+ 0.078\theta^5$ in., $v = 24.3\theta^2 - 15.5\theta^3 + 2.45\theta^4$ in/s,

$a = 305\theta - 292\theta^2 + 61.6\theta^3$ in/s^2, $j = 1920 - 3670\theta + 1160\theta^2$ in/s^3,

For $\pi \le \theta \le 2\pi$, $s = 1.29(2\pi - \theta)^3 - 0.616(2\pi - \theta)^4 + 0.078(2\pi - \theta)^5$ in.,

$v = -24.3(2\pi - \theta)^2 + 15.5(2\pi - \theta)^3 - 2.45(2\pi - \theta)^4$ in/s,

$a = 305(2\pi - \theta) - 292(2\pi - \theta)^2 + 61.6(2\pi - \theta)^3$ in/s^2,

$j = -1920 + 3670(2\pi - \theta) - 1160(2\pi - \theta)^2$ in/s^3,

B) v_{max} occurs at $\theta = \frac{\pi}{2}$ and $\frac{3\pi}{2}$ · $v(\theta = \frac{\pi}{2}) = 14.8$ in/s, $v(\theta = \frac{3\pi}{2}) = -14.8$ in/s,

a_{max} occurs when $j = 0$, $\theta^2 - 3.16\theta + 1.66 = 0$,

$$\theta = \frac{3.16 \pm\sqrt{(3.16)^2 - 4(1.66)}}{2} = 0.664, \quad 2.48, \quad a_{max} = \pm 92.4 \text{ in/s}^2,$$

C) For $\theta = 120^\circ = \frac{2\pi}{3}$ rad, $s = 3.14$ in., $v = 11.3$ in/s, $a = -76.1$ in/s^2,

$j = -678$ in/s^3, D) By trial and error, $s = 3$ when $\theta = 2.02$,

Then $v = 12.2$ in/s, $a = -67.6$ in/s^2, $j = -760$ in/s^3

5.32 A) For $0 \le \theta \le \frac{2\pi}{3}$, $s = \frac{h}{\alpha}\theta$ where $h = 50$ mm and $\alpha = \frac{2\pi}{3}(120^\circ)$, Then $s = \frac{75}{\pi}\theta$,

$v = \frac{75\omega}{\pi} = 150$ mm/s, $a = j = 0$, The acceleration and jerk are infinite at $\theta = 0$ and 120° B) For $0 \le \theta \le \frac{2\pi}{3}$, $v_{max} = 150$ mm/s and $a \to \infty$ at $\theta = 0$ and $\frac{2\pi}{3}$

5.33 A) For $0 \le \theta \le \frac{2\pi}{3}$, $s = \frac{h}{2} - \frac{h}{2}\cos\frac{\pi}{\alpha}\theta$ where $h = 50$ mm and $\alpha = \frac{2\pi}{3}$,

Then $s = 25 - 25\cos\frac{3\theta}{2}$ mm, $v = 75\pi \sin\frac{3\theta}{2}$ mm/s, $a = 225\pi^2 \cos\frac{3\theta}{2}$ mm/s^2,

$j = -675\pi^3 \sin\frac{3\theta}{2}$ mm/s^3 ($j \to \infty$ at $\theta = 0$ and $\frac{2\pi}{3}$) B) For $0 \le \theta \le \frac{2\pi}{3}$,

$v_{max} = 236$ mm/s at $\theta = \frac{\pi}{3} = 60^\circ$, and $a_{max} = 2220$ mm/s^2 at $\theta = 0$ and -2220 mm/s^2 at $\theta = \frac{2\pi}{3}$

5.34 A) For $0 \le \theta \le \frac{2\pi}{3}$, $s = \frac{h}{\alpha}\theta - \frac{h}{2\pi}\sin\frac{2\pi\theta}{\alpha}$ where $h = 50$ mm and $\alpha = \frac{2\pi}{3}$,

Then $s = \frac{75}{\pi}\theta - \frac{25}{\pi}\sin 3\theta$ mm, $v = 150 - 150\cos 3\theta$ mm/s, $a = 900\pi \sin 3\theta$ mm/s^2,

$j = 5400\pi^2 \cos 3\theta$ mm/s^3, B) For $0 \le \theta \le \frac{2\pi}{3}$, $v_{max} = 300$ mm/s at $\theta = \frac{\pi}{3}$,

and $a_{max} = 2830$ mm/s^2 at $\theta = \frac{\pi}{6}$ and -2830 mm/s^2 at $\theta = \frac{\pi}{2}$

5.35 A) For $0 \le \theta \le \frac{\pi}{3}$, $s = \frac{2h}{\alpha^2}\theta^2$ where $h = 50$ mm and $\alpha = \frac{2\pi}{3}$,

Then $s = \frac{225}{\pi^2}\theta^2$ mm, $v = \frac{900}{\pi}\theta$ mm/s, $a = 1800$ mm/s^2, $j = 0$

($j \to \infty$ at $\theta = 0$ and $\frac{\pi}{3}$), For $\frac{\pi}{3} \le \theta \le \frac{2\pi}{3}$, $s = -h + \frac{4h}{\alpha}\theta - \frac{2h}{\alpha^2}\theta^2$, where

$h = 50$ mm and $\alpha = \frac{2\pi}{3}$, Then $s = -50 + \frac{300}{\pi}\theta - \frac{225}{\pi^2}\theta^2$ mm, $v = 600 - \frac{900}{\pi}\theta$ mm/s,

$a = -1800$ mm/s^2, $j = 0$ ($j \to \infty$ at $\theta = \frac{\pi}{3}$ and $\frac{2\pi}{3}$), B) For $0 \le \theta \le \frac{2\pi}{3}$,

$v_{max} = 300$ mm/s at $\theta = \frac{\pi}{3}$, $a_{max} = 1800$ mm/s^2 for $0 \le \theta \le \frac{\pi}{3}$ and -1800 mm/s^2 for

$\frac{\pi}{3} \le \theta \le \frac{2\pi}{3}$

5.36 A) For $0 \le \theta \le \frac{2\pi}{3}$, $s = \frac{10h}{\alpha^3}\theta^3 - \frac{15h}{\alpha^4}\theta^4 + \frac{6h}{\alpha^5}\theta^5$ where $h = 50$ mm and $\alpha = \frac{2\pi}{3}$,

Then $s = 500\left(\frac{3}{2\pi}\right)^3\theta^3 - 750\left(\frac{3}{2\pi}\right)^4\theta^4 + 300\left(\frac{3}{2\pi}\right)^5\theta^5$ mm,

$v = 4500\left(\frac{3}{2\pi}\right)^2\theta^2 - 9000\left(\frac{3}{2\pi}\right)^3\theta^3 + 4500\left(\frac{3}{2\pi}\right)^4\theta^4$ mm/s,

$a = 27{,}000\left(\frac{3}{2\pi}\right)\theta - 81{,}000\left(\frac{3}{2\pi}\right)^2\theta^2 + 54{,}000\left(\frac{3}{2\pi}\right)^3\theta^3$ mm/s^2,

$j = 81{,}000 - 486{,}000\left(\frac{3}{2\pi}\right)\theta + 486{,}000\left(\frac{3}{2\pi}\right)^2\theta^2$ mm/s^3, B) For $0 \le \theta \le \frac{2\pi}{3}$,

$v_{max} = 281$ mm/s at $\theta = \frac{\pi}{3}$, a_{max} occurs when $j = 0$. Rearranging,

$27\theta^2 - 18\pi\theta + 2\pi^2 = 0$, $\theta = \frac{18\pi \pm \sqrt{(18\pi)^2 - 4(27)(2\pi^2)}}{2(27)} = 0.442, \quad 1.65$

$a(\theta = 0.442) = 2600$ mm/s^2 and $a(\theta = 1.65) = -2600$ mm/s^2

5.37 The maximum velocity magnitude during the rise will equal that during the return. For the rise, with h = 22mm, $\theta_i = 0$, and $\alpha = \frac{4\pi}{9}$,

$s = 220\left(\frac{9}{4\pi}\right)^3\theta^3 - 330\left(\frac{9}{4\pi}\right)^4\theta^4 + 132\left(\frac{9}{4\pi}\right)^5\theta^5 = 80.82\theta^3 - 86.83\theta^4 + 24.87\theta^5$,

$v = \omega\,(242.5\theta^2 - 347.3\theta^3 + 124.4\theta^4)$ where $\omega = \frac{3200}{60}(2\pi) = 335.1$ rad/s

The maximum velocity occurs when $\theta = 40^\circ = \frac{2\pi}{9}$ rad $= 0.6981$ rad, with $v_{max} = 335.1\left[242.5(.6981)^2 - 347.3(.6981)^3 + 124.4(.6981)^4\right] = 9909$ mm/s

5.38 Let θ_c refer to the cam angle where the transition from constant acceleration to constant deceleration takes place. Then $s = C\theta^2$, for $0 \le \theta \le \theta_c$, $s = C_1 + C_2\theta + C_3\theta^2$ for $\theta_c \le \theta \le \pi$, Differentiating, $v = 2\omega C\theta$ and $a = 2\omega^2 C$ for $0 \le \theta \le \theta_c$, $v = \omega(C_2 + 2C_3\theta)$ and $a = 2\omega^2 C_3$ for $\theta_c \le \theta \le \pi$,

From the given information, $2\omega^2 C = -3(2\omega^2 C_3)$ or $C_3 = -C/3$, Since $V = 0$ when $\theta = \pi$, $C_2 = -2C_3\pi = \frac{2\pi}{3}C$. The velocities are equal at $\theta = \theta_c$, $2\omega C\theta_c = \omega(C_2 + 2C_3\theta_c)$, from which $2C\theta_c = C_2 + 2C_3\theta_c = \frac{2\pi}{3}C - \frac{2C}{3}\theta_c$, $\theta_c = \frac{\pi}{4}$,

The displacements at $\theta = \theta_c$ are equal, from which $C(\frac{\pi}{4})^2 = C_1 + C_2(\frac{\pi}{4}) + C_3(\frac{\pi}{4})^2$, leading to $C_1 = (\frac{\pi}{4})^2 C - (\frac{\pi}{4})\frac{2\pi}{3}C + (\frac{\pi}{4})^2\frac{C}{3} = -\frac{\pi^2}{12}C$, The deflection at $\theta = \pi$ is 2 in:

$2 = C_1 + C_2\pi + C_3\pi^2 = -\frac{\pi^2}{12}C + \frac{2\pi^2}{3}C - \frac{\pi^2}{3}C = \frac{\pi^2}{4}C$, $C = \frac{8}{\pi^2}$, and

$a = 2C\omega^2 = \frac{16}{\pi^2}(\frac{300}{60}2\pi)^2 = 1600 \text{ in/s}^2$

5.39 Parabolic motion with $h = 1.5$ in and $\alpha = 8\pi/9$, $\omega = 375(\frac{2\pi}{60}) = 39.27$ rad/s,

$a = \frac{4h\omega^2}{\alpha^2} = \frac{4(1.5)(39.27)^2}{(8\pi/9)^2} = 1186 \text{ in/s}^2$, $v_{max} = \frac{2h\omega}{\alpha} = \frac{2(1.5)(39.27)}{(8\pi/9)} = 42.19$ in/s

5.40 During constant acceleration motion, $v = at$ and $s = \frac{1}{2}at^2$, Let t_1 refer to the time at the end of the constant acceleration portion of the rise, at which point the maximum velocity has been attained. Then,

$t_1 = \frac{v_{max}}{a}$ and $s = \frac{1}{2}\frac{v_{max}^2}{a}$ or $a = \frac{v_{max}^2}{2s} = \frac{(35)^2}{2(0.5)} = 1225 \text{ in/s}^2$, From symmetry, the deceleration is -1225 in/s^2. The elapsed time during the acceleration of the follower is equal to that during the deceleration and is $t_1 = \frac{v_{max}}{a} = \frac{35}{1225} = 0.0286$s

The elapsed time t_2 during the constant velocity motion is $t_2 = \frac{s}{v} = \frac{0.75}{35}$ $= 0.0214$s and the total time is $t = 2t_1 + t_2 = 2(0.0286) + 0.0214 = 0.0786$ s.

The degrees of cam rotation will, therefore, be $\theta = \frac{200 \text{ rev/m}}{60 \text{ s/m}} (360 \frac{\text{deg}}{\text{rev}})(0.0786\text{s})$

$= 94.32$ deg

5.41 Eight conditions. Therefore, $s = C_o + C_1\theta + C_2\theta^2 + C_3\theta^3 + C_4\theta^4 + C_5\theta^5 + C_6\theta 6$

$+ C_7\theta^7$, 1. $s = 0$ when $\theta = \theta_i = 0$; $C_o = 0$, 2. $v = 0$ when $\theta = 0$; $C_1 = 0$,

3. $a = 0$ when $\theta = 0$; $C_2 = 0$, 4. $j = 0$ when $\theta = 0$; $C_3 = 0$,

5. $s = h$ when $\theta = \alpha$; $h = C_4\alpha^4 + C_5\alpha^5 + C_6\alpha^6 + C_7\alpha^7$, 6. $v = 0$ when $\theta = \alpha$;

$0 = \omega(4C_4\alpha^3 + 5C_5\alpha^4 + 6C_6\alpha^5 + 7C_7\alpha^6)$, 7. $a = 0$ when $\theta = \alpha$;

$0 = \omega^2(12C_4\alpha^2 + 20C_5\alpha^3 + 30C_6\alpha^4 + 42C_7\alpha^5)$, 8. $j = 0$ when $\theta = \alpha$;

$0 = \omega^3(24C_4\alpha + 60C_5\alpha^2 + 120C_6\alpha^3 + 210C_7\alpha^4)$, Solving, $C_4 = 35h/\alpha^4$,

$C_5 = -84h/\alpha^5$, $C_6 = 70h/\alpha^6$, $C_7 = -20h/\alpha^7$

5.47 Let the parameter λ be defined such that the coordinates of the center of any circle in the family of circles are $x = \lambda$ and $y = -\lambda$. Then

$$F(x,y,z) = (x - \lambda)^2 + (y + \lambda)^2 - \frac{\lambda 2}{2} = 0 \quad (1)$$

$\frac{\partial F}{\partial \lambda}(x,y,z) = -2(x - \lambda) + 2(y + \lambda) - \lambda = 0$ (2), a) From Eq. (2),

$-2x + 2\lambda + 2y + 2\lambda - \lambda = 0$ or $\lambda = \frac{2}{3}x - \frac{2}{3}y$, Substituting into Eq. (1),

$(\frac{1}{3}x + \frac{2}{3}y)^2 + (\frac{2}{3}x + \frac{1}{3}y)^2 - \frac{1}{2}(\frac{2}{3}x - \frac{2}{3}y)^2 = 0$, Rearranging,

$y^2 + 4xy + x^2 = 0$, Solving, $y = (-2 \pm\sqrt{3})x$ (two straight lines), b) From Part(a).

$\lambda = \frac{2}{3}x - \frac{2}{3}(-2 \pm\sqrt{3})x = (2 \mp \frac{2}{\sqrt{3}})x$, $x = (\frac{3 \pm\sqrt{3}}{4})\lambda$ and $y = x - \frac{3}{2}\lambda = (\frac{-3 \pm\sqrt{3})}{4}\lambda$

5.48 $s = \frac{h}{\pi}\theta - \frac{h}{2\pi}\sin 2\theta$, $s' = \frac{h}{\pi} - \frac{h}{\pi}\cos 2\theta$, $x = -(r_b + s)\sin\theta - s'\cos\theta = -\sin\theta$

$(r_b + \frac{h}{\pi}\theta) - \frac{h}{2\pi}\cos\theta\,(1 - \cos 2\theta) = -\sin\theta(100+15.9\theta) - 7.96\cos\theta(1 - \cos 2\theta)$

$y = (r_b + s)\cos\theta - s'\sin\theta = \cos\theta\,(r_b + \frac{h}{\pi}\theta) - \frac{h}{2\pi}\sin\theta\,(3 - \cos 2\theta)$

$= \cos\theta\,(100 + 15.9\theta) - 7.96\sin\theta\,(3 - \cos 2\theta)$, Minimum face width $= 2s'_{max}$

$= \frac{4h}{\pi} = 63.7$ mm

5.49 $s = \frac{h}{\pi}\theta$, $s' = \frac{h}{\pi}$, $x = -(r_b + s)\sin\theta - s'\cos\theta = -\sin\theta\,(r_b + \frac{h}{\pi}\theta) - \frac{h}{\pi}\cos\theta$

$= -\sin\theta\,(100 + 15.9\theta) - 15.9\cos\theta$, $y = (r_b + s)\cos\theta - s'\sin\theta = \cos\theta\,(r_b + \frac{h}{\pi}\theta)$

$- \frac{h}{\pi}\sin\theta = \cos\theta\,(100 + 15.9\theta) - 15.9\sin\theta$, Minimum face width $= 2s'_{max}$

$= \frac{2h}{\pi} = 31.8$ mm

5.50 For $0 \le \theta \le \frac{\pi}{2}$, $s = \frac{2h}{\pi^2}\theta^2$, $s' = \frac{4h}{\pi^2}\theta$, $x = -(r_b + s)\sin\theta - s'\cos\theta$

$= -\sin\theta\,(r_b + \frac{2h}{\pi^2}\theta^2) - \frac{4h}{\pi^2}\theta\cos\theta = -\sin\theta\,(100 + 10.1\theta^2) - 20.3\theta\cos\theta$,

$y = (r_b + s)\cos\theta - s'\sin\theta = \cos\theta\,(r_b + \frac{2h}{\pi^2}\theta^2) - \frac{4h}{\pi^2}\theta\sin\theta = \cos\theta\,(100 + 10.1\theta^2)$

$- 20.3\theta\sin\theta$, For $\frac{\pi}{2} \le \theta \le \pi$, $s = -h + \frac{4h}{\pi}\theta - \frac{2h}{\pi^2}\theta^2$, $s' = \frac{4h}{\pi} - \frac{4h}{\pi^2}\theta$,

$x = -\sin\theta\,(r_b - h + \frac{4h}{\pi}\theta - \frac{2h}{\pi^2}\theta^2) - \cos\theta\,(\frac{4h}{\pi} - \frac{4h}{\pi^2}\theta) = -\sin\theta\,(50 + 63.7\theta - 10.1\theta^2)$

$- \cos\theta\,(63.7 - 20.3\theta)$, $y = \cos\theta\,(r_b - h + \frac{4h}{\pi}\theta - \frac{2h}{\pi^2}\theta^2) - \sin\theta\,(\frac{4h}{\pi} - \frac{4h}{\pi^2}\theta)$

$= \cos\theta\,(50 + 63.7\theta - 10.1\theta^2) - \sin\theta\,(63.7 - 20.3\theta)$, Minimum face width

$= 2s'_{max} = 2(\frac{2h}{\pi}) = 63.7$ mm

5.51 For $0 \le \theta \le 2\pi$, $s = \frac{h}{2} - \frac{h}{2}\cos\theta = 25 - 25\cos\theta$, $s' = \frac{h}{2}\sin\theta = 25\sin\theta$

$x_c = -(r_b + r_f)\sin\theta - s\sin\theta = -(r_b + r_f + \frac{h}{2})\sin\theta + \frac{h}{2}\sin\theta\cos\theta$,

$x_c' = -(r_b + r_f + \frac{h}{2})\cos\theta + \frac{h}{2}\cos 2\theta$, $y_c = (r_b + r_f)\cos\theta + s\cos\theta = (r_b + r_f + \frac{h}{2})$

$\cos\theta - \frac{h}{2}\cos^2\theta$, $y_c' = -(r_b + r_f + \frac{h}{2})\sin\theta + \frac{h}{2}\sin 2\theta$,

$x_c'^2 + y_c'^2 = (r_b + r_f + \frac{h}{2})^2 + \frac{h^2}{4} - h\,(r_b + r_f + \frac{h}{2})\cos\theta$,

$$x = -R\sin\theta + \frac{h}{2}\sin\theta\cos\theta \pm \left[\frac{-r_f R\sin\theta + r_f \frac{h}{2}\sin 2\theta}{\sqrt{R^2 + \frac{h^2}{4} - hR\cos\theta}}\right]$$

$$y = R\cos\theta - \frac{h}{2}\cos^2\theta + \left[\frac{-r_f R\cos\theta + r_f \frac{h}{2}\cos 2\theta}{\sqrt{R^2 + \frac{h^2}{4} - hR\cos\theta}}\right], \quad \text{where } R = r_b + r_f + \frac{h}{2}$$

Pressure angle: $\tan\phi = \dfrac{s'}{s + r_b + r_f} = \dfrac{25\sin\theta}{150 - 25\cos\theta} = \dfrac{\sin\theta}{6 - \cos\theta}$,

$\theta = 30^\circ$: $\tan\phi = 0.0974$, $\phi = 5.6^\circ$, $\theta = 60^\circ$: $\tan\phi = 0.1574$, $\phi = 8.9^\circ$,

$\theta = 90^\circ$: $\tan\phi = 0.1667$, $\phi = 9.5^\circ$

5.52 $s = \frac{h}{\pi}\theta - \frac{h}{2\pi}\sin 2\theta$, $s' = \frac{h}{\pi} - \frac{h}{\pi}\cos 2\theta$, $s'' = \frac{2h}{\pi}\sin 2\theta$,

$\rho = r_b + s + s'' = r_b + \frac{h}{\pi}\theta + \frac{3h}{2\pi}\sin 2\theta = 100 + 15.9\theta + 23.9\sin 2\theta$,

$\rho_{min} = \rho(\theta = 0) = 100$ mm, $\rho_{max} = \rho(\theta = 180^\circ) = 150$ mm

5.53 $s = \frac{h}{\pi}\theta$, $s' = \frac{h}{\pi}$, $s'' = 0$, $\rho = r_b + s + s'' = r_b + \frac{h}{\pi}\theta = 100 + 15.9\theta$,

$\rho_{min} = \rho(\theta = 0) = 100$ mm, $\rho_{max} = \rho(\theta = 180^\circ) = 150$ mm

5.54 For $0<\theta<\frac{\pi}{2}$, $s = \frac{2h}{\pi^2}\theta^2$, $s' = \frac{4h}{\pi^2}\theta$, $s'' = \frac{4h}{\pi^2}$,

$\rho = r_b + s + s'' = r_b + \frac{4h}{\pi^2} + \frac{2h}{\pi^2}\theta^2 = 120.3 + 10.1\theta^2$,

$\rho_{min} = \rho(\theta = 0) = 120$ mm, $\rho_{max} = \rho(\theta = \frac{\pi}{2}) = 145$ mm,

For $\frac{\pi}{2}<\theta<2\pi$, $s = -h + \frac{4h}{\pi}\theta - \frac{2h}{\pi^2}\theta^2$, $s' = \frac{4h}{\pi} - \frac{4h}{\pi^2}\theta$, $s'' = -\frac{4h}{\pi^2}$,

$\rho = r_b - \frac{4h}{\pi^2} - h + \frac{4h}{\pi}\theta - \frac{2h}{\pi^2}\theta^2 = 29.7 + 63.7\theta - 10.1\theta^2$,

$\rho_{min} = \rho(\theta = \frac{\pi}{2}) = 105$ mm, $\rho_{max} = \rho(\theta = \pi) = 130$ mm

5.55 $s = \frac{h}{2} - \frac{h}{2}\cos\theta$, $s' = \frac{h}{2}\sin\theta$, $s'' = \frac{h}{2}\cos\theta$,

$$\rho_p = \frac{\left[(r_b + r_f + \frac{h}{2} - \frac{h}{2}\cos\theta)^2 + \frac{h^2}{4}\sin^2\theta\right]^{3/2}}{(r_b + r_f + \frac{h}{2} - \frac{h}{2}\cos\theta)(r_b + r_f + \frac{h}{2} - \frac{h}{2}\cos\theta - \frac{h}{2}\cos\theta) + \frac{h^2}{2}\sin^2\theta}$$

Rearranging, $\rho_p = \dfrac{(R^2 + \frac{h^2}{4} - hR\cos\theta)^{3/2}}{R^2 + \frac{h^2}{2} - \frac{3hR}{2}\cos\theta}$, where $R = r_b + r_f + \frac{h}{2}$,

Substituting numerical values, $\rho_p = \dfrac{25(37 - 12\cos\theta)^{3/2}}{38 - 18\cos\theta}$,

$$\frac{d\rho_p}{d\theta} = \frac{450 \sin\theta(1 - 6\cos\theta)(37 - 12 \cos\theta)^{1/2}}{(38 - 18 \cos\theta)^2}$$

$\frac{d\rho_p}{d\theta} = 0$ when $\theta = 0, \quad 80.4^\circ, \quad 180^\circ, \quad 279.6^\circ, \quad 360^\circ$

$\rho_{p\ min} = 147.9$ mm at $\theta = 80.4^\circ, \quad 279.6^\circ$; $\quad \rho_{p\ max} = 156.2$ mm at $\theta = 0^\circ, \quad 360^\circ$

CHAPTER 6

SPUR GEARS: DESIGN AND ANALYSIS

6.5 $d = \frac{pN}{\pi} = \frac{(1.5708)(40)}{\pi} = 20$ in

6.6 $d = \frac{pN}{\pi} = \frac{(.7854)(28)}{\pi} = 7.0$ in, $\omega = \frac{V_p}{r} = \frac{2V_p}{d} = \frac{2(12)(12)}{7.0} = 41.1$ rad/s,

$n = \frac{60\omega}{2\pi} = \frac{60(41.1)}{2\pi} = 392$ rpm

6.7 $d = mN = 2(40) = 80$ mm, $\omega = \frac{V_p}{r} = \frac{2V_p}{d} = \frac{2(2500)}{80} = 62.5$ rad/s,

$n = \frac{60\omega}{2\pi} = \frac{60(62.5)}{2\pi} = 597$ rpm

6.8 (A) $r_v = \frac{n_2}{n_1} = \frac{520}{350} = 1.48$, (B) $\frac{N_1}{N_2} = \frac{n_2}{n_1}$, $N_2 = \frac{N_1 n_1}{n_2} = \frac{35(350)}{520} = 23.6$

Let $N_2 = 24$. Note: exact speed ratio cannot be obtained.

6.9 $r_v = \frac{N_1}{N_2}$, $N_1 = r_v N_2 = \frac{1}{3}(75) = 25$, $c = \frac{1}{2}(d_2 - d_1) = \frac{1}{2P}(N_2 - N_1) =$

$\frac{1}{2(3)}(75-25) = 8.33$ in

6.10 $P = \frac{\pi}{p} = \frac{\pi}{.7854} = 4$, $d = \frac{N}{P} = \frac{24}{4} = 6$ in, (A) Working depth = whole depth - clearance

$= \frac{2.25}{P} - \frac{0.25}{P} = \frac{2}{P} = \frac{2}{4} = 0.5$ in, (B) $d_b = d\cos\phi = 6\cos 20^{\circ} = 5.64$ in,

(C) $a = \frac{1}{P} = \frac{1}{4} = 0.25$ in, $d_a = d + 2a = 6 + 2(0.25) = 6.5$ in,

(D) Tooth thickness at pitch circle, $t = \frac{1.5708}{P} = \frac{1.5708}{4} = 0.393$ in,

and involute angle given by $\cos\psi = \frac{r_b}{r} = \frac{2.82}{3} = 0.94$,

$\psi = 20^\circ = 0.349$ rad, $\quad \text{inv}\,\psi = \tan\psi - \psi = 0.364 - 0.349 = 0.015$

At base circle, $\cos\psi_b = \dfrac{r_b}{r_b} = 1, \quad \psi_b = 0, \quad \text{inv}\,\psi_b = 0,$

$$\frac{t_b}{2r_b} - \frac{t}{2r} = \text{inv}\,\psi - \text{inv}\,\psi_b, \quad t_b = 2(2.82)\left[\frac{0.393}{2(3)} + .015 - 0\right] = 0.454 \text{ in},$$

(E) At addendum circle, $r_a = 3.25, \quad \cos\psi_a = \dfrac{r_b}{r_a} = \dfrac{2.82}{3.25} = .868,$

$\psi_a = 29.8^\circ = 0.520$ rad, $\quad \text{inv}\,\psi_a = \tan\psi_a - \psi_a = 0.573 - 0.520 = 0.053,$

$$t_a = 2r_a\left(\frac{t}{2r} + \text{inv}\,\psi - \text{inv}\,\psi_a\right) = 2(3.25)\left[\frac{.393}{2(3)} + .015 - .053\right] = 0.179 \text{ in}$$

6.11 $\quad \omega_1 = \dfrac{(300 \text{ rpm})\ 2\pi}{60} = 31.4$ rad/s, $\quad r_v = \dfrac{N_1}{N_2}, \quad N_1 = r_v N_2 = \dfrac{1}{3}(60) = 20,$

$d_1 = \dfrac{N_1}{P} = \dfrac{20}{4} = 5$ in, $\quad V_p = r_1\omega_1 = (\tfrac{5}{2})(31.4) = 78.5$ in/s

6.12 $\quad r_v = \dfrac{1}{5} = \dfrac{N_1}{N_2}, \quad N_2 = 5N_1, \quad C = 15 = \dfrac{1}{2P}(N_1+N_2) = \dfrac{N_1+5N_1}{2(4)}, \quad 6N_1 = 120,$

$N_1 = 20, \quad N_2 = 100$

6.13 $\quad r_v = \dfrac{N_1}{N_2} = 0.4, \quad N_1 = 0.4\,N_2, \quad c = 87.5 = \dfrac{m}{2}(N_1+N_2) = \dfrac{1.25}{2}(1.4\,N_2),$

$N_2 = 100, \quad N_1 = 0.4(100) = 40, \quad \omega_1 = \dfrac{(1000 \text{ rpm})2\pi}{60} = 104.7$ rad/s,

$d_1 = mN_1 = 1.25\ (40) = 50$ mm, $\quad V_p = r_1\omega_1 = (\tfrac{50}{2})(104.7) = 2618$ mm/s

6.14 $\quad N_2 = N_1+84, \ m = \dfrac{p}{\pi} = \dfrac{15.708}{\pi} = 5\text{mm}, \quad c = 350 = \dfrac{m}{2}(N_1+N_2) = \dfrac{5}{2}(2N_1+84),$

$N_1 = \dfrac{1}{5}\left[350 - \dfrac{5(84)}{2}\right] = 28, \quad N_2 = 112, \quad d_1 = mN_1 = 5(28) = 140$ mm,

$\omega_1 = \dfrac{V_p}{r_1} = \dfrac{4000}{(140/2)} = 57.1$ rad/s, $\quad n_1 = \dfrac{60\omega_1}{2\pi} = 545$ rpm, $\quad n_2 = n_1(\dfrac{N_1}{N_2}) =$
$545\ (\dfrac{28}{112}) = 136$ rpm

6.15 $\quad$ (A) $N_2 = \dfrac{N_1}{r_v} = \dfrac{32}{1/3} = 96, \quad c = \dfrac{1}{2P}(N_1+N_2) = \dfrac{1}{2(4)}(32+96) = 16$ in,

(B) $d_2 = \dfrac{N_2}{P} = \dfrac{96}{4} = 24$ in, $\quad r_{b2} = r_2\cos\phi = \dfrac{24}{2}\cos 20^\circ = 11.3$ in

6.16 $\quad$ (A) $N_2 = \dfrac{N_1}{r_v} = \dfrac{30}{0.3} = 100, \quad c = \dfrac{m}{2}(N_1+N_2) = \dfrac{6}{2}(30+100) = 390$ mm,

(B) $d_1 = mN_1 = 6(30) = 180$ mm, $d_2 = mN_2 = 6(100) = 600$ mm,

$r_{b1} = r_1 \cos\phi = \frac{180}{2} \cos 20^\circ = 84.6$ mm, $r_{b2} = r_2 \cos\phi = \frac{600}{2} \cos 20^\circ = 282$ mm,

(C) Tooth thickness at pitch circle, $t = 1.5708\, m = 1.5708(6) = 9.42$ mm

and involute angle $\psi = \phi = 20^\circ = 0.349$ rad, $\text{inv}\,\psi = \tan\psi - \psi = 0.364 - 0.349 =$

0.015 rad, At base circles, $\cos\psi_b = \frac{r_b}{r_b} = 1$, $\psi_b = \text{inv}\,\psi_b = 0$,

and $t_b = 2r_b\left(\frac{t}{2r} + \text{inv}\,\psi - \text{inv}\,\psi_b\right)$, $t_{b1} = 2(84.6)\left[\frac{9.42}{2(90)} + .015 - 0\right] = 11.39$ mm,

$t_{b2} = 2(282)\left[\frac{9.42}{2(300)} + .015 - 0\right] = 17.31$ mm

6.17 (A) $C = \frac{1}{2P}(N_1 + N_2)$, $N_2 = 2(5)(15) - 35 = 115$, (B) $d_1 = \frac{N_1}{P} = \frac{35}{5} = 7.0$ in,

$r_1 = 3.5$ in, $d_2 = \frac{N_2}{P} = \frac{115}{5} = 23.0$ in, $r_2 = 11.5$ in,

$r_{b1} = r_1 \cos\phi = 3.5 \cos 20^\circ = 3.29$ in, $r_{b2} = r_2 \cos\phi = 11.5 \cos 20^\circ = 10.81$ in,

$r_1' = \frac{r_{b1}}{\cos\phi'} = \frac{3.29}{\cos 24^\circ} = 3.60$ in, $r_2' = \frac{r_{b2}}{\cos\phi'} = \frac{10.81}{\cos 24^\circ} = 11.83$ in,

$c' = r_1' + r_2' = 3.60 + 11.83 = 15.43$ in

6.18 $(r_{b1} + r_{b2})$ is unchanged, $r_{b1} + r_{b2} = r_1 \cos\phi + r_2 \cos\phi = c \cos\phi =$

$r_1' \cos\phi' + r_2' \cos\phi' = c' \cos\phi'$, $\therefore c' \cos\phi' = c \cos\phi$,

$c' = c\frac{\cos\phi}{\cos\phi'} = 200 \frac{\cos 20^\circ}{\cos 23^\circ} = 204.2$ mm

6.19 $a_1 = a_2 = \frac{1}{P} = 0.333$, $P = 3$, $\cos\phi = 0.940$, $\sin\phi = 0.342$,

$d_1 = \frac{N_1}{P} = \frac{24}{3} = 8$ in, $r_1 = 4$ in, $r_2 = \frac{r_1}{r_v} = \frac{4}{1/2} = 8$ in,

$p_b = p\cos\phi = \frac{\pi\cos\phi}{P} = \frac{\pi\cos 20^\circ}{3} = .984$ in, Using Eq. (6-17),

$$CR = \frac{\sqrt{(8.333)^2 - 64(.940)^2} - 8(.342)}{.984} + \frac{\sqrt{(4.333)^2 - 16(.940)^2} - 4(.342)}{.984}, \quad CR = 1.67$$

6.20 $p_b = p\cos\phi = \frac{\pi}{P}\cos\phi$, $P = \frac{\pi}{p_b}\cos\phi = \frac{\pi}{.712}\cos 25^\circ = 4$, $a_1 = a_2 = \frac{1}{P} = 0.25$,

$\cos\phi = 0.906$, $\sin\phi = 0.423$, $d_1 = \frac{N_1}{P} = \frac{32}{4} = 8$ in, $r_1 = 4$ in,

$d_2 = \frac{N_2}{P} = \frac{48}{4} = 12$ in, $r_2 = 6$ in, From Eq. (6-17),

$$CR = \frac{\sqrt{(6.25)^2 - 36(.906)^2} - 6(.423)}{.712} + \frac{\sqrt{(4.25)^2 - 16(.906)^2} - 4(.423)}{.712}, \quad CR = 1.51$$

6.21 $r_v = \frac{1}{3} = \frac{N_1}{N_2}$, $N_2 = 3N_1$, $C = \frac{N_1 + N_2}{2P} = \frac{N_1 + 3N_1}{2(6)} = \frac{N_1}{3}$, $\therefore$ Find minimum N_1,

Minimum number of teeth: $N_1 \geq 20$, Contact ratio: $\phi = 20^\circ$,

$a_1 = a_2 = \frac{1}{P} = 0.167$ in, $p_b = p\cos\phi = \frac{\pi\cos\phi}{P} = \frac{\pi\cos 20^\circ}{6} = 0.492$ in,

$r_1 = \frac{N_1}{2P} = \frac{N_1}{12}$, $r_2 = 3r_1 = \frac{N_1}{4}$, Using Eq. (6-17) for different values of N_1,

N_1	20	22	24	26
CR	1.66	1.68	1.70	1.71

$\therefore N_{1\,min} = 24$ and $c_{min} = \frac{24}{3} = 8$ in

6.22 $a_1 = a_2 = 0.8\,m = 4.0$ mm, $\cos\phi = 0.940$, $\sin\phi = 0.342$,

$p_b = p\cos\phi = m\pi\cos\phi = 5\pi\cos 20^\circ = 14.76$ mm, $d_1 = mN_1 = 5(32) = 160$ mm,

$r_1 = 80$ mm, $r_2 = c - r_1 = 200 - 80 = 120$ mm, From Eq. (6-17), CR = 1.39

6.23 $a_1 = a_2 = m = 4$ mm, $\cos\phi = 0.940$, $\sin\phi = 0.342$,

$p_b = p\cos\phi = m\pi\cos\phi = 4\pi\cos 20^\circ = 11.81$ mm, $d_2 = mN_2 = 4(60) = 240$ mm,

$r_2 = 120$ mm, $r_1 = r_v r_2 = \frac{1}{3}(120) = 40$ mm, Using Eq. (6-17),

$$CR = \frac{\sqrt{(124)^2 - (120)^2 (.940)^2} - 120 (.342)}{11.81} + \frac{\sqrt{(44)^2 - (40)^2 (.940)^2} - 40 (.342)}{11.81},$$

$CR = 1.66$

6.24 $d_1 = mN_1 = 5(32) = 160$ mm, $r_1 = 80$ mm, $r_2 = c - r_1 = 200 - 80 = 120$ mm,

$a_1 = a_2 = m = 5$ mm, $\cos\phi = 0.940$, $\sin\phi = 0.342$,

$n_2 = \frac{r_1}{r_2} n_1 = \frac{80}{120} (1000) = 667$ rpm,

$|\omega_2 - \omega_1| = \frac{2\pi}{60} |n_2 - n_1| = \frac{2\pi}{60} (667 + 1000) = 174$ rad/s, Initial tooth contact:

$EP = \sqrt{(r_2 + a_2)^2 - r_2^2 \cos^2\phi} - r_2 \sin\phi = \sqrt{(125)^2 - (120)^2 (.940)^2} - 120 (.342)$

$= 12.8$ mm, $v_s = |\omega_2 - \omega_1| EP = (174)(12.8) = 2230$ mm/s, Final tooth contact:

$PF = \sqrt{(r_1 + a_1)^2 - r_1^2 \cos^2\phi} - r_1 \sin\phi = \sqrt{(85)^2 - (80)^2 (.940)^2} - 80 (.342)$

$= 12.3$ mm, $v_s = |\omega_2 - \omega_1| PF = (174)(12.3) = 2140$ mm/s

6.25 $n_2 = r_v n_1 = \frac{1}{2} (200) = 100$ rpm, $|\omega_2 - \omega_1| = \frac{2\pi}{60} |n_2 - n_1| = \frac{2\pi}{60} (100 + 200)$

$= 31.4$ rad/s, $d_1 = \frac{N_1}{P} = \frac{32}{4} = 8$ in, $r_1 = 4$ in, $r_2 = r_1/r_v = \frac{4}{1/2} = 8$ in,

$a_1 = a_2 = \frac{1}{P} = 0.25$ in, $\cos\phi = 0.906$, $\sin\phi = 0.423$,

Initial tooth contact: $EP = \sqrt{(r_2 + a_2)^2 - r_2^2 \cos^2\phi} - r_2 \sin\phi$

$= \sqrt{(8.25)^2 - 64(.906)^2} - 8(.423) = 0.557$ in,

$v_s = |\omega_2 - \omega_1| EP = (31.4)(.557) = 17.5$ in/s, Final tooth contact:

$PF = \sqrt{(r_1 + a_1)^2 - r_1^2 \cos^2\phi} - r_1 \sin\phi = \sqrt{(4.25)^2 - 16\ (.906)^2} - 4\ (.423)$

$= 0.528$ in, $\quad v_s = |\omega_2 - \omega_1| PF = (31.4)\ (.528) = 16.6$ in/s

6.26 $d_1 = \frac{N_1}{P} = \frac{12}{3} = 4$ in, $\quad r_1 = 2$ in, $\quad r_2 = \frac{r_1}{r_v} = \frac{2}{1/2} = 4$ in, $\quad c = r_1 + r_2 = 6$ in,

$a_1 = a_2 = \frac{1}{P} = 0.333$ in, $\quad$ For the gear, $r_{a2} = r_2 + a_2 = 4.333$ in,

and $r_{a2}\ (max) = \sqrt{r_2^2 \cos^2\phi + c^2 \sin^2\phi} = \sqrt{16(.940)^2 + 36(.342)^2} = 4.283$ in,

$\therefore$ Remove $4.333 - 4.283 = 0.05$ in, $\quad$ For the pinion, $r_{a1} = r_1 + a_1 = 2.333$ in,

and $r_{a1}\ (max) = \sqrt{r_1^2 \cos^2\phi + c^2 \sin^2\phi} = \sqrt{4(.940)^2 + 36(.342)^2} = 2.783$ in,

Since $r_{a1} < r_{a1}$ (max), pinion is OK

6.27 $d_1 = \frac{N_1}{P} = \frac{12}{3} = 4$ in, $\quad r_1 = 2$ in, $\quad r_2 = \frac{r_1}{r_v} = \frac{2}{1/2} = 4$ in, $\quad c = r_1 + r_2 = 6$ in,

$a_2 = \frac{1}{P} = 0.333$ in, $\quad r_{a2} = r_2 + a_2 = 4.333$ in,

$\sin\phi = \sqrt{\frac{r_{a2}^2 - r_2^2}{c^2 - r_2^2}} = \sqrt{\frac{(4.333)^2 - (4)^2}{(6)^2 - (4)^2}}, \quad \sin\phi = 0.372, \quad \phi = 21.9^\circ$

6.28 Standard gears, $k = 1$, $\quad 4kN_2 + 4k^2 \leq N_1^2 \sin^2\phi + 2N_1N_2 \sin^2\phi$,

$4(1)(45) + 4(1)^2 \leq N_1^2(.423)^2 + 2N_1(45)(.423)^2, \quad N_1^2 + 90N_1 - 1028 \geq 0$

From equality, $\quad N_1 = \frac{-90 \pm \sqrt{(90)^2 + 4\ (1028)}}{2} = -100.2, \quad 10.2,$ $\quad$ Therefore,

$N_1\ (min) = 11$

6.29 $p_b = \dfrac{2\pi r_{b1}}{N_1} = \dfrac{2\pi(1.81)}{10} = 1.14$ in, (A), (B), (C):

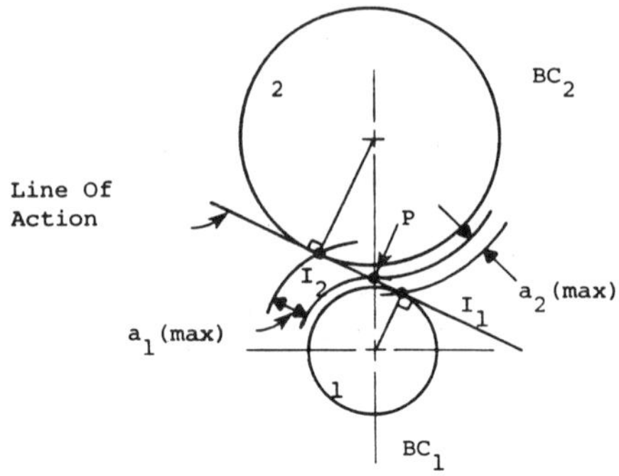

(D) $CR = \dfrac{I_1 I_2}{p_b} = \dfrac{2.55}{1.14} = 2.2$

6.30 (A) $N_2 = Pd_2 = 4(16) = 64$, (B) $a = \dfrac{1}{P} = \dfrac{1}{4} = 0.25$ in, (C) $d_1 = \dfrac{N_1}{P} = \dfrac{32}{4} = 8$ in,

$r_1 = 4$ in, $c = r_1 + r_2 = 4 + 8 = 12$ in,

$r_{a1}\text{ (max)} = \sqrt{r_1^2 \cos^2\phi + c^2 \sin^2\phi} = \sqrt{16(.968)^2 + 144(.250)^2} = 4.90$ in,

$a_1\text{ (max)} = r_{a1}\text{ (max)} - r_1 = 0.90$ in

6.31 $d_1 = mN_1 = 3(13) = 39$ mm, $r_1 = 19.5$ mm, $r_2 = \dfrac{r_1}{r_v} = \dfrac{19.5}{1/2} = 39$ mm,

$c = r_1 + r_2 = 58.5$ mm, $a_1 = a_2 = m = 3$ mm, Gear 2: $r_{a2} = r_2 + a_2 = 42$ mm,

$r_{a2}\text{ (max)} = \sqrt{r_2^2 \cos^2\phi + c^2 \sin^2\phi} = \sqrt{(39)^2 (.940)^2 + (58.5)^2 (.342)^2} = 41.76$ mm,

$\therefore$ Remove $42 - 41.76 = 0.24$ mm, Pinion 1: $r_{a1} = r_1 + a_1 = 22.5$ mm,

r_{a1} (max) $= \sqrt{r_1^2 \cos^2\phi + c^2 \sin^2\phi} = \sqrt{(19.5 \times .940)^2 + (68.5 \times .342)^2} = 27.13$ mm,

Pinion is OK

<u>6.32</u> $N_2 = 40, \quad k = 1, \quad 4k\,(N_2 + k) \le N_1 \sin^2\phi\,(N_1 + 2N_2),$

$4(1)(40 + 1) \le N_1\,(.342)^2\,(N_1 + 2 \times 40), \quad N_1^2 + 80N_1 - 1402 \ge 0,$

$N_1 > \dfrac{-80 + \sqrt{(80)^2 + 4\,(1402)}}{2} = 14.8, \quad \therefore N_1\,(\text{min}) = 15$

<u>6.33</u> $N_1 = 15, \quad k = 1, \quad 4k\,N_2 + 4k^2 \le N_1^2 \sin^2\phi + 2N_1 N_2 \sin^2\phi$

$[4(1) - 2(15)(.342)^2]\,N_2 \le (15)^2(.342)^2 - 4(1)^2, \quad N_2 \le 45.44, \quad \therefore N_2(\text{max}) = 45$

<u>6.34</u> $d_1 = mN_1 = 4(14) = 56$ mm, $\quad r_1 = 28$ mm, $\quad d_2 = mN_2 = 4(45) = 180$ mm,

$r_2 = 90$ mm, $\quad r_{b1} = r_1 \cos\phi = 26.3$ mm, $\quad r_{b2} = r_2 \cos\phi = 84.6$ mm,

$a_1 = a_2 = m = 4$ mm, $\quad c = r_1 + r_2 = 118$ mm, $\quad r_{a2} = r_2 + a_2 = 90 + 4 = 94$ mm,

r_{a2} (max) $= \sqrt{r_{b2}^2 + c^2 \sin^2\phi} = \sqrt{(84.6)^2 + (118 \times .342)^2} = 93.7$ mm,

$\therefore$ Increase center distance so that $\sqrt{r_{b2}^2 + c'^2 \sin^2\phi'} = r_{a2}$,

where c' is new center distance, and ϕ' is new pressure angle,

Then, $\quad c'^2 \sin^2\phi' = r_{a2}^2 - r_{b2}^2 = (94)^2 - (84.6)^2,$

But $c' = r_1' + r_2' = \dfrac{r_{b1} + r_{b2}}{\cos\phi'} = \dfrac{110.9}{\cos\phi'}$, $\quad$ Substituting, $(110.9)^2 \tan^2\phi'$

$= 1678.8, \quad \tan\phi' = 0.369, \quad \phi' = 20.3^\circ$

6.35 $d_1 = \frac{N_1}{P} = \frac{20}{5} = 4$ in, $r_1 = 2$ in, $r_2 = c - r_1 = 6 - 2 = 4$ in,

$r_v = \frac{r_1}{r_2} = \frac{1}{2}$, n_2 = speed of spur gear = $r_v n_1 = \frac{1}{2}(2000) = 1000$ rpm,

n_3 = speed of worm = n_2 = 1000 rpm, $r_v = \frac{n_4}{n_3} = \frac{N_3}{N_4} = \frac{2}{56} = \frac{1}{28}$ for worm,

n_4 = speed of wormwheel = $r_v n_3 = \frac{1000}{28}$, n_4 = speed of hoisting drum = 35.7 rpm,

$v = r\omega = \frac{8}{2}(35.7)(2\pi) = 897$ in/min, Load moves 897 in. in one minute

6.36 $d_b = d\cos\phi = 6\cos 14.5^\circ = 5.81$ in,

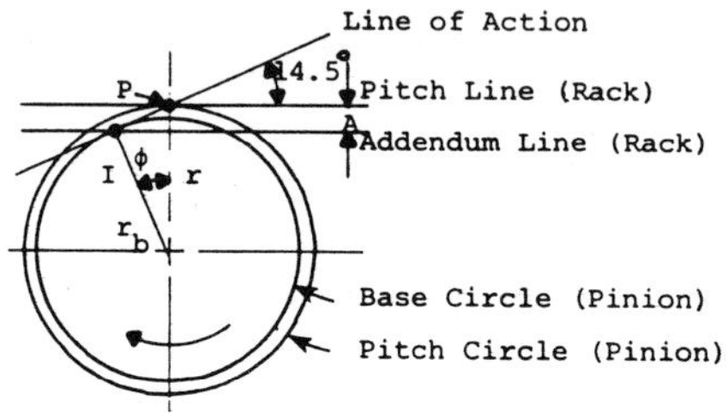

$A = r - r_b\cos\phi = \frac{6}{2} - \frac{5.81}{2}\cos 14.5^\circ = 0.188$ in

6.37 $P = \frac{N_1}{d_1} = \frac{15}{6} = 2.5$, $a_1 = \frac{1}{P} = 0.4$ in

(A)

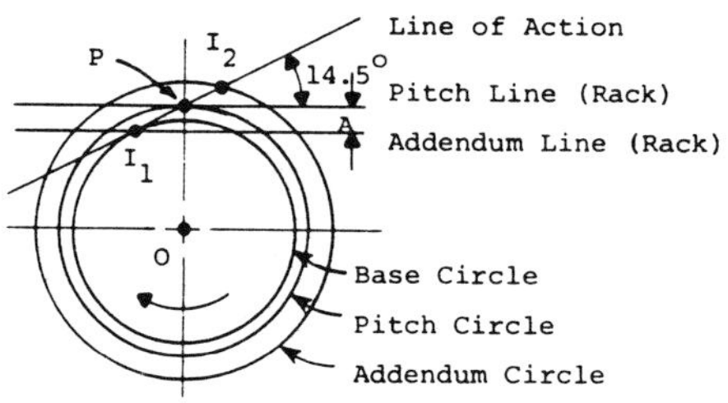

Not to Scale

(B) $CR = \dfrac{I_1 I_2}{p_b} = \dfrac{\sqrt{OI_2^{\,2} - OI_1^{\,2}}}{p_b}$, $\quad I_1 I_2 = \sqrt{(3.4)^2 - \left(\dfrac{5.81}{2}\right)^2} = 1.77$ in,

$p_b = \dfrac{\pi d_{b1}}{N_1} = \dfrac{\pi(5.81)}{15} = 1.22$, $\quad CR = 1.77/1.22 = 1.45$, $\quad$ (C) Adequate

6.38 From Problem 6.36, $\quad A = r - r_b \cos\phi = r - r\cos^2\phi = r\sin^2\phi$,

For standard addendum, $\quad A = \dfrac{1}{P} = \dfrac{2r}{N}$ and $\sin^2\phi = \dfrac{A}{r} = \dfrac{2}{N}$, $\quad \sin\phi = \sqrt{\dfrac{2}{N}}$, For N = 15,

$\sin\phi = 0.365$, $\quad \phi = 21.4^{\circ}$

6.39 $r_v = \dfrac{N_1}{N_2} = \dfrac{1}{5}$, $\quad N_2 = 5N_1$, $\quad C = 10 = \dfrac{1}{2}(d_2 - d_1) = \dfrac{1}{2P}(N_2 - N_1) = \dfrac{4N_1}{2(4)}$,

$N_1 = 20$, $\quad N_2 = 100$, $\quad d_1 = \dfrac{N_1}{P} = \dfrac{20}{4} = 5$ in, $\quad r_1 = 2.5$ in,

$v_p = r_1\omega_1 = 2.5(250)\dfrac{2\pi}{60} = 65.4$ in/s

6.40 (A) $C_{AB} = \dfrac{N_1 + N_2}{2P} = \dfrac{20 + 60}{2(4)} = 10$ in, $\quad C_{BC} = \dfrac{N_2 + N_3}{2P} = \dfrac{60 + 40}{2(4)} = 12.5$ in,

(B) $n_1 = 800$ rpm ccω, $\quad n_2 = \dfrac{N_1}{N_2} n_1 = \dfrac{20}{60}(800) = 267$ rpm cω,

$n_3 = \dfrac{N_2}{N_3} n_2 = \dfrac{N_1 \times N_2}{N_2 \times N_3} n_1 = \dfrac{20}{40}(800) = 400$ rpm ccω, (C) $d_1 = \dfrac{N_1}{P} = \dfrac{20}{4} = 5$ in,

$r_1 = 2.5$ in, $\quad d_2 = \dfrac{N_2}{P} = \dfrac{60}{4} = 15$ in, $\quad r_2 = 7.5$ in, $\quad d_3 = \dfrac{N_3}{P} = \dfrac{40}{4} = 10$ in,

$r_3 = 5.0$ in, $\quad v_p = r_1\omega_1 = r_2\omega_2 = r_3\omega_3 = 2.5(800)\dfrac{2\pi}{60} = 209$ in/s,

(D) $T_A = \dfrac{63,025\ hp}{n_1} = \dfrac{63,025(10)}{800} = 788$ in.lb, $\quad T_B = 0$ (gear 2 is an idler gear),

$T_C = \dfrac{63,025\ hp}{n_3} = \dfrac{63,025(10)}{400} = 1576$ in.lb

(E)

$$F_t = F_{Ax} = \frac{T_A}{r_1} = \frac{788}{2.5} = 315 \text{ lb.}$$

$$F_r = F_{Ay} = F_t \tan\phi = 115 \text{ lb.}$$

$$F_{Bx} = F_{By} = F_t + F_r = 430 \text{ lb.}$$

$$F_{Cx} = F_r = 115 \text{ lb.}, \; F_{Cy} = F_t = 315 \text{ lb.}$$

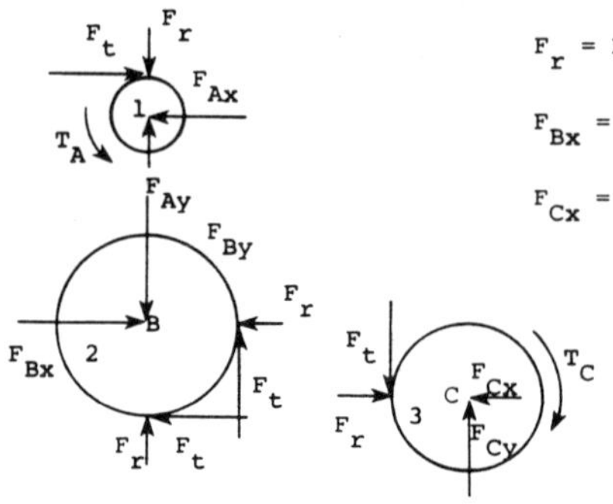

<u>6.41</u> $d_1 = mN_1 = 8(20) = 160$ mm, $r_1 = 80$ mm, $d_2 = mN_2 = 8(40) = 320$ mm,

$r_2 = 160$ mm, $d_3 = mN_3 = 8(100) = 800$ mm, $r_3 = 400$ mm,

(A) $C_{AB} = r_1 + r_2 = 80 + 160 = 240$ mm, $C_{BC} = r_3 - r_2 = 400 - 160 = 240$ mm,

(B) $n_1 = 1000$ rpm, $\omega_1 = 105$ rad/s cω, $n_2 = \frac{N_1}{N_2} n_1 = \frac{20}{40}(1000) = 500$ rpm ccω,

$n_3 = \frac{N_2}{N_3} n_2 = \frac{40}{100}(500) = 200$ rpm ccω, (C) $T_A = \frac{1000 \text{ kw}}{\omega_1} = \frac{1000(20)}{105} = 191$ N·m,

$T_B = 0$ (gear 2 is an idler), $T_C = \frac{1000 \text{ kw}}{\omega_3} = \frac{1000(20)}{20.9} = 955$ N·m,

(D) $F_t = F_{Ay} = F_{Cy} = \frac{T_A}{r_1} = \frac{191}{80/1000} = 2388$ N, $F_r = F_{Ax} = F_{Cx} = F_t \tan\phi = 869$ N,

$F_{Bx} = 0$, $F_{By} = 2F_t = 4776$ N

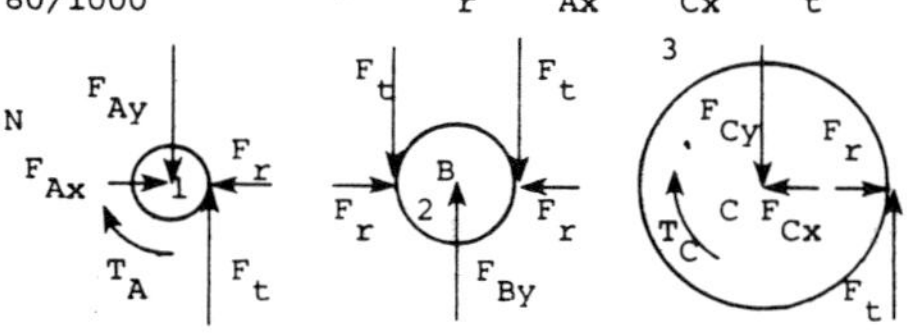

6.42 $d_1 = \frac{N_1}{P} = \frac{24}{5} = 4.8$ in, $\quad d_2 = \frac{N_2}{P} = \frac{60}{5} = 12$ in, $\quad d_3 = \frac{N_3}{P} = \frac{20}{4} = 5$ in,

$d_4 = \frac{N_4}{P} = \frac{60}{4} = 15$ in, $\quad n_A = 1000$ rpm cω,

$n_B = \frac{N_1}{N_2} n_A = \frac{24}{60}(1000) = 400$ rpm ccω, $\quad n_C = \frac{N_3}{N_4} n_B = \frac{20}{60}(400) = 133$ rpm cω,

$T_A = \frac{63{,}025 \text{ hp}}{n_A} = \frac{63{,}025(20)}{1000} = 1260$ in.lb, $\quad T_B = \frac{63{,}025 \text{ hp}}{n_B} = \frac{63{,}025(20)}{400} = 3151$ in.lb,

$T_C = \frac{63{,}025 \text{ hp}}{n_C} = \frac{63{,}025(20)}{133} = 9477$ in.lb, $\quad F_{t12} = \frac{T_A}{d_1/2} = \frac{1260}{2.4} = 525$ lb,

$F_{t34} = \frac{T_C}{d_4/2} = \frac{9477}{7.5} = 1264$ lb, $\quad F_{r34} = F_{t34} \tan 20^\circ = 1264\,(.364) = 460$ lb,

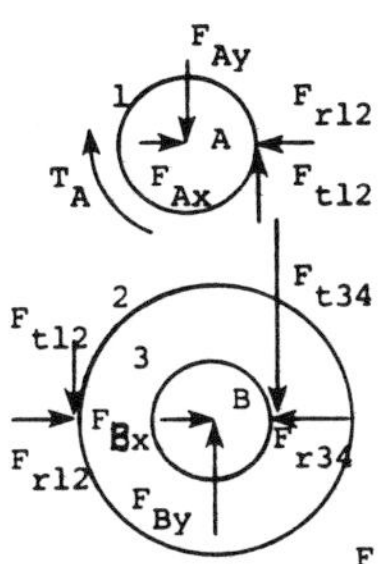

$F_{r12} = F_{t12} \tan 25^\circ = 525\,(.466) = 245$ lb.

$F_{Ax} = F_{r12} = 245$ lb.
$F_{Ay} = F_{t12} = 525$ lb. $\Big\}\; F_A = 579$ lb.

$F_{Bx} = F_{r34} - F_{r12} = 215$ lb.

$F_{By} = F_{t12} + F_{t34} = 1789$ lb.

$F_B = 1802$ lb.

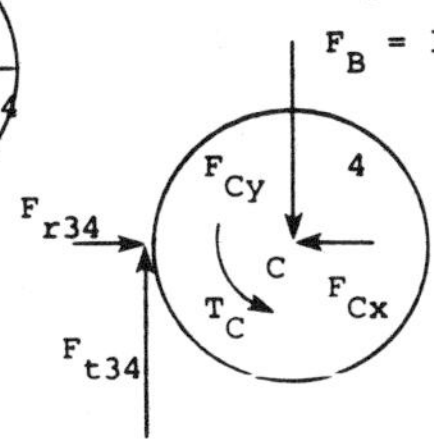

$F_{Cx} = F_{r34} = 460$

$F_{Cy} = F_{t34} = 1264$

$F_C = 1345$ lb.

6.43 $r_1 = \frac{m}{2} N_1 = \frac{4}{2}(32) = 64$ mm, $\quad r_2 = \frac{m}{2} N_2 = 160$ mm, $\quad r_3 = \frac{m}{2} N_3 = \frac{6}{2}(24) = 72$ mm,

$r_4 = \frac{m}{2} N_4 = 150$ mm, $\quad \omega_A = \frac{1200(2\pi)}{60} = 126$ rad/s ccω,

$\omega_B = \frac{N_1}{N_2}\omega_A = \frac{32}{80}(126) = 50.4$ rad/s cω, $\quad \omega_C = \frac{N_3}{N_4}\omega_B = \frac{24}{50}(50.4) = 24.2$ rad/s ccω,

$T_A = \frac{1000 \text{ kw}}{\omega_A} = \frac{1000(30)}{126} = 238$ N·m, $\quad T_B = \frac{1000 \text{ kw}}{\omega_B} = \frac{1000(30)}{50.4} = 595$ N·m,

$$T_C = \frac{1000\ kw}{\omega_C} = \frac{1000(30)}{24.2} = 1240\ N\cdot m$$

$$F_{t12} = \frac{T_A}{r_1} = \frac{238}{64}(1000) = 3720\ N, \qquad F_{r12} = F_{t12}\tan 20^{\circ} = 1350\ N,$$

$$F_{t34} = \frac{T_C}{r_4} = \frac{1240}{150}(1000) = 8270\ N, \qquad F_{r34} = F_{t34}\tan 25^{\circ} = 3850\ N,$$

$$\left.\begin{array}{l} F_{Ax} = F_{r12} = 1350\ N \\ F_{Ay} = F_{t12} = 3720\ N \end{array}\right\} \quad F_A = 3960\ N,$$

$$\left.\begin{array}{l} F_{Bx} = F_{r34} - F_{r12} = 2500\ N \\ F_{By} = F_{t12} + F_{t34} = 11{,}990\ N \end{array}\right\} \quad F_B = 12{,}250\ N$$

$$\left.\begin{array}{l} F_{cx} = F_{r34} = 3850 \\ F_{Cy} = F_{t34} = 8270 \end{array}\right\} \quad F_C = 9120\ N$$

CHAPTER 7

HELICAL, WORM, AND BEVEL GEARS: DESIGN AND ANALYSIS

7.1 $p^n = p\cos\psi = \frac{\pi}{P}\cos\psi \qquad p^n = \frac{\pi}{5}\cos 40^\circ = 0.481$ in., $\qquad N_f = \frac{N}{\cos^3\psi}$,

$N_{f1} = \frac{20}{(.766)^3} = 44.5, \quad N_{f2} = 66.7$

7.2 $\cos\psi = \frac{\tan\phi^n}{\tan\phi} = \frac{\tan 20^\circ}{\tan 25^\circ} = 0.781, \qquad N_{f1} = \frac{N_1}{\cos^3\psi} = \frac{15}{(.781)^3} = 31.5,$

$N_{f2} = \frac{45}{(.781)^3} = 94.5, \qquad c = \frac{1}{2P}(N_{f1} + N_{f2}) = \frac{31.5 + 94.5}{2(10)} = 6.3$ in

7.3 $c = 5 = \frac{N_1 + N_2}{2P^n\cos\psi}, \qquad N_1 + N_2 = 10\,(6)\cos\psi = 60\cos\psi,$

From text, minimum $\psi = 15^\circ$, $\quad \therefore N_1 + N_2 = 60\cos 15^\circ = 58, \quad$ But $N_1 = 20$,

$\therefore N_2 = 58 - 20 = 38, \qquad$ and $r_v = \frac{N_1}{N_2} = \frac{20}{38} = 0.526,$

$\tan\phi = \frac{\tan\phi^n}{\cos\psi} = \frac{60}{58}\tan 20^\circ = 0.376, \qquad \phi = 20.6^\circ$

7.4 $r_v = \frac{1}{4} = \frac{N_1}{N_2}, \qquad N_2 = 4N_1 = 80, \qquad \cos\psi = \frac{\tan\phi^n}{\tan\phi} = \frac{\tan 20^\circ}{\tan 25^\circ} = 0.781,$

$c = 10 = \frac{N_1 + N_2}{2P^n\cos\psi} = \frac{100}{2P^n(.781)}, \qquad P^n = 6.40, \qquad P = P^n\cos\psi = 6.40(.781) = 5.0,$

$d_1 = \frac{N_1}{P} = \frac{20}{5} = 4.0$ in, $\qquad d_2 = \frac{N_2}{P} = 16.0$ in

7.6 $r_v = \frac{N_1}{N_2} = 0.4, \qquad N_1 = 0.4\,N_2, \qquad \cos\psi = \frac{\tan\phi^n}{\tan\phi} = \frac{\tan 20^\circ}{\tan 23^\circ} = 0.857,$

$$c = 8 = \frac{p^n(N_1 + N_2)}{2\pi\cos\psi} = \frac{0.6\ (1.4\ N_2)}{2\pi(.857)}, \quad N_2 = 51.3, \quad \text{Use } N_2 = 51,$$

$$N_1 = 0.4\ N_2 = 20.5, \quad \text{Use } N_1 = 21$$

7.7 $p = \pi m = \pi(6) = 18.8$ mm, $p^n = p\cos\psi = 18.8 \cos 30^\circ = 16.3$ mm,

$$r_v = \frac{n_2}{n_1} = \frac{400}{1000} = \frac{N_1}{N_2}, \quad N_1 = 0.4\ N_2, \quad c = 252 = \frac{m}{2}(N_1 + N_2) = \frac{6}{2}(1.4\ N_2),$$

$$N_2 = 60, \quad N_{f2} = \frac{N_2}{\cos^3\psi} = \frac{60}{(.866)^3} = 92.4, \quad N_1 = 0.4(60) = 24,$$

$$N_{f1} = \frac{N_1}{\cos^3\psi} = 37.0$$

7.8 $v_p = \frac{d_1}{2}\omega_1$, $d_1 = \frac{2V_p}{\omega_1}$, $d_1 = 2(4500)/(\frac{2\pi}{60} \times 800) = 107$ mm,

$$p = \frac{\pi d_1}{N_1} = \frac{\pi(107)}{20} = 16.8, \quad \cos\psi = \frac{p^n}{p} = \frac{15}{16.8} = .893, \quad \psi = 26.8^\circ$$

7.9 $\cos\psi = \frac{\tan\phi^n}{\tan\phi} = \frac{\tan 20^\circ}{\tan 24^\circ} = 0.818$, $c = \frac{m^n(N_1 + N_2)}{2\cos\psi} = \frac{5(30 + 60)}{2(.818)} = 275$ mm,

$$N_{f1} = \frac{N_1}{\cos^3\psi} = \frac{30}{(.818)^3} = 54.8, \quad N_{f2} = \frac{N_2}{\cos^3\psi} = \frac{60}{(.818)^3} = 109.6$$

7.10 $P = P^n \cos\psi = 4 \cos 20^\circ = 3.76$, $d_1 = \frac{N_1}{P} = \frac{30}{3.76} = 7.98$,

$$V_p = \frac{d_1}{2}\omega_1 = \frac{7.98}{2(12)}(800 \times 2\pi) = 1671 \text{ ft/min}, \quad F_t = \frac{33{,}000\ hp}{v_p} = \frac{33{,}000(100)}{1671} = 1975 \text{ lb.}$$

$$\tan\phi = \frac{\tan\phi^n}{\cos\psi} = \frac{\tan 25^\circ}{\cos 20^\circ} = 0.496, \quad F_r = F_t \tan\phi = 1975(.496) = 980 \text{ lb.}$$

$$F_a = F_t \tan\psi = 1975 \tan 20^\circ = 719 \text{ lb.}$$

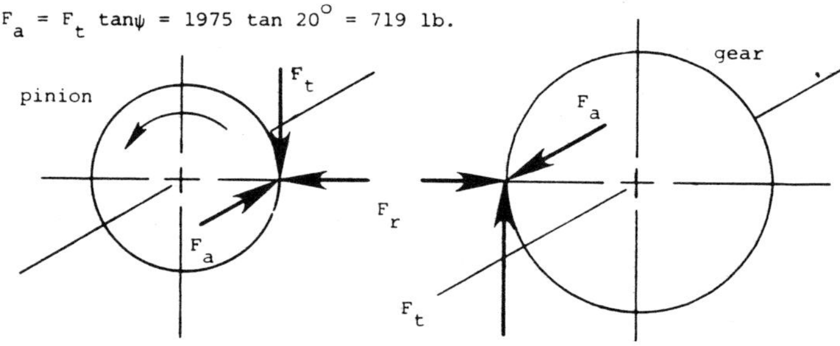

7.11 $c = 200 = \frac{m}{2}(N_1 + N_2) = \frac{m}{2}(20 + 40) \qquad m = 6.67,$

$$d_1 = mN_1 = 6.67\ (20) = 133\ \text{mm}, \qquad v_p = \frac{d_1}{2}\omega_1 = \frac{133}{2}\left(\frac{1200 \times 2\pi}{60}\right) = 8360\ \text{mm/s},$$

$$F_t = \frac{kW(10)^6}{v_p} = \frac{50(10)^6}{8360} = 5980\ \text{N}, \qquad \cos\psi = \frac{m^n(N_1 + N_2)}{2c} = \frac{6(60)}{2(200)} = 0.9,$$

$$\psi = 25.8^\circ, \qquad \tan\phi = \frac{\tan\phi^n}{\cos\psi} = \tan 20^\circ/.9 = 0.404, \qquad F_r = F_t\tan\phi = 5980(.404) = 2410\text{N}$$

$$F_a = F_t\tan\psi = 5980\ \tan 25.8^\circ = 2890\ \text{N}$$

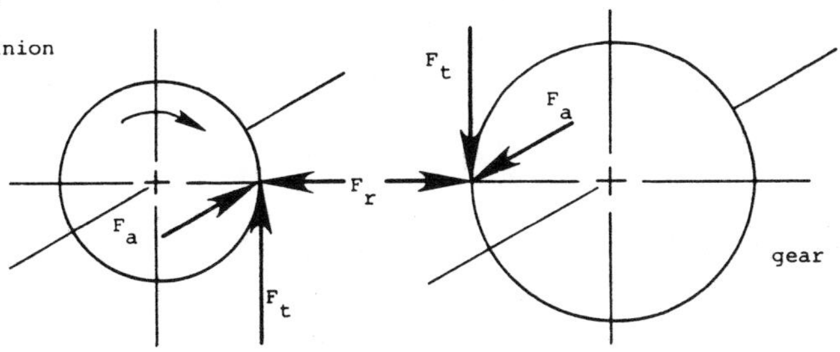

7.12 $\omega_A = \frac{2\pi n_A}{60} = \frac{2\pi(1000)}{60} = 104.7\ \text{rad/s}$

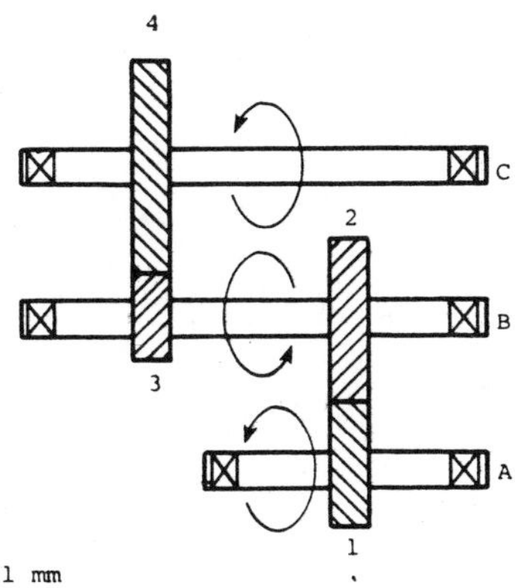

$$\omega_B = \frac{N_1}{N_2}\omega_A = \frac{30}{40}(104.7) = 78.54\ \text{rad/s}$$

$$\omega_C = \frac{N_3}{N_4}\omega_B = \frac{20}{50}(78.54) = 31.42\ \text{rad/s}$$

$$T = \frac{1000\ kW}{\omega} = \frac{20,000}{\omega}$$

$$T_A = \frac{20,000}{104.7} = 191.0\ \text{N·m}$$

$$T_B = 254.6\ \text{N·m}, \qquad T_C = 636.5\ \text{N·m}$$

$$m = \frac{m^n}{\cos\psi} = \frac{8}{\cos 30^\circ} = 9.238, \qquad d_1 = mN_1 = 277.1\ \text{mm}$$

$$T_A = \frac{d_1}{2}F_{t1} = \frac{d_1}{2}F_{n1}\cos\phi^n\cos\psi, \qquad F_{n1} = F_{n2} = \frac{2T_A}{d_1\cos\phi^n\cos\psi}$$

$$= \frac{2(191.0)(1000)}{(277.1)\cos 20^\circ\cos 30^\circ} = 1694\text{N}, \qquad d_4 = mN_4 = 461.9\ \text{mm},$$

$$F_{n3} = F_{n4} = \frac{2T_c}{d_4 \cos\phi^n \cos\psi} = \frac{2(636.5)(1000)}{(461.9)\cos 20^\circ \cos 30^\circ} = 3388\text{N},$$

$F_{aA} = F_{a1} = F_{n1} \cos\phi^n \sin\psi = 1694 \cos 20^\circ \sin 30^\circ = 796$ N to left

$F_{aB} = F_{a3} - F_{a2} = F_{n3}\cos\phi^n \sin\psi - 796 = 1592 - 796 = 796$ N to left,

$F_{ac} = F_{a4} = 1592$ N to right

7.13 $d_1 = \dfrac{N_1 p^n}{\pi \cos\psi_1} = \dfrac{24(.7854)}{\pi \cos 35^\circ} = 7.32$ in, $r_v = \dfrac{N_1}{N_2} = \dfrac{1}{2}$, $N_2 = 2N_1 = 48$,

$\psi_2 = \Sigma - \psi_1 = 90^\circ - 35^\circ = 55^\circ$, $d_2 = \dfrac{N_2 p^n}{\pi \cos\psi_2} = \dfrac{48(.7854)}{\pi \cos 55^\circ} = 20.92$ in,

$c = \frac{1}{2}(d_1 + d_2) = \frac{1}{2}(7.32 + 20.92) = 14.12$ in.

7.14 $d_1 = 7.32$ in., $N_2 = 48$, $\psi_2 = \psi_1 - \Sigma = 35^\circ - 10^\circ = 25^\circ$,

$d_2 = \dfrac{N_2 p^n}{\pi \cos\psi_2} = \dfrac{48(.7854)}{\pi \cos 25^\circ} = 13.24$ in., $c = \frac{1}{2}(d_1 + d_2) = \frac{1}{2}(7.32 + 13.24) = 10.28$ in.

7.15 $\psi_1 = \psi_2 = 30^\circ$, $r_v = 0.4 = \dfrac{N_1}{N_2}$, $N_1 = 0.4 N_2$,

$c = 12 = \dfrac{1}{2P^n}\left(\dfrac{N_1}{\cos\psi_1} + \dfrac{N_2}{\cos\psi_2}\right) = \dfrac{1.4N_2}{2(4)\cos 30^\circ}$, $N_2 = 59.38$, Use $N_2 = 60$,

$N_1 = 0.4(60) = 24$

7.16 $r_v = \dfrac{N_1}{N_2} = \dfrac{1}{2}$, $N_2 = 2N_1 = 2(32) = 64$,

$c = 220 = \dfrac{m^n}{2}\left(\dfrac{N_1}{\cos\psi_1} + \dfrac{N_2}{\cos\psi_2}\right) = \dfrac{4}{2}\left(\dfrac{32}{\cos\psi_1} + \dfrac{64}{\cos\psi_2}\right)$, $\dfrac{1}{\cos\psi_1} + \dfrac{2}{\cos\psi_2} = \dfrac{220}{64}$ (1),

$\psi_1 + \psi_2 = \Sigma = 60^\circ$ (2), Solving Eqs. (1) and (2) by trial and error:

$\psi_1 = 33^\circ$, $\psi_2 = 27^\circ$

7.17 $p_w = p_g = \dfrac{\ell}{N_W} = \dfrac{1}{3}$ in., $r_v = \dfrac{\ell}{\pi d_g}$, $d_g = \dfrac{\ell}{\pi r_v} = \dfrac{1}{\pi(.1)} = 3.18$ in,

$c = \dfrac{d_w + d_g}{2}$, $d_w = 2c - d_g = 5.0 - 3.18 = 1.82$ in.

7.18 $r_v = \dfrac{\ell}{\pi d_g}$, $d_g = \dfrac{\ell}{\pi r_v} = \dfrac{3}{\pi(.05)} = 19.1$ in, $\tan\lambda = \dfrac{\ell}{\pi d_w} = \dfrac{3}{\pi(1.5)} = 0.637$,

Assuming 90° shafts, $\psi_g = \lambda = 32.5^{\circ}$

<u>7.19</u> $\ell = p_w N_w = 1.5(3) = 4.5$ in., $\tan\lambda = \frac{\ell}{\pi d_w}$, $d_w = \frac{\ell}{\pi \tan\lambda} = \frac{4.5}{\pi \tan 20^{\circ}} = 3.94$ in.,

$d_g = \frac{\ell}{\pi r_v} = \frac{4.5}{\pi(.05)} = 28.65$ in., $c = \frac{1}{2}(d_w + d_g) = \frac{1}{2}(3.94 + 28.65) = 16.3$ in.

<u>7.20</u> $\ell = p_w N_w = .5(3) = 1.5$ in., $d_g = \frac{\ell}{\pi r_v} = \frac{1.5}{\pi(.04)} = 11.94$ in.

$c = \frac{1}{2}(d_w + d_g) = \frac{1}{2}(2.5 + 11.94) = 7.2$ in.

<u>7.21</u> $\tan\lambda = \frac{\ell}{\pi d_w}$, $\ell = \pi d_w \tan\lambda$, $r_v = \frac{\ell}{\pi d_g}$, $\ell = \pi d_g r_v$,

$\pi d_g r_v = \pi d_w \tan\lambda$, $d_g = \frac{\tan\lambda}{r_v} d_w = \frac{\tan 20^{\circ}}{0.05} d_w = 7.28\, d_w$,

$2c = 20 = d_w + d_g = 8.28\, d_w$, $d_w = \frac{20}{8.28} = 2.42$ in., $d_g = 2c - d_w = 17.58$ in.

<u>7.22</u> (A) $\ell = p_w N_w = 20(1) = 20$ mm, $\tan\lambda = \frac{\ell}{\pi d_w} = \frac{20}{\pi(50)} = 0.127$, $\lambda = 7.3^{\circ}$,

(B) $r_v = \frac{N_w}{N_g} = \frac{1}{40} = \frac{\ell}{\pi d_g}$, $d_g = \frac{40\ell}{\pi} = \frac{40(20)}{\pi} = 254.6$ mm,

$c = \frac{1}{2}(d_w + d_g) = \frac{1}{2}(50 + 254.6) = 152$ mm,

(C) $\omega_g = \frac{N_w}{N_g}\omega_w = \frac{1}{40}(\frac{2\pi}{60} \times 500) = 1.31$ rad/s, $T_g = \frac{d_g}{2} F_{tg} = \frac{kW(1000)}{\omega_g} = \frac{500}{1.31} =$

382 N·m, $F_{tg} = \frac{2T_g}{d_g} = \frac{2(382)}{255/1000} = 3000$ N, $F_{rg} = F_{tg} \tan\phi = 3000 \tan 25^{\circ} =$

1400 N, $F_{ag} = F_{tg} \tan\lambda = 3000(.127) = 381$ N

<u>7.23</u> $T_A = \frac{kW(1000)}{\omega_A} = \frac{1000(60)}{(2\pi \times 1000)} = 9.55$ N·m, $T_B = \frac{N_2}{N_1} T_A = \frac{48}{24}(9.55) = 19.1$ N·m,

$T = \frac{N_4}{N_3} T_B = \frac{40}{1}(19.1) = 764$ N·m,

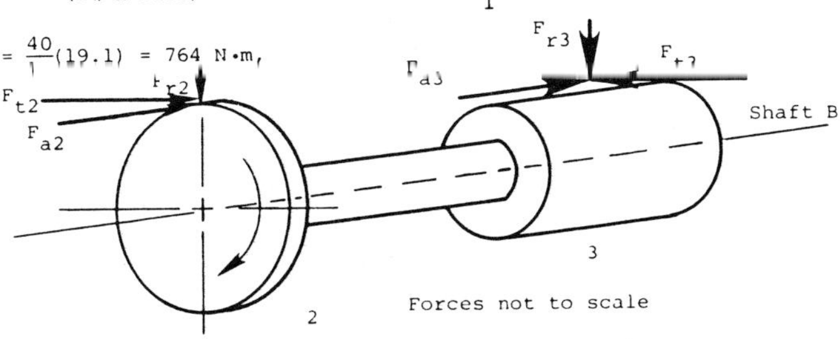

Helical gears (Nos. 1 and 2): $\frac{d_2}{d_1} = \frac{N_2}{N_1}$, $d_2 = \frac{N_2}{N_1} d_1 = \frac{48}{24} d_1 = 2d_1$,

$2c = 240 = d_1 + d_2 = 3d_1$, $d_1 = 80\text{mm}$, $F_t = \frac{2T_A}{d_1} = \frac{2(9.55)}{80/1000} = 239$ N,

$F_r = F_t \tan\phi = \frac{F_t \tan\phi^n}{\cos\psi} = 92.6$ N, $F_a = F_{aA} = F_t \tan\psi = 239 \tan 20^\circ = 87.0$ N,

Worm and worm gear (Nos. 3 and 4): $\ell = p_3 N_3 = 20(1) = 20$ mm,

$r_v = N_3/N_4 = 1/40$, $d_4 = \frac{\ell}{\pi r_v} = \frac{20}{\pi(1/40)} = 255$ mm,

$F_{t4} = F_{a3} = \frac{2T_C}{d_4} = \frac{2(764)}{255/1000} = 5990$ N, $F_{r4} = F_{r3} = F_{t4}\tan\phi = 5990 \tan 25^\circ = 2790$ N

$\tan\lambda = \ell/\pi d_w = \frac{20}{\pi(300 - 255)} = 0.141$, $F_{aC} = F_{a4} = F_{t3} = F_{t4}\tan\lambda = 5990(.141) = 844\text{N}$,

$F_{aB} = F_{a2} + F_{a3} = 87 + 5990 = 6077$ N

7.24 (A) $\tan\gamma = r_v = 0.4$, $\gamma = 21.8^\circ$, $\tan\Gamma = \frac{1}{r_v} = 2.5$, $\Gamma = 68.2^\circ$,

(B) $r_{p1} = \frac{N_1}{2P} = \frac{18}{12} = 1.5$ in, $r_{p2} = \frac{1.5}{r_v} = 3.75$ in., $r_{b1} = \frac{r_{p1}}{\cos\gamma} = 1.62$ in,

$r_{b2} = \frac{r_{p2}}{\cos\Gamma} = \frac{3.75}{\cos 68.2^\circ} = 10.1$ in., (C) $N_{f2} = \frac{2Pr_{p2}}{\cos\Gamma} = \frac{2(6)(3.75)}{\cos 68.2^\circ} = 121.2$

7.25 $\Sigma = 60^\circ$, $r_v = 0.4$, (A) $\tan\gamma = \frac{\sin\Sigma}{(1/r_v) + \cos\Sigma} = 0.289$, $\gamma = 16.1^\circ$,

$\tan\Gamma = \frac{\sin\Sigma}{r_v + \cos\Sigma} = 0.962$, $\Gamma = 43.9^\circ$, (B) $r_{b1} = \frac{r_{p1}}{\cos\gamma} = \frac{1.5}{\cos 16.1^\circ} = 1.56$ in.,

$r_{b2} = \frac{r_{p2}}{\cos\Gamma} = \frac{3.75}{\cos 43.9^\circ} = 5.20$ in., (C) $N_{f2} = 2Pr_{b2} = 2(6)(5.20) = 62.4$

7.26 $\Sigma = 110^\circ$, $r_v = 0.4$, (A) $\tan\gamma = \frac{\sin\Sigma}{(1/r_v) + \cos\Sigma} = 0.435$, $\gamma = 23.5^\circ$

$\tan\Gamma = \frac{\sin\Sigma}{r_v + \cos\Sigma} = 16.2$, $\Gamma = 86.5^\circ$, (B) $r_{b1} = \frac{r_{p1}}{\cos\gamma} = \frac{1.5}{\cos 23.5^\circ} = 1.64$ in.,

$r_{b2} = \frac{r_{p2}}{\cos\Gamma} = \frac{3.75}{\cos 86.5^\circ} = 61.4$ in., (C) $N_{f2} = 2Pr_{b2} = 2(6)(61.4) = 737$

7.27 $r_v = \frac{N_p}{N_g} = \frac{24}{48} = 0.5$, $\tan\gamma = \frac{N_p}{N_g} = 0.5$, $\gamma = 26.6^\circ$, $\Gamma = 63.4^\circ$,

$d_{op} = \frac{N_p}{P} = \frac{24}{4} = 6$ in., $d_{ip} = d_{op} - 2w\sin\gamma = 6 - 2(2)\sin 26.6^\circ = 4.2$ in.,

$r_{mp} = \frac{1}{4}(d_{ip} + d_{op}) = \frac{1}{4}(4.2 + 6) = 2.55$ in.,

$$T_p = \frac{63{,}025 \text{ hp}}{n_p} = \frac{63{,}025(50)}{1000} = 3150 \text{ lb in}, \quad T_p = F_{tp} r_{mp},$$

$$F_{tp} = F_{tg} = \frac{T_p}{r_{mp}} = \frac{3150}{2.55} = 1240 \text{ lb.}, \quad F_{rp} = F_{ag} = F_{tp} \tan\phi\cos\gamma$$

$$= 1240 \tan 20^\circ \cos 26.6^\circ = 404 \text{ lb.}, \quad F_{ap} = F_{rg} = F_{tp} \tan\phi\sin\gamma$$

$$= 1240 \tan 20^\circ \sin 26.6^\circ = 202 \text{ lb.}$$

8.1 If 1 and 2 have the same DP as 3 and 4, $N_1 + N_2 = N_3 + N_4$ from which $N_2 = N_3 + 20$.

Then $\frac{n_4}{n_1} = \frac{N_1N_3}{N_2N_4} = \frac{20N_3}{40(N_3+20)}$

If $N_3 = 20$ (and $N_2 = 40$), $N_4/n_1 = 1/4$. If $N_3 = 40$ (and $N_2 = 60$), $n_4/n_1 = 1/3$.

The lowest and highest speeds obtainable are, therefore, $n_4 = 500$ and 667 RPM. Otherwise n_4 may range from 333 RPM to 3000 RPM.

8.2 $n_4 = n_1N_1N_3/(N_2N_4) = 0.5714$ RPM.

8.3 $n_3/n_1 = N_1/N_3$, etc. $N_3 = \frac{N_1}{n_3/n_1}$, etc. Therefore, N_1 should be a whole number multiple of each ratio. Let $N_1 = 24$. Then, $N_3 = 24/.8 = 30$; $N_4 = 24/.75 = 32$; $N_5 = 24/.6 = 40$; $N_6 = 24/.5 = 48$; $N_7 = 24/.4 = 60$. The sum $D_1/2 + D_2 + D_3/2$ must exceed the distance between input and output shafts and the shaft distance must be sufficient to allow gears 1 and 7 to clear. To satisfy the latter condition, let the shaft distance be $(N_1/2 + N_7/2 + 4)$ divided by the diametral pitch or 46/DP. Then, $N_1/2 + N_2 + N_3/2$ must exceed 46 or N_2 must exceed 19; say $N_2 = 26$. Diametral pitch will be based on transmitted torque.

8.4 There are many solutions to this problem.

a) See sketch.

b) Let $N_1 = N_4 = 20$; $N_2 = N_8 = 40$; $N_3 = N_7 = 30$; $N_5 = N_6 = 18$; $N_9 = 36$.

8.5 Factors of 2000 are $2^4.5^3$.

" " 1131 " 3.13.29.

For $n_4/n_1 = (N_1N_3)/(N_2N_4) = 1131/2000 = (3.13.29)/(2^4.5^3)$

$= (29.39)/(40.50)$ we may let $N_1 = 29$, $N_2 = 40$, $N_3 = 39$ and $N_4 = 50$ for approximately equal reductions in each stage. This is not the only solution.

8.6

	S_1	S_2	S_3	P	C
I	1	1	1	1	1
II	-1	-50/49	-50/51	50/30	0
Sum	0	-1/49	1/51	8/3	1
Sum X 100	0	-2.04	1.96	233	100 (CW positive)

ANS. (under S_2 and S_3)

8.7

	S	P_1	P_2	R	C
I	1	1	1	1	1
II	-1	NS/N_{P1}	$-N_S/N_{P2}$	$-N_S/N_r$	0
Sum	0	$1 + \frac{N_S}{N_{P1}}$	$1 - \frac{N_S}{N_{P2}}$	$1 - \frac{N_S}{N_R}$	1
Sum X 200	0	$200(1 + \frac{N_S}{N_{P1}})$	$200(1 - \frac{N_S}{N_{P2}})$	$200(1 - \frac{N_3}{N_R})$	200 Answer (CW positive)

Not to scale

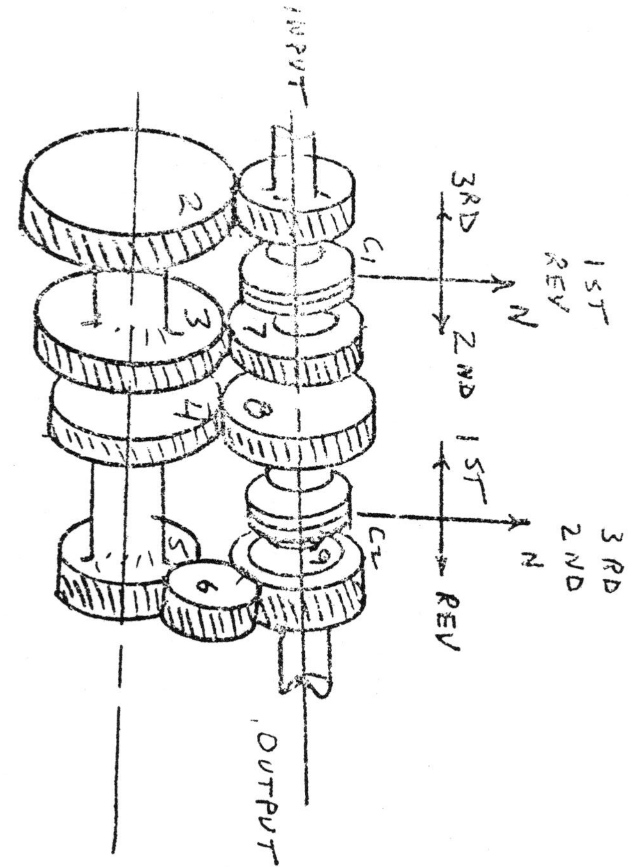

8.4

Countershaft gears 2,3,4,5 mesh with gears 1,7,8 and 6 respectively at all times.
Gear 1 is driven by **input shaft**.
Gears 7,8 and 9 turn freely with respect to the output shaft except when engaged by synchromest clutches C_1 or C_2 which are splined to the output shaft.

Position	Speed ratio n_{out}/n_{in}
3rd	$+\ 1:1$
2nd	$+\ \frac{N_1 N_3}{N_2 N_7}$
1st	$+\ \frac{N_1 N_4}{N_2 N_8}$
Rev	$-\ \frac{N_1 N_5}{N_2 N_9}$

8.8

	S	P_1	P_2	R	C
I	1	1	1	1	1
II	$-N_R/N_S$	N_R/N_{P1}	$-N_R/N_{P2}$	-1	0
Sum	$-(N_R/N_S-1)$	$N_R/N_{P1} + 1$ -	$(\frac{N_R}{N_{P2}} - 1)$	0	1
Sum X 200	$-200(\frac{N_R}{N_S}-1)$	$200(\frac{N_R}{N_{P1}} + 1)$	$-200(\frac{N_R}{N_{P2}} - 1)$	0	200 Answer (CW positive)

8.9 a) From table, $$\frac{n_{R2}}{n_S} = \frac{1 - \frac{N_{R1} N_{P2}}{N_{P1} N_{R2}}}{1 + \frac{N_{R1}}{N_S}} = \frac{1 - \frac{95.21}{20.100}}{1 + \frac{95}{55}} = .000915$$

n_{R2} = 100(0.000915) = 0.0915 RPM CW.

b) A speed increase of this magnitude is impossible due to friction.

8.10 Let the sun gear be the input and the carrier the output. Then

$$n_o/n_i = 1000/3500 = n_c/n_s = \frac{1}{1 + \frac{N_R}{N_S}}$$

(by the tabular method). From this equation, we find

$N_R = 2.5\ N_S$. Also, for this type train, $N_R = 2N_P + N_S$.

a) Thus, if we let $N_P = 20$, then $N_R = 2(20) + N_S = 2.5\ N_S$.

from which $N_S = 40/1.5$, say 27 teeth. Then, $N_R = 27 + 2(20) = 67$

b) $n_c = n_S \left(\frac{1}{1 + \frac{N_R}{N_S}}\right) = 3500 \left(\frac{1}{1 + \frac{67}{27}}\right) = 1007$ RPM.

c) $(N_S + N_R)$ / number of planets = an integer. $N_S + N_R = 27 + 67 =$

94 = 2(47). Thus, two equally spaced planets may be used.

8.11

	S_1	P_1&P_2	S_2	C
I	1	1	1	1
II	-1	51/20	$\frac{-(51)(19)}{-(20)(50)}$	0
Sum	0	71/20	31/1000	1
Sum X 1000	0	3550	31	1000 RPM (Ans.) all CW.

8.12

	S_1	P_1&P_2	S_2	C
I	1	1	1	1
II	-1000/969	50/19	-1	0
Sum	-31/969	69/19	0	1
Sum X 1000	-31.99	3632	0	1000 RPM (Ans.) (CW positive)

8.13 The train may be separated into two tandem trains, the first made up of C_1, S_1, S_2, P_1, P_2 and the second made up of C_2, S_3, S_4, P_3, P_4. Considering the first set only, we have:

	S_1	P_1&$_2$P	S_2&C_2	C_1
I	1	1	1	1
II	-1	101/100	$\frac{-101(99)}{100(100)}$	0
Sum	0	$\frac{201}{100}$	$\frac{1}{10,000}$	1

for a speed ratio of n_{S2}/n_{C1} = 1/10,000. Similarly, for the second set, n_{S4}/n_{C2} = 1/10,000. Since S_2 and C_2 turn at the same speed, the speed ratio for the entire train is the product of the two ratios: n_{S4}/n_{C1} = 1/100,000,000. 60 RPM corresponds to 1 sec/rev of C_1. Then, S_4 makes 1 rev. in 100,000,000 sec (or about 3 years and 2 months).

8.14

	S	P_1	P_2	R	C
I Lock Train	u	u	u	u	u
II Lock carrier	v	$-v\frac{N_s}{N_{p1}}$	$v\frac{N_s}{N_{p2}}$	$v\frac{N_s}{N_R}$	0
Sum	u + v			$u + v\frac{N_s}{N_R}$	u

(a) $u + v = 400$ RPM (CW positive) (b) $u + v\frac{N_s}{N_R} = u + 0.4\,v = 300$ RPM

Multiply (a) by -0.4 and add to (b) to obtain 0.6 u = 140 RPM or carrier speed N_c = 233 RPM CW.

8.15

	S_1	P_1&P_2	S_2	C
I Lock Train	u	u	u	u
II Lock carrier	v	$-\frac{N_{s_1}}{N_{P_1}}$	$v\frac{N_{s1}N_{p2}}{N_{p1}N_{s2}}$	0
Sum	u + v		$u + \frac{N_{s1}N_{p2}}{N_{p1}N_{s2}}v$	u

$n_C = u$; $n_{S1} = u + v$ then, $v = n_{S1} - u = n_{S1} - n_C$ and $n_{S2} = u + \frac{N_{s1}N_{p2}}{N_{p1}N_{s2}}\,v = n_C + \frac{N_{s1}N_{p2}}{N_{p1}N_{s2}}\,(n_{S1} - n_C)$

Solutions 8.16 through 8.33 are based on the formula method:

$$r^* = n_Y^*/n_X^* = \frac{n_Y - n_C}{n_X - n_C}$$ and refer to figure 8.7

8.16 $r^* = n_R^*/n_S^* = (n_R - n_C)/(n_S - n_C)$ (fig. 8.7B)

$$r^* = \frac{-N_S N_{P2}}{N_{P1} N_R} = \frac{0 - n_C}{n_S - n_C} = \frac{-n_C/n_S}{1 - n_C/n_S}$$

$$r^* - r^* n_C/n_S = -n_C/n_S$$

$$r^* = n_C/n_S(r^* - 1)$$

$$n_C/n_S = r^*/(r^* - 1) = -r^*/(-r^* + 1)$$

$$n_C/n_S = \frac{N_S N_{P2}}{N_S N_{P2} + N_{P1} N_R} \quad \text{(ans)}$$

8.17 $r^* = n^*_{S2}/n^*_{S1} = (n_{S2} - n_C)/(n_{S1} - n_C)$ (fig. 8.7C)

$$r^* = \frac{N_{S1} N_{P2}}{N_{P1} N_{S2}} = \frac{0 - n_C}{n_{S1} - n_C}$$

$$n_{S1} r^* - n_C r^* = -n_C$$

$$n_C = \frac{r^*}{r^* - 1} n_{S1} = \frac{N_{S1} N_{P2} n_{S1}}{N_{S1} N_{P2} - N_{P1} N_{S2}} \quad \text{(ans)}$$

8.18 (Fig. 8-7D)

$$r^* = n_{R2}^*/n_{R1}^* = (n_{R2} - n_C)/(n_{R1} - n_C)$$

$$r^* = N_{R1} N_{P2}/(N_{P1} N_{R2}) = (0 - n_C)/(n_{R1} - n_C)$$

$$r^* n_{R1} - r^* n_C = -n_C$$

$$n_C = \frac{r^* n_{R1}}{r^* - 1} = \frac{N_{R1} N_{P2} n_{R1}}{N_{R1} N_{P2} - N_{P1} N_{R2}} \quad \text{(ans)}$$

8.19 (Fig. 8.7F)

$$r^* = n_{R2}^*/n_{R1}^* = (n_{R2} - n_C)/(n_{R1} - n_C)$$

$$r^* = \frac{-N_{R1} N_{P2} N_{P4}}{N_{P1} N_{P3} N_{R2}} = \frac{0 - n_C}{n_{R1} - n_C}$$

$$r^* n_{R1} - r^* n_C = -n_C$$

$$n_C = \frac{r^* n_{R1}}{r^* - 1} = \frac{N_{R1} N_{P2} N_{P4} n_{R1}}{N_{R1} N_{P2} N_{P4} + N_{P1} N_{P3} N_{R2}} \quad \text{(ans)}$$

8.20 (Fig. 8.7G)

$$r^* = n_{S2}^*/n_{S1}^* = (n_{S2} - n_C)/(n_{S1} - n_C)$$

$$r^* = \frac{-N_{S1} N_{P2} N_{P4}}{N_{P1} N_{P3} N_{S2}} = \frac{0 - n_C}{n_{S1} - n_C}$$

8.20 (Cont'd.)

$r^* n_{S1} - r^* n_C = -n_C \qquad n_C = \dfrac{-r^* n_{S1}}{-r^*+1}$

$$n_C = \frac{N_{S1}N_{P2}N_{P4}n_{S1}}{N_{S1}N_{P2}N_{P4}+N_{P1}N_{P3}N_{S2}} \qquad \text{(ans)}$$

8.21 (Fig. 8.7H)

$r^* = n_{R2}^*/n_{R1}^* = (n_{R2}-n_C)/(n_{R1}-n_C)$

$$r^* = \frac{-N_{R1}N_{P2}}{N_{P1}N_{R2}} = \frac{0-n_C}{n_{R1}-n_C}$$

$r^* n_{R1} - r^* n_C = -n_C \qquad n_C = \dfrac{-r^* n_{R1}}{1-r^*}$

$$n_C = \frac{N_{R1}N_{P2}n_{R1}}{N_{P1}N_{R2}+N_{R1}N_{P2}} \qquad \text{(ans)}$$

8.22 (Fig. 8.7I)

$r^* = n_R^*/n_S^* = (n_R - n_C)/(n_S-n_C)$

$r^* = N_S/N_R = (0-n_C)/(n_S-n_C)$

$n_S r^* - n_C r^* = -n_C \qquad n_C = \dfrac{n_S r^*}{r^*-1}$

$$n_C = \frac{N_S n_S}{N_S-N_R} \qquad \text{(ans)}$$

8.23 (Fig. 8.7J)

$r^* = n_R^*/n_S^* = (n_R - n_C)/(n_S - n_C)$

$$r^* = \frac{N_S N_{P2}}{N_{P1}N_R} = \frac{0-n_C}{n_S-n_C}$$

$n_S r^* - n_C r^* = -n_C \qquad n_C = \dfrac{n_S r^*}{r^*-1}$

$$n_C = \frac{N_S N_{P2} n_S}{N_S N_{P2}-N_{P1}N_R} \qquad \text{(ans)}$$

8.24 (Fig. 8.7K)

$r^* = n_R^*/n_S^* = (n_R-n_C)/(n_S-n_C)$

$r^* = \dfrac{N_S N_{P3}}{N_{P2}N_R} = \dfrac{0-n_C}{n_S-n_C} \qquad n_C = \dfrac{n_S r^*}{r^*-1}$

$$n_C = \frac{N_S N_{P3} n_S}{N_S N_{P3}-N_{P2}N_R} \qquad \text{(ans)}$$

8.25 (Fig. 8-7L)

$$r^* - n_R^*/n_S^* = (n_R - n_C)/(n_S - n_C)$$

$$r^* = \frac{N_S N_{P2} N_{P4}}{N_{P1} N_{P3} N_R} = \frac{0 - n_C}{n_S - n_C} \qquad n_C = \frac{n_S r^*}{r^* - 1}$$

$$n_C = \frac{N_S N_{P2} N_{P4} n_S}{N_S N_{P2} - N_{P1} N_{P3} N_R} \qquad \text{(ans)}$$

8.26 (Fig. 8-7B)

$$r^* = n_R^*/n_S^* = (n_R - n_C)/(n_S - n_C)$$

$$r^* = \frac{-N_S N_{P2}}{N_{P1} N_R} = \frac{n_R - n_C}{0 - n_C} \qquad -r^* n_C = n_R - n_C$$

$$n_C = n_R/(1 - r^*) = \frac{n_R}{1 + \dfrac{N_S N_{P2}}{N_{P1} N_R}} \qquad \text{(ans)}$$

8.27 (Fig. 8.7E)

$$r^* = n_{S2}^*/n_{S1}^* = (n_{S2} - n_C)/(n_{S1} - n_C)$$

$$r^* = \frac{-N_{S1} N_{P3}}{N_{P2} N_{S2}} = \frac{0 - n_C}{n_{S1} - n_C} \qquad r^* n_{S1} - r^* n_C = -n_C$$

$$n_C = -r^* n_{S1}/(1 - r^*)$$

$$n_C = \frac{N_{S1} N_{P3} n_{S1}}{N_{S1} N_{P3} + N_{P2} N_{S2}} \qquad \text{(ans)}$$

(Check) Tabular (superposition) analysis of the gear train.
(Fig. 8.7E)

Gear	Input S_1	P_1	P_2 & P_3	Fixed S_2	Carrier C
Step I Rotations with train locked	+1	+1	+1	+1	+1
Step II Rotations with planet carrier fixed	$+\frac{N_{S2} N_{P2}}{N_{P3} N_{S1}}$	$-\frac{N_{S2} N_{P2}}{N_{P3} N_{P1}}$	$+\frac{N_{S2}}{N_{P3}}$	-1	0
Total rotations	$1 + \frac{N_{S2} N_{P2}}{N_{P3} N_{S1}}$			0	+1

Output speed $n_C = n_{S1}/[1 + \frac{N_{S2} N_{P2}}{N_{P3} N_{S1}}]$

$n_C = (n_{S1} N_{P3} N_{S1})/(N_{P3} N_{S1} + N_{S2} N_{P2})$ (ans checks)

8.28 (Fig. 8.7E)

$$r^* = n_{S2}^*/n_{S1}^* = (n_{S2}-n_C)/(n_{S1}-n_C)$$

$$r^* = \frac{-N_{S1}N_{P3}}{N_{P2}N_{S2}} = \frac{n_{S2}-n_C}{0-n_C} \qquad -r^* n_C = n_{S2} - n_C$$

$$n_C = n_{S2}/(1-r^*)$$

$$n_C = \frac{n_{S2}}{1 + \dfrac{N_{S1}N_{P3}}{N_{P2}N_{S2}}} \qquad \text{(ans)}$$

8.29 (Fig. 8-7H)

$$r^* = n_{R2}^*/n_{R1}^* = (n_{R2}-n_C)/(n_{R1}-n_C)$$

$$r^* = \frac{-N_{R1}N_{P2}}{N_{P1}N_{R2}} = \frac{n_{R2}-n_C}{0-n_C}$$

$$n_{R2} = -n_C r^* + n_C \qquad n_C = n_{R2}/(1-r^*)$$

$$n_C = n_{R2}/(1 + \frac{N_{R1}N_{P2}}{N_{P1}N_{R2}}) \qquad \text{(ans)}$$

8.30 (Fig. 8.7I)

$$r^* = (n_R^*/n_S^*) = (n_R-n_C)/(n_S-n_C)$$

$$r^* = N_S/N_R = (n_R-n_C/(0-n_C)$$

$$-n_C r^* = n_R - n_C \qquad n_R = n_C(1-r^*)$$

$$n_C = n_R/(1 - N_S/N_R)$$

8.31 (Fig. 8.7J)

$$r^* = n_R^*/n_S^* = (n_R-n_C)/(n_S-n_C)$$

$$r^* = \frac{N_S N_{P2}}{N_{P1}N_R} = \frac{n_R-n_C}{0-n_C}$$

$$n_C = n_R/(1-r^*) = n_R/(1 - \frac{N_S N_{P2}}{N_{P1}N_R}) \qquad \text{(ans)}$$

8.32 (Fig. 8.7K)

$$r^* = n_R^*/n_S^* = (n_R-n_C)/(n_S-n_C)$$

$$r^* = \frac{N_S N_{P3}}{N_{P2}N_R} = \frac{n_R-n_C}{0-n_C} \qquad n_R = n_C(1-r^*)$$

$$n_C = n_R/(1 - \frac{N_S N_{P3}}{N_{P2}N_R}) \qquad \text{(ans)}$$

8.33 (Fig. 8.7L)

$$r^* = n_R^*/n_S^* = (n_R - n_C)/(n_S - n_C)$$

$$r^* = \frac{N_S N_{P2} N_{P4}}{N_{P1} N_{P3} N_R} = \frac{n_R - n_C}{0 - n_C}$$

$$n_C = n_R/(1 - r^*)$$

$$n_C = n_R/(1 - \frac{N_S N_{P2} N_{P4}}{N_{P1} N_{P3} N_R}) \qquad \text{(ans)}$$

8.34 Tabular (superposition) analysis of the gear train. (Fig. 8-7B)

Gear	Input S	P_1	P_2	Fixed R	Carrier C
Step I Rotations with train locked	+1	+1	+1	+1	+1
Step II Rotations with planet carrier fixed	$+\frac{N_R}{N_{P2}}\frac{N_{P1}}{N_S}$	$-\frac{N_R}{N_{P2}}$	$-\frac{N_R}{N_{P2}}$	-1	0
Total rotations	$1 + \frac{N_R N_{P1}}{N_{P2} N_S}$			0	+1

$$\text{Output speed } n_C = n_S/[1 + \frac{N_R N_{P1}}{N_{P2} N_S}] = \frac{n_S N_{P2} N_S}{N_{P2} N_S + N_R N_{P1}} \qquad \text{(ans)}$$

8.35 Tabular (superposition) analysis of the gear train. (Fig. 8-7C)

Gear	Input S_1	P_1	P_2	Fixed S_2	Carrier C
Step I Rotations with train locked	+1	+1	+1	+1	+1
Step II Rotations with planet carrier fixed	$-\frac{N_{S2} N_{P1}}{N_{P2} N_{S1}}$	$+\frac{N_{S2}}{N_{P2}}$	$+\frac{N_{S2}}{N_{P2}}$	-1	0
Total rotations	$1 - \frac{N_{S2} N_{P1}}{N_{P2} N_{S1}}$			0	+1

$$\text{Output speed } n_C = n_{S1}/(1 - \frac{N_{S2} N_{P1}}{N_{P2} N_{S1}}) = \frac{n_{S1} N_{P2} N_{S1}}{N_{P2} N_{S1} - N_{S2} N_{P1}} \qquad \text{(ans)}$$

8.36 Tabular (superposition analysis of the gear train. (Fig. 8-7D)

Gear	Input R_1	P_1	P_2	Fixed R_2	Carrier C
Step I Rotations with train locked	$+1$	$+1$	$+1$	$+1$	$+1$
Step II Rotations with planet carrier fixed	$-\frac{N_{R2}N_{P1}}{N_{P2}N_{R1}}$	$-\frac{N_{R2}}{N_{P2}}$	$-\frac{N_{R2}}{N_{P2}}$	-1	0
Total rotations	$1-\frac{N_{R2}N_{P1}}{N_{P2}N_{R1}}$			0	$+1$

Output speed $n_C = n_{R1}/[1 - \frac{N_{R2}N_{P1}}{N_{P2}N_{R1}}]$

$$n_C = (n_{R1}N_{P2}N_{R1})/(N_{P2}N_{R1} - N_{R2}N_{P1}) \quad \text{(ans)}$$

8.37 Tabular (superposition) analysis of the gear train. (Fig. 8-7F)

Gear	Input R_1	P_1 & P_2	P_3 & P_4	Fixed R_2	Carrier C
Step I Rotations with train locked	$+1$	$+1$	$+1$	$+1$	$+1$
Step II Rotations with planet carrier fixed	$+\frac{N_{R2}N_{P3}N_{P1}}{N_{P4}N_{P2}N_{R1}}$	$+\frac{N_{R2}N_{P3}}{N_{P4}N_{P2}}$	$-\frac{N_{R2}}{N_{P4}}$	-1	0
Total rotations	$1+\frac{N_{R2}N_{P3}N_{P1}}{N_{P4}N_{P2}N_{R1}}$			0	$+1$

Output speed $n_C = n_{R1}/[1 + \frac{N_{R2}N_{P3}N_{P1}}{N_{P4}N_{P2}N_{R1}}]$

$$n_C = (n_{R1}N_{P4}N_{P2}N_{R1})/(N_{P4}N_{P2}N_{R1} + N_{R2}N_{P3}N_{P1}) \quad \text{(ans)}$$

8.38 Tabular (superposition) analysis of the gear train. (Fig. 8.7G)

Gear	Input S_1	P_1 & P_2	P_3 & P_4	Fixed S_2	Carrier C
Step I Rotations with train locked	$+1$	$+1$	$+1$	$+1$	$+1$
Step II Rotations with planet carrier fixed	$+\frac{N_{S2}N_{P3}N_{P1}}{N_{P4}N_{P2}N_{S1}}$	$-\frac{N_{S2}N_{P3}}{N_{P4}N_{P2}}$	$+\frac{N_{S2}}{N_{P4}}$	-1	0
Total rotations	$1+\frac{N_{S2}N_{P3}N_{P1}}{N_{P4}N_{P2}N_{S1}}$			0	$--1$

Output speed $n_C = n_{S1}/[1 + \frac{N_{S2}N_{P3}N_{P1}}{N_{P4}N_{P2}N_{S1}}]$

$$n_C = (n_{S1}N_{P4}N_{P2}N_{S1})/(N_{P4}N_{P2}N_{S1} + N_{S2}N_{P3}N_{P1}) \quad \text{(ans)}$$

8.39 Tabular (superposition) analysis of the gear train. (Fig. 8.7H)

Gear	Input R_1	P_1 & P_2	P_3	Fixed R_2	Carrier C
Step I Rotations with train locked	+1	+1	+1	+1	+1
Step II Rotations with planet carrier fixed	$+\frac{N_{R2}N_{P1}}{N_{P2}N_{R1}}$	$+\frac{N_{R2}N_{P3}}{N_{P3}N_{P2}}$	$-\frac{N_{R2}}{N_{P3}}$	-1	0
Total rotations	$1+\frac{N_{R2}N_{P1}}{N_{P2}N_{R1}}$			0	+1

Output speed $n_C = n_{R1}/[1 + \frac{N_{R2}N_{P1}}{N_{P2}N_{R1}}]$

$n_C = (n_{R1}N_{P2}N_{R1})/(N_{P2}N_{R1} + N_{R2}N_{P1})$ (ans)

8.40 Tabular (superposition) analysis of the gear train. (Fig. 8.7I)

Gear	Input S	P_1	P_2	Fixed R	Carrier C
Step I Rotations with train locked	+1	+1	+1	+1	+1
Step II Rotations with planet carrier fixed	$-\frac{N_R}{N_S}$	$+\frac{N_R}{N_{P1}}$	$-\frac{N_R}{N_{P2}}$	-1	0
Total rotations	$1-\frac{N_R}{N_S}$	$1+\frac{N_R}{N_{P1}}$	$1-\frac{N_R}{N_{P2}}$	0	+1

Output speed $n_C = n_S/(1 - N_R/N_S)$

$n_C = n_S N_S/(N_S - N_R)$ or

$n_C = -n_S N_S/(N_R - N_S)$ (ans)

8.41 Tabular (superposition) analysis of the gear train. (Fig. 8.7J)

Gear	Input S	P_1 & P_1	P_3	Fixed R	Carrier C
Step I Rotations with train locked	+1	+1	+1	+1	+1
Step II Rotations with planet carrier fixed	$-\frac{N_R N_{P1}}{N_{P2}N_S}$	$+\frac{N_R}{N_{P2}}$	$-\frac{N_R}{N_{P3}}$	-1	0
Total rotations	$1-\frac{N_R N_{P1}}{N_{P2}N_S}$			0	+1

Output speed $n_C = n_S/[1 - \frac{N_R N_{P1}}{N_{P2}N_S}]$

$n_C = (n_S N_{P2}N_S)/(N_{P2}N_S - N_R N_{P1})$ (ans)

8.42 Tabular (superposition) analysis of the gear train. (Fig. 8.7K)

Gear	Input S	P_1	P_2 & P_3	Fixed R	Carrier C
Step I Rotations with train locked	+1	+1	+1	+1	+1
Step II Rotations with planet carrier fixed	$-\frac{N_R N_{P2}}{N_{P3} N_S}$	$+\frac{N_R N_{P2}}{N_{P3} N_{P1}}$	$-\frac{N_R}{N_{P3}}$	-1	0
Total rotations	$1-\frac{N_R N_{P2}}{N_{P3} N_S}$			0	+1

Output speed $n_C = n_S/[1 - \frac{N_R N_{P2}}{N_{P3} N_S}]$

$n_C = (n_S N_{P3} N_S)/(N_{P3} N_S - N_R N_{P2})$ (ans)

8.43 Tabular (superposition) analysis of the gear train. Fig. 8.7L)

Gear	Input S	P_1 & P_2	P_3 & P_4	Fixed R	Carrier C
Step I Rotations with train locked	+1	+1	+1	+1	+1
Step II Rotations with planet carrier fixed	$-\frac{N_R N_{P3} N_{P1}}{N_{P4} N_{P2} N_S}$	$+\frac{N_R N_{P3}}{N_{P4} N_{P2}}$	$-\frac{N_R}{N_{P4}}$	-1	0
Total rotations	$1-\frac{N_R N_{P3} N_{P1}}{N_{P4} N_{P2} N_S}$			0	+1

Output speed $n_C = n_S/[1 - \frac{N_R N_{P3} N_{P1}}{N_{P4} N_{P2} N_S}]$

$n_C = (n_S N_S N_{P2} N_{P4})/(N_S N_{P2} N_{P4} - N_R N_{P1} N_{P3})$ (ans)

8.44 Tabular (superposition) analysis of the gear train. (Fig. 8.7B)

Gear	Fixed S	P_1	P_2	Input R	Carrier C
Step I Rotations with train locked	+1	+1	+1	+1	+1
Step II Rotations with planet carrier fixed	-1	$+\frac{N_S}{N_{P1}}$	$+\frac{N_S}{N_{P1}}$	$+\frac{N_S N_{P2}}{N_{P1} N_R}$	0
Total rotations	0			$1+\frac{N_S N_{P2}}{N_{P1} N_R}$	+1

Output speed $n_C = n_R/[1 + \frac{N_S N_{P2}}{N_{P1} N_R}]$

$n_C = (n_R N_{P1} N_R)/(N_{P1} N_R + N_S N_{P2})$ (ans)

8.45 Tabular (superposition) analysis of the gear train. (Fig. 8.7H)

Gear	Fixed R_1	P_1 & P_2	P_3	Input R_2	Carrier C
Step I Rotations with train locked	+1	+1	+1	+1	+1
Step II Rotations with planet carrier fixed	-1	$-\frac{N_{R1}}{N_{P1}}$	$+\frac{N_{R1}N_{P2}}{N_{P1}N_{P3}}$	$+\frac{N_{R1}N_{P2}}{N_{P1}N_{R2}}$	0
Total rotations	0			$1+\frac{N_{R1}N_{P2}}{N_{P1}N_{R2}}$	+1

Output speed $n_C = n_{R2}/[1 + \frac{N_{R1}N_{P2}}{N_{P1}N_{R2}}]$

$n_C = (n_{R2}N_{P1}N_{R2})/(N_{P1}N_{R2} + N_{R1}N_{P2})$ (ans)

8.46 Tabular (superposition) analysis of the gear train. (Fig. 8.7I)

Gear	Fixed S	P_1	P_2	Input R	Carrier C
Step I Rotations with train locked	+1	+1	+1	+1	+1
Step II Rotations with planet carrier fixed	-1	$+\frac{N_S}{N_{P1}}$	$-\frac{N_S}{N_{P2}}$	$-\frac{N_S}{N_R}$	0
Total rotations	0			$1-\frac{N_S}{N_R}$	+1

Output speed $n_C = n_R/[1 - N_S N_R]$

$n_C = (n_R N_R)/(N_R - N_S)$ (ans)

8.47 Tabular (superposition) analysis of the gear train. (Fig. 8.7J)

Gear	Fixed S	P_1 & P_2	P_3	Input R	Carrier C
Step I Rotations with train locked	+1	+1	+1	+1	+1
Step II Rotations with planet carrier fixed	-1	$+\frac{N_S}{N_{P1}}$	$-\frac{N_S N_{P2}}{N_{P1}N_{P3}}$	$-\frac{N_S N_{P2}}{N_{P1}N_R}$	0
Total rotations	0			$1-\frac{N_S N_{P2}}{N_{P1}N_R}$	+1

Output speed $n_C = n_R/[1 - \frac{N_S N_{P2}}{N_{P1}N_R}]$

$n_C = (n_R N_{P1} N_R)/(N_{P1}N_R - N_S N_{P2})$ (ans)

8.48 Tabular (superposition) analysis of the gear train. (Fig. 8.7K)

Gear	Fixed S	P_1	P_2 & P_3	Input R	Carrier C
Step I Rotations with train locked	+1	+1	+1	+1	+1
Step II Rotations with planet carrier fixed	-1	$+\frac{N_S}{N_{P1}}$	$-\frac{N_S}{N_{P2}}$	$-\frac{N_S N_{P3}}{N_{P2} N_R}$	0
Total rotations	0			$1-\frac{N_S N_{P3}}{N_{P2} N_R}$	+1

Output speed $n_C = n_R/[1 - \frac{N_S N_{P3}}{N_{P2} N_R}]$

$n_C = (n_R N_{P2} N_R)/(N_{P2} N_R - N_S N_{P3})$ (ans)

8.49 Tabular (superposition) analysis of the gear train. (Fig. 8.7L)

Gear	Fixed S	P_1 & P_2	P_3 & P_4	Input R	Carrier C
Step I Rotations with train locked	+1	+1	+1	+1	+1
Step II Rotations with planet carrier fixed	-1	$+\frac{N_S}{N_{P1}}$	$-\frac{N_S N_{P2}}{N_{P1} N_{P3}}$	$-\frac{N_S N_{P2} N_{P4}}{N_{P1} N_{P3} N_R}$	0
Total rotations	0			$1-\frac{N_S N_{P2} N_{P4}}{N_{P1} N_{P3} N_R}$	+1

Output speed $n_C = n_R/[1 - \frac{N_S N_{P2} N_{P4}}{N_{P1} N_{P3} N_R}]$

$n_C = (n_R N_{P1} N_{P3} N_R)/(N_{P1} N_{P3} N_R - N_S N_{P2} N_{P4})$ (ans)

8.50 To produce the required speed reduction, it will be necessary to fix one of the sun gears, drive the carrier, and use the other sun gear as the output. The result is

$n_{S2}/n_C = 1 - \frac{N_{S1} N_{P2}}{N_{P1} N_{S2}}$ which is set equal to 0.0625/100.

One solution which gives the exact ratio is $N_{S1}=82$, $N_{P1}=40$, $N_{P2}=39$, $N_{S2}=80$. Then, the module for gears S_1 and P_1 is 4 mm which is standard. The module for the other two is not a standard value. The solution is shown below, using the tabular method.

8.50 (Cont'd.)

Gear	Stationary S_1	P_1	P_2	Output S_2	Input C
No. of teeth	82	40	39	80	--
I Lock train	+1	+1	+1	+1	+1
II Correct S_1 Fix C	-1	$+\frac{N_{S1}}{N_{P1}}$	$+\frac{N_{S1}}{N_{P1}}$	$-\frac{N_{S1}N_{P2}}{N_{P1}N_{S2}}$	0
Σ Rotations	0	$1+\frac{N_{S1}}{N_{P1}}$	$1+\frac{N_{S1}}{N_{P1}}$	$1-\frac{N_{S1}N_{P2}}{N_{P1}N_{S2}}$	+1
	0	+3.05	+3.05	+.000625	+1
n (rpm)	0	305	305	0.0625	100
module m (mm)	4	4	4.1008	4.1008	--
$m_1 = \frac{2C}{N_{S1}+N_{P1}}$ (S_1 and P_1)	(std)	(std)	(non std)	(non std)	

$m_2 = \frac{2C}{N_{S2}+N_{P2}}$ (S_2 & P_2)

8.51 a)

Gear	S	P_1	P_2	R	C
No. of teeth	N_S	N_{P1}	N_{P2}	N_R	
I (train locked)	+1	+1	+1	+1	+1
II (carrier fixed)	$-N_R/N_S$	$+N_R/N_{P1}$	$-N_R/N_{P2}$	-1	0
Σ (Sum)	$1-N_R/N_S$	$1+N_R/N_{P1}$	$1-N_R/N_{P2}$	0	+1

From the above table $\omega_C/\omega_S = \frac{1}{1-N_R/N_S} = -100/300$

From which $N_R = 4N_S$

$N_S = d_S/m = 80/4 = 20$ teeth $\quad N_R = 80 \quad d_R = 320$ mm

For standard teeth (full depth) addendum = m

Equal planets

$r_S + 2r_P + m < r_R - m$ for planet 1 to clear ring

$40 + 2r_P + 4 < 160 - 4 \quad d_P < 112$

To mesh, $d_S + 2d_P + 2d_P \geq d_R \quad 80 + 4d_P \geq 320$

$d_P \geq 60 \quad 60 \leq d_P < 112 \quad 15 \leq N_P < 28$

i.e. $15 \leq N_P \leq 27$

$N_{P1} = N_{P2} = 20$ is satisfactory. Then, with $n_R=0$, $n_C=-100$ and $n_S=+300$ the result is $n_{P1}=-500 \quad n_{P2}=+300$ (ans)

8.51 (Cont'd.)

b) $n_R^*/n_S^* = (n_R - n_C)/(n_S - n_C) \quad +N_S/N_R = (0 - n_C)/(n_S - n_C)$

$20/80 = +100/(300+100)$ checks

8.52

Gear	Input S_1	P_1	P_2	Output S_2	Fixed R	C
I	1	1		1	1	1
II	$\frac{N_R}{N_{P2}}\frac{N_{P1}}{N_{S1}}$	$-N_R/N_{P2}$		$+N_R/N_{S2}$	-1	0
Σ	$1 + \frac{N_R N_{P1}}{N_{P2} N_{S1}}$	$1 - N_R/N_{P2}$		$1 + N_R/N_{S2}$	0	1

$$n_{S2}/n_{S1} = (1 + N_R/N_{S2})/(1 + \frac{N_R N_{P1}}{N_{P2} N_{S1}}) \quad \text{(ans)}$$

8.53 $N = d/m = d/5$

$N_{S1} = 20,\ N_{P1} = 20,\ N_{P2} = 19,\ N_{S2} = 21,\ N_R = 59$

$$n_{R1}^*/n_{S1}^* = -\frac{N_{S2}N_{P2}}{N_{P1}N_R} = \frac{n_R - n_C}{n_{S1} - n_C} = -\frac{20x19}{20x59}$$

$$-0.32203 = \frac{0 - n_C}{n_{S1} - n_C}$$

$$n_C/n_{S1} = 0.32203/1.32203 = 0.24358$$

$$n_{S2}^*/n_{S1}^* = \frac{N_{S1}N_{P2}}{N_{P1}N_{S2}} = \frac{n_{S2} - n_C}{n_{S1} - n_C}$$

$$\frac{20x19}{20x21} = \frac{n_{S2}/n_{S1} - 0.24358}{1 - 0.24358}$$

$$n_{S2}/n_{S1} = 0.92795$$

8.54 Using the results of prob. 8.52:

$$\frac{n_{S2}}{n_{S1}} = \frac{1 + 59/21}{1 + \frac{59x20}{19x20}} = 0.92796$$

8.55 $N = d/m = d/2 \qquad N_S = 60/2 = 30$, etc.

$$\frac{n_{R2}^*}{n_S^*} = \frac{-N_S N_{P3}}{N_{P1}N_{R2}} = \frac{n_{R2} - n_C}{n_S - n_C}$$

$$-\frac{30x22}{30x82} = \frac{0 - n_C}{n_S - n_C} = \frac{-n_C/n_S}{1 - n_C/n_S}$$

$$-\frac{22}{82} + \frac{22}{82}\frac{n_C}{n_S} = -\frac{n_C}{n_S}$$

8.55 (Cont'd.)

$$\frac{n_C}{n_S} = \frac{82}{104} \times \frac{22}{82} = \frac{22}{104}$$

$$\frac{n_{R1}^*}{n_S^*} = -\frac{N_S N_{P2}}{N_{P1} N_{R1}} = \frac{n_{R1} - n_C}{n_S - n_C}$$

$$-\frac{30\text{x}20}{30\text{x}80} = -\frac{1}{4} = \frac{n_{R1}/n_S - n_C/n_S}{1 - n_C/n_S} = \frac{n_{R1}/n_S - 22/104}{1 - 22/104}$$

$$n_{R1}/n_S = 0.014423 = \omega_{R1}/\omega_S$$

8.56 Using the last row of table 8.6:

$$\frac{\omega_{R1}}{\omega_S} = \frac{1 - \dfrac{N_{R2}N_{P2}}{N_{P3}N_{R1}}}{1 + \dfrac{N_{R2}N_{P1}}{N_{P3}N_S}} = \frac{1 - \dfrac{82\text{x}20}{22\text{x}80}}{1 + \dfrac{82\text{x}30}{22\text{x}30}} = 0.014423$$

8.57 The train may be assembled if d_P = 32 mm. Then, $d_S + 2d_P = d_R$. Also $(N_S+N_R)/4$ is an integer.

Gear	equation/procedure	S (input)	P	R(fixed)	C(output)
Diameter	--	36	32	100	--
N	N = d/m = d/2	18	16	50	--
I	Lock train	+1	+1	+1	+1
II	Correct R with C fixed	$\frac{50}{18}$	$-\frac{50}{10}$	-1	0
sum	I & II	+3.7777	-2.125	0	+1
ω(rad/s)	$\frac{400}{3.777}$ x sum	400 cw	225 ccw	0	105.88 cw

The carrier turns at 105.88 rad/s cw

8.58 Using the diameters given in the above solution:

$$\frac{n_R^*}{n_S^*} = \frac{n_R - n_C}{n_S - n_C}$$ or, in terms of angular velocity (rad/s):

$$-\frac{N_S}{N_R} = \frac{0 - \omega_C}{\omega_S - \omega_C} \qquad \frac{-36}{100} = \frac{0 - \omega_C}{400 - \omega_C}$$

$$\omega_C = 4\text{x}36/1.36 = 105.88 \text{ rad/s cw}$$

$$\frac{\omega_P^*}{\omega_S^*} = -\frac{N_S}{N_P} = -\frac{36}{32} = \frac{\omega_P - \omega_C}{\omega_S - \omega_C} = \frac{\omega_P - 105.88}{400 - 105.88}$$

$$\omega_P = -\frac{36}{32}(400 - 105.88) + 105.88 = -225 \qquad \omega_P = 225 \text{ rad/s ccw}$$

8.59 Using the results of Table 10,

$$v = (x + y)/2 \quad \text{or} \quad \omega_C = (\omega_{S1} + \omega_{S2})/2$$

- Assume the figure represents a view from the rear of the vehicle. Take angular velocity as positive if CW as viewed from the right. A speedometer reading of 20 MPH indicates a carrier speed corresponding to 20 MPH on a straightaway (or 352 in/sec, since 1 MPH = 17.6 in/sec). Dividing by tire radius, we have

$$\omega_C = \frac{352 \text{ in/sec}}{13 \text{ in}} = 27.08 \text{ rad/sec.}$$

The path of the right wheel has a radius of 360+56 = 416 in. If there is no slipping, the ratio of angular velocities of the wheels is given by the ratio of path radii from which

$$\frac{\omega_{S_2}}{\omega_{S_1}} = 360/416. \quad \text{Substituting in} \quad \omega_C = \frac{\omega_{S_1} + \omega_{S_2}}{2} \quad \text{we have}$$

$$27.08 + \omega_{S_1} + \frac{360}{416}\omega_{S_1} \quad \text{from which } \omega_{S_1} = 29.02 \text{ rad/sec.}$$

Then $\omega_{S_2} = \frac{360}{416}\omega_{S_1} = 25.14$ rad/sec.

The speed of S_1 relative to the carrier is $\omega_{S_{1C}} = \omega_{S_1} - \omega_C$ = 1.94 rad/sec. The speed of the planet about its axis is the speed of the planet with respect to the carrier:

$$\omega_{PC} = \frac{N_{S1}}{N_P}\omega_{S_{1C}} = \frac{30}{20} * 1.94 = 2.91 \text{ rad/sec}$$

(CW in the upper diagram).

8.60 Using gear S_1 as gear x and gear S_2 as gear y in eqn. 9, we have

$$r^* = \frac{n_{S2}^*}{n_{S1}^*} = \frac{n_{S2} - n_C}{n_{S1} - n_C} = \frac{\omega_{S2} - \omega_C}{\omega_{S1} - \omega_C}$$

Let CW be positive as viewed from the right.

$$r^* = -\left(\frac{N_{S1}N_P}{N_P N_{S2}}\right) = -1 \qquad -1 = \frac{\omega_{S2} - \omega_C}{\omega_{S1} - \omega_C} \qquad -\omega_{S1} + \omega_C = \omega_{S2} + \omega_C$$

from which $\omega_C = (\omega_{S1} + \omega_{S2})/2$. The remainder of the solution is identical to that of prob. 8.59.

8.61 a)

Gear	S_1	S_2	C
I Rotations with train locked	V	V	V
II Rotations with carrier fixed	W	$-\frac{35}{25}W = -1.4W$	0
III Total rotations	X = V+W = 40 (CW)	y = V - 1.4W = -15 (CC)	V

8.61 a) (Cont'd.)

Simultaneous solution of the equations in row III yields $2.4W = 55$ $W = 22.917$ $V = 40 - W =$ <u>17.083 carrier rotations</u> (CW)

Rotations relative to carrier:

$n_{S1C} = n_{S1} - n_C = W = 22.917$ rotations

$n_{PC} = n_{S1C}N_{S1}/N_P = 22.917 \times 35/22$

$\underline{n_{PC} = 36.458 \text{ rotations}}$ (ans)

b) From row III, simultaneous solution yields

$1.4\,x + y = 1.4\,V + V = 2.4\,V$

Carrier rotations $V = (1.4\,x + y)/2.4$

$V = (7/12)x + (5/12)y$ (ans)

8.62 a) Let the train be viewed from the left end, with clockwise rotation positive.

$$\frac{n_{S2}^*}{n_{S1}^*} = -\frac{N_{S1}}{N_{S2}} = \frac{n_{S2} - n_C}{n_{S1} - n_C}$$

$$-\frac{35}{25} = \frac{-15 - n_C}{40 - n_C} \qquad n_C = 17.083 \text{ rotations cw.}$$

$n_{S1}^* = n_{S1} - n_C$

$n_P^* = n_{S1}^* N_{S1}/N_P = (40 - 17.083)35/22$

$= 36.458$ cw rotations viewed from the outside.

b) From part a):

$$-\frac{35}{25} = \frac{y - n_C}{x - n_C}$$

$n_C = (y + 1.4x)/2.4$

8.63 Speed ratio $n_4/n_1 = d_1d_3/(d_2d_4)$

<u>For $n_4 = 400$,</u> using $d_3 = 5$,

$400/1800 = 4 \times 5/(10 \times d_{2max})$ from which $d_{2max} = 9$

<u>For $n_4 = 1400$,</u> using $d_2 = 5$

$1400/1800 = 4d_{3max}/5 \times 10$ from which $d_{3max} = 7.92$

Thus, $5 \le d_2 \le 9$ in and $5 \le d_3 \le 7.92$ in

8.64 Speed ratio $n_4/n_1 = d_1d_3/(d_2d_4)$

For $n_4 = 400$, using $d_3 = 100$

$400/1800 = 80 \times 100/(200d_2)$ from which $d_2 = 180$

8.64 (Cont'd.)

For $n_4 = 1400$, using $d_2 = 100$

$1400/1800 = 80d_3/(100x200)$ from which $d_3 = 194.44$

Thus, $100 \le d_2 \le 180$ and $100 \le d_3 \le 194.44$

8.65 $n_2/n_1 = r_1/r_2 = r_1/(2-r_1) = 16/90$, etc. from which $r_1 = 0.3019$ in for 16 RPM, 0.5406 in. for 33 1/3 RPM, 0.6667 in. for 45 RPM and 0.9286 in. for 78 RPM. (ans)

8.66 $n_2/n_1 = r_1/r_2 = r_1/(38-r_1)$ from which $r_1 = (38\ n_2/n_1)/(n_2/n_1 + 1)$

For 16 r/min: $r_1 = (38x16/90)/(16/90 + 1) = 5.736$ mm.

For 33 1/3 r/min: $r_1 = 10.270$ mm.

For 45 r/min: $r_1 = 12.667$ mm

For 78 r/min: $r_1 = 17.643$ mm

8.67 Same as Fig. 31, but disk 2 drives. Output to input speed ratio $n_1/n_2 = r_2/r_1$ to vary from 2000/1000 through zero to -2000/1000. Disk 1 is designed to shift axially along a diameter of disk 2. It must move from $r_2 = 2r_1$ on one side of disk 2, through the center, to a distance $2r_1$ on the other side. Thus, the diameter of disk 2 must exceed twice the diameter of disk 1.

8.68 Carrier C is the input and ring R_2 the output. From table 8.11, speed ratio $n_{R2}/n_C = 1 - r_{R1}N_{P2}/(r_{P1}N_{R2})$ from which

$r_{P1} = r_{R1}(N_{P2}/N_{R2})/(1 - n_{R2}/n_C) = 2.5(15/45)/(1 - n_{R2}/1800)$

For 180 r/min cc,

$r_{P1} = 2.5(15/45)/(1+180/1800) = 0.7575$ $d_{P1} = 1.515$ in.

at point of contact near small end of cone.

For 450 r/min cw,

$r_{P1} = 2.5(15/45)/(1-450/1800) = 1.111$

$d_{P1} = 2.222$ in. at point of contact near large end of cone.

8.69 As in the above solution,

$r_{P1} = r_{R1}(N_{P2}/N_{R2})/(1-n_{R2}/n_C) = 100(15/45)/(1-n_{R2}/1800)$

For 180 r/min cc

$r_{P1} = 100(15/45)/(1+180/1800) = 30.303$

$d_{P1} = 60.606$ mm at the point of contact near the small end of the cone.

For 450 r/min cw

$r_{P1} = 100(15/45)/(1-450/1800) = 44.444$

$d_{P1} = 88.889$ mm at the point of contact near the large end of the cone.

8.70 Center distance $C = 112.5 \text{ mm} = \frac{d_1+d_2}{2} = \frac{d_3+d_4}{2}$

$$d_1 + d_2 = d_3 + d_4 = 225 \text{ mm}$$

$$N = d/m \quad N_1 + N_2 = N_3 + N_4 = \frac{2C}{m} = \frac{225}{5} = 45$$

$$\omega_{out}/\omega_{in} = \frac{N_1 N_3}{N_2 N_4}$$

The 6 possible combinations of N_1 and N_2 are as follows:

N_1	N_2
20	25
21	24
22	23
23	22
24	21
25	20

and the same pairs of values are possible for N_3 and N_4. Thus

$$\omega_{out}/\omega_{in}\text{ (min)} = \frac{20\text{x}20}{25\text{x}25} = \underline{0.64} \quad \text{and}$$

$$\omega_{out}/\omega_{in}\text{ (max)} = \frac{25\text{x}25}{20\text{x}20} = \underline{1.5625} \quad \text{(ans)}$$

$$d = mN = 5N$$

<u>For minimum output speed</u>

$$d_1 = d_3 = 5\text{x}20 = 100 \text{ mm}$$

$$d_2 = d_4 = 5\text{x}25 = 125 \text{ mm}$$

<u>For maximum output speed</u>

$$d_1 = d_2 = 5\text{x}25 = 125 \text{ mm}$$

$$d_2 = d_4 = 5\text{x}20 = 100 \text{ mm}$$

8.71 Since each of the 6 N_1, N_2 combinations corresponds to 6 N_3, N_4 combinations, 6x6=36 different gear trains are possible. Due to duplication of ratios, however, only <u>19 different ratios are possible</u>. (ans)

They are given in the following table.

8.71 (Cont'd.)

Possible speed ratios $\text{out}/\text{in} = \frac{N_1 N_3}{N_2 N_4}$

N_1	N_2	N_3 = 20, N_4 = 25	N_3 = 21, N_4 = 24	N_3 = 22, N_4 = 23	N_3 = 23, N_4 = 22	N_3 = 24, N_4 = 21	N_3 = 25, N_4 = 21
20	25	$(\frac{20}{25})^2 = 0.64$	$\frac{20x21}{24x25} = 0.7$	$\frac{20x22}{23x25} =$ 0.765217	$\frac{20x23}{22x25} =$ 0.8364	$\frac{20x24}{21x25} =$ 0.914286	1
21	24	$\frac{20x21}{24x25}$	$(\frac{21}{24})^2 =$ 0.76563	$\frac{21x22}{23x24} =$ 0.8370	$\frac{21x23}{22x24} =$ 0.914773	1	$\frac{21x25}{20x24} =$ 1.09375
22	23	$\frac{20x23}{22x25}$	$\frac{21x22}{23x24}$	$(\frac{22}{23})^2$ =0.914933	1	$\frac{22x24}{21x22}$ =1.09317	$\frac{22x25}{20x22}$ 1.19565
23	22	$\frac{20x23}{22x25}$	$\frac{21x23}{22x24}$	1	$(\frac{23}{22})^2$ =1.09298	$\frac{23x24}{21x22}$ =1.194805	$\frac{23x25}{20x22}$ 1.30682
24	21	$\frac{20x24}{21x25}$	1	$\frac{22x24}{21x23}$	$\frac{23x24}{21x22}$	$(\frac{24}{21})^2$ =1.30612	$\frac{24x25}{20x21}$ =1.42857
25	20	1	$\frac{21x25}{20x24}$	$\frac{22x25}{20x23}$	$\frac{23x25}{20x22}$	$\frac{24x25}{20x21}$	$(\frac{25}{20})^2$ =1.5625

8.72

Component	Sun S (fixed)	Rotor P (Internal gear acting as planet)	Crankshaft C (carrier)
Teeth	N_S	N_P	
Step I	+1	+1	+1
II	-1	$-N_S/N_P$	0
Sum	0	$1 - N_S/N_P$	1

From which the ratio is: $n_C/n_P = 1/(1-N_S/N_P)$

$$= 1/(1-1/1.5)$$

$$n_C/n_P = 3$$

8.73 $r^* = n_P^*/n_S^* = N_S/N_P = 1/1.5$

$$r^* = \frac{n_P^*}{n_S^*} = \frac{n_P - n_C}{n_S - n_C} = \frac{1 - n_C/n_P}{n_S/n_P - n_C/n_P} = \frac{1 - n_C/n_P}{0 - n_C/n_P}$$

From which the ratio is:

$$n_C/n_P = 1/(1-r^*) = 1/(1-1/1.5) = 3$$

8.74 See free body diagrams.

Gear	S(input)	P	R(fixed)	C(output)
N	25	20	65	--
$d(in)=N/P_d$	5	4	13	$r=r_S+r_P=4.5$
Step I	+1	+1	+1	+1
Step II	$+N_R/N_S$	$-N_R/N_P$	-1	0
Sum	$1+N_R/N_S = 3.6$	$1-N_R/N_P = -2.25$	0	+1
n(r/min) = 630xsum/3.6	630 cw	394 ccw	0	175 cw
T (in lb)	2000. cw	0	5200 cw	7200 ccw
HP	20	0	0	20

Torque $T = 63{,}025 \times HP/n_S$

$T_S = 63025 \times 20/630 = 2000$ in.lb $= 2\ F^t d_S/2$

Tangential force $F_t = 2000/5 = 400$ lb

Radial force $F_r = F_t \tan 20^o = 146$ lb

$T_C = 2 \times 2\ F_t r_C = 2 \times 2 \times 400 \times 4.5 = 7200$ in.lb

8.75 Using the formula method:

$$\frac{n_{S2}^*}{n_{S1}^*} = r^* = + \frac{N_{S1}N_{P2}}{N_{P1}N_{S2}} = \frac{n_{S2} - n_C}{n_{S1} - n_C} = \frac{n_{S2} - n_C}{0 - n_C}$$

$(1-r^*)n_C = n_{S2}$ $\quad n_{S2}/n_C = 1-r^* = 1 -(N_{S1}N_{P2})/(N_{P1}N_{S2})$

$$\frac{n_{S2}}{n_C} = \frac{10}{1960} = 1 - \frac{N_{S1}N_{P2}}{N_{P1}N_{S2}} = 1 - \frac{25 \times 39}{N_{P1} \times 49} \qquad \underline{N_{P1} = 20}$$

Alternatively, the speed ratio may be determined by the tabular method.

Gear	S_1	P_1	P_2	S_2	C
# of Teeth	N_{S1}	N_{P1}	N_{P2}	N_{S2}	
I	+1	+1	+1	+1	+1
II	-1	$+N_{S1}/N_{P1}$	$+N_{S1}/N_{P1}$	$-\frac{N_{S1}N_{P2}}{N_{P1}N_{S2}}$	0
Total	0	$1+N_{S1}/N_{P1}$	$1+N_{S1}/N_{P1}$	$1-\frac{N_{S1}N_{P2}}{N_{P1}N_{S2}}$	+1

$$n_{S2}/n_C = \frac{1 -(N_{S1}N_{P2})/(N_{P1}N_{S2})}{1} = 1 -(N_{S1}N_{P2})/(N_{P1}N_{S2})$$

resulting in $N_{P1} = 20$ as above

8.74
(Not to scale)

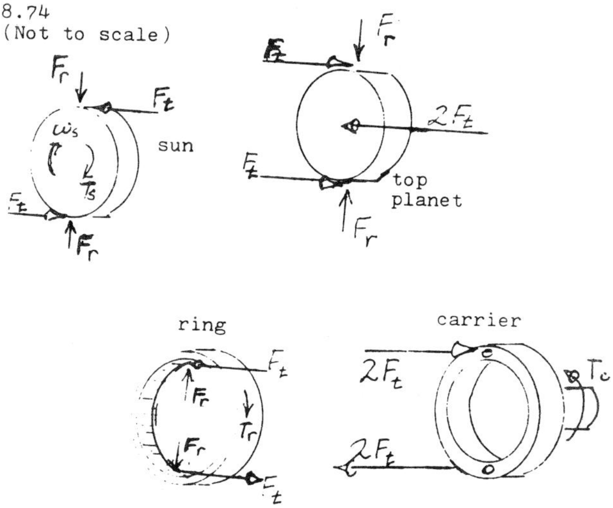

8.76
(Not to scale)

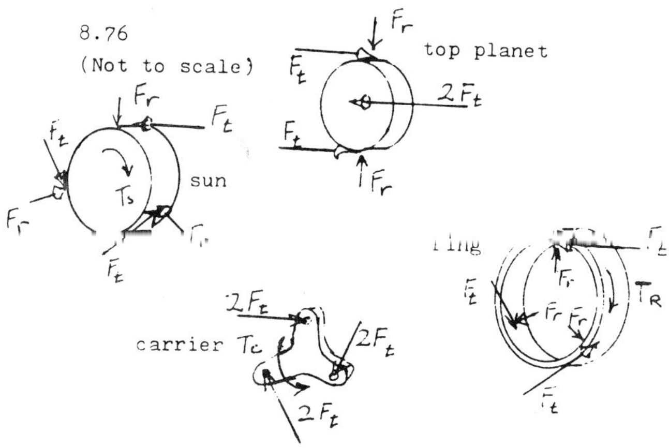

8.76 See free body diagrams.

Radius: $r_C = r_S + r_P = 20 + 10 = 30$; $r_R = r_S + 2\ r_P = 20+2\times10=40$

<u>Module</u>: $m = d_P/N_P = 2\times10/20 = 1$ mm

<u>Teeth</u>: $N_S = d_S/m = 2\times20 = 40$ etc.

<u>Formula method</u> $r^* = n_R^*/n_S^* = -N_S/N_R = \dfrac{n_R-n_C}{n_S-n_C} - \dfrac{40}{80} = \dfrac{200-n_C}{0-n_C}$

$\underline{n_C = 133.3}$

<u>Tangential force</u>: $3F_t = T_R/r_R = 12000/40 = 300$ $F_t = 100$ N

<u>Radial force</u>: $F_r = F_t \tan\phi = 100 \tan 25^o = 46.6$N

<u>Torque</u>: $T_S = 3F_t r_S = 3\times100\times20 = \underline{6000\ N_{mm}\ cw}$

$T_C = 3\times2F_t \times r_C = 3\times200\times30 = \underline{18000\ N_{mm}\ cc}$

<u>Power</u>: $P(kw) = T\omega/10^6$ $P_R = 12000\times20.94/10^6 = \underline{.2513\ kw}$

$\underline{P_C = 0.2513\ kw}$

Alternate solution with tabulation of results

Gear	Fixed S	P	Input R	Output C	
Radius r	20	10	40	30	mm
Teeth N	40	20	80	--	
Step I	+1	+1	+1	+1	
Step II	-1	+2	+.5	0	
Sum	0	+3	+1.5	+1	
Speed n_{rpm}	0	400 cw	200 cw	133.3 cw	
Torque T(N•mm)	6000 cw	0	12,000 cw	18,000 cc	
Power (kw)	0	0	0.2513	0.2513	
Angular velocity ω rad/s	0	41.89	20.94	13.96	

8.77 Let the sun gear be the input and the planet carrier the output. A clutch engaging the sun gear to the planet carrier (with the ring free to rotate) produces a 1 : 1 speed ratio. If the clutch is disengaged and the ring held stationary with a band brake, there is a speed reduction. The desired speed reduction is produced (approximately) if the sun, planet and ring gears have 19, 21 and 61 teeth, respectively. See spreadsheet table S8.77 . The train may be balanced with four planets.

8.78 Let the sun gear be the input and the planet carrier the output. A clutch engaging the sun gear to the planet carrier (with the ring free to rotate) produces a 1 : 1 speed ratio. If the clutch is disengaged and the ring held stationary with a band brake, there is a speed reduction. The desired speed reduction is produced (approximately) if the sun, planet and ring gears have 23, 21 and 65 teeth, respectively. See spreadsheet table S8.78. The train may be balanced with four planets.

8.79 Let the sun gear be the input and the planet carrier the output. A clutch engaging the sun gear to the planet carrier (with the ring free to rotate) produces a 1 : 1 speed ratio. If the clutch is disengaged and the ring held stationary with a band brake, there is a speed reduction.

58.77

Planetary gear train with stationary ring

Superposition method. Check by formula method.

	SUN	PLANET	RING	CARRIER	Planets 3 Balanced?	4
Teeth	18	19	56		NO	NO
Speed n/n(c)	4.111111	-1.94737	0	1		
Speed n/n(s)	1	-.473684	0	.2432432		
Speed check	1		0	.2432432		
Teeth	18	20	58		NO	YES
Speed n/n(c)	4.222222	-1.9	0	1		
Speed n/n(s)	1	-.45	0	.2368421		
Speed check	1		0	.2368421		
Teeth	18	21	60		YES	NO
Speed n/n(c)	4.333333	-1.85714	0	1		
Speed n/n(s)	1	-.428571	0	.2307692		
Speed check	1		0	.2307692		
Teeth	19	19	57		NO	YES
Speed n/n(c)	4	-2	0	1		
Speed n/n(s)	1	-.5	0	.25		
Speed check	1		0	.25		
Teeth	19	20	59		YES	NO
Speed n/n(c)	4.105263	-1.95	0	1		
Speed n/n(s)	1	-.475	0	.2435897		
Speed check	1		0	.2435897		
→ Teeth	19	21	61		NO	YES
Speed n/n(c)	4.210526	-1.90476	0	1		
Speed n/n(s)	1	-.452381	0	.2375		
Speed check	1		0	.2375		
Teeth	20	19	58		YES	NO
Speed n/n(c)	3.9	-2.05263	0	1		
Speed n/n(s)	1	-.526316	0	.2564103		
Speed check	1		0	.2564103		
Teeth	20	20	60		NO	YES
Speed n/n(c)	4	-2	0	1		
Speed n/n(s)	1	-.5	0	.25		
Speed check	1		0	.25		
Teeth	20	21	62		NO	NO
Speed n/n(c)	4.1	-1.95238	0	1		
Speed n/n(s)	1	-.47619	0	.2439024		
Speed check	1		0	.2439024		

58.78

Planetary gear train with stationary ring
Superposition method. Check by formula method.

	SUN	PLANET	RING	CARRIER	Planets 3 Balanced?	4
Teeth	21	20	61		NO	NO
Speed n/n(c)	3.904762	-2.05	0	1		
Speed n/n(s)	1	-.525	0	.2560976		
Speed check	1		0	.2560976		
Teeth	21	21	63		YES	YES
Speed n/n(c)	4	-2	0	1		
Speed n/n(s)	1	-.5	0	.25		
Speed check	1		0	.25		
Teeth	21	22	65		NO	NO
Speed n/n(c)	4.095238	-1.95455	0	1		
Speed n/n(s)	1	-.477273	0	.244186		
Speed check	1		0	.244186		
Teeth	22	20	62		YES	YES
Speed n/n(c)	3.818182	-2.1	0	1		
Speed n/n(s)	1	-.55	0	.2619048		
Speed check	1		0	.2619048		
Teeth	22	21	64		NO	NO
Speed n/n(c)	3.909091	-2.04762	0	1		
Speed n/n(s)	1	-.52381	0	.255814		
Speed check	1		0	.255814		
Teeth	22	22	66		NO	YES
Speed n/n(c)	4	-2	0	1		
Speed n/n(s)	1	-.5	0	.25		
Speed check	1		0	.25		
Teeth	23	20	63		NO	NO
Speed n/n(c)	3.73913	-2.15	0	1		
Speed n/n(s)	1	-.575	0	.2674419		
Speed check	1		0	.2674419		
→ Teeth	23	21	65		NO	YES
Speed n/n(c)	3.826087	-2.09524	0	1		
Speed n/n(s)	1	-.547619	0	.2613636		
Speed check	1		0	.2613636		
Teeth	23	22	67		YES	NO
Speed n/n(c)	3.913043	-2.04545	0	1		
Speed n/n(s)	1	-.522727	0	.2555556		
Speed check	1		0	.2555556		

58.79

Planetary gear train with stationary ring

Superposition method. Check by formula method.

	SUN	PLANET	RING	CARRIER	Planets 3 Balanced?	Planets 4 Balanced?
Teeth	18	18	54		YES	YES
Relative speed	4	-2	0	1		
Speed (rad/s)	1	-.5	0	.25		
Speed check	1		0	.25		
Teeth	19	18	55		NO	NO
Relative speed	3.894737	-2.05556	0	1		
Speed (rad/s)	1	-.527778	0	.2567568		
Speed check	1		0	.2567568		
Teeth	20	18	56		NO	YES
Relative speed	3.8	-2.11111	0	1		
Speed (rad/s)	1	-.555556	0	.2631579		
Speed check	1		0	.2631579		
Teeth	21	18	57		YES	NO
Relative speed	3.714286	-2.16667	0	1		
Speed (rad/s)	1	-.583333	0	.2692308		
Speed check	1		0	.2692308		
Teeth	22	18	58		NO	YES
Relative speed	3.636364	-2.22222	0	1		
Speed (rad/s)	1	-.611111	0	.275		
Speed check	1		0	.275		
Teeth	23	18	59		NO	NO
Relative speed	3.565217	-2.27778	0	1		
Speed (rad/s)	1	-.638889	0	.2804878		
Speed check	1		0	.2804878		
Teeth	24	18	60		YES	YES
Relative speed	3.5	-2.33333	0	1		
Speed (rad/s)	1	-.666667	0	.2857143		
Speed check	1		0	.2857143		
Teeth	25	18	61		NO	NO
Relative speed	3.44	-2.38889	0	1		
Speed (rad/s)	1	-.694444	0	.2906977		
Speed check	1		0	.2906977		
→ Teeth	26	18	62		NO	YES
Relative speed	3.384615	-2.44444	0	1		
Speed (rad/s)	1	-.722222	0	.2954545		
Speed check	1		0	.2954545		

The desired speed reduction is produced (approximately) if the sun, planet and ring gears have 26, 18 and 62 teeth, respectively. See spreadsheet table S8.79. The train may be balanced with four planets.

8.80 Let the sun gear be the input and the planet carrier the output. A clutch engaging the sun gear to the planet carrier (with the ring free to rotate) produces a 1 : 1 speed ratio. If the clutch is disengaged and the ring held stationary with a band brake, there is a speed reduction. The results are shown in spreadsheet table S8.80.
Four equally-spaced planets may be used. It can be seen that the proposed design meets the kinematic specifications, and that sun, ring and carrier torques balance (a partial check of results). Note that the error is insignificant, and results from truncation errors. The design was not evaluated for tooth stress, etc.

8.81 Let the sun gear be the input and the planet carrier the output. A clutch engaging the sun gear to the planet carrier (with the ring free to rotate) produces a 1 : 1 speed ratio. If the clutch is disengaged and the ring held stationary with a band brake, there is a speed reduction. Three equally-spaced planets may be used. The results are shown in spreadsheet table S8.81. It can be seen that the

Planetary gear train with stationary ring

Module: 1.5 Power: 2.8 kw Pressure angle: 20

Superposition method. Check by formula method.

Shaft torque in N mm

	SUN	PLANET	RING	CARRIER	Planets 3	4
					Balanced?	
Teeth	20	18	56		NO	YES
Radius (mm)	15	13.5	42	28.5		
Relative speed	3.8	-2.11111	0	1		
Speed (rad/s)	220	-122.222	0	57.89474		
Speed check	220		0	57.89474		
Shaft torque	12727.27	0	35636.36	-48363.6		
Tangential force (N)					282.8283	212.1212
Radial force (N)					102.9411	77.20581
Shaft power (kw)	2.8	0	0	-2.8		
Torque balance (N mm)					-1.8E-12	ERROR

S 8.81

Planetary gear train with stationary ring

Module: 5 Power: 12 kw Pressure angle: 20

Superposition method. Check by formula method.

Shaft torque in N mm

	SUN	PLANET	RING	CARRIER	Planets 3	4
					Balanced?	
Teeth	24	21	66		YES	NO
Radius (mm)	60	52.5	165	112.5		
Relative speed	3.75	-2.14286	0	1		
Speed (rad/s)	400	-228.571	0	106.6667		
Speed check	400		0	106.6667		
Shaft torque	30000	0	82500	-112500		
Tangential force (N)					166.6667	125
Radial force (N)					60.66171	45.49628
Shaft power (kw)	12	0	0	-12		
Torque balance (N mm)					0 OK	

proposed design meets the kinematic specifications, and that sun, ring and carrier torques balance (a partial check of results). The design was not evaluated for tooth stress, etc.

8.82 Let the sun gear be the input and the planet carrier the output. A clutch engaging the sun gear to the planet carrier (with the ring free to rotate) produces a 1 : 1 speed ratio. If the clutch is disengaged and the ring held stationary with a band brake, there is a speed reduction. Either three or four equally-spaced planets may be used. The results are shown in spreadsheet table. S8.82 It can be seen that the proposed design meets the kinematic specifications, and that sun, ring and carrier torques balance (a partial check of results). The design was not evaluated for tooth stress, etc.

8.83 Let the sun gear be the input and the planet carrier the output. A clutch engaging the sun gear to the planet carrier (with the ring free to rotate) produces a 1 : 1 speed ratio. If the clutch is disengaged and the ring held stationary with a band brake, there is a speed reduction. The results are shown in spreadsheet table S8.83. It can be seen that the proposed design meets the kinematic specifications including a speed ratio of precisely 1 : 4.

58.82

Planetary gear train with stationary ring
Module: 2.5 Power: 3 kw Pressure angle: 20
Superposition method. Check by formula method.

Shaft torque in N mm

	SUN	PLANET	RING	CARRIER	Planets 3	4
					Balanced?	
Teeth	22	20	62		YES	YES
Radius (mm)	27.5	25	77.5	52.5		
Relative speed	3.818182	-2.1	0	1		
Speed (rad/s)	400	-220	0	104.7619		
Speed check	400		0	104.7619		
Shaft torque	7500	0	21136.36	-28636.4		
Tangential force (N)					90.90909	68.18182
Radial force (N)					33.0882	24.81615
Shaft power (kw)	3	0	0	-3		
Torque balance (N mm)					0	OK

S8.83

Planetary gear train with stationary ring

Module: 1.5 Power: 1.75 kw Pressure angle: 20

Superposition method. Check by formula method.

Shaft torque in N mm

	SUN	PLANET	RING	CARRIER	Planets 3	4
					Balanced?	
					NO	YES
Teeth	20	20	60			
Radius (mm)	15	15	45	30		
Relative speed	4	-2	0	1		
Speed (rad/s)	400	-200	0	100		
Speed check	400		0	100		
Shaft torque	4375	0	13125	-17500		
Tangential force (N)					97.22222	72.91667
Radial force (N)					35.38599	26.5395
Shaft power (kw)	1.75	0	0	-1.75		
Torque balance (N mm)					0 OK	

58,83 contd

Ring and carrier torques balance (a partial check of results). The design was not evaluated for tooth stress, etc.

CHAPTER 9

STATIC-FORCE ANALYSIS

9.1 A)

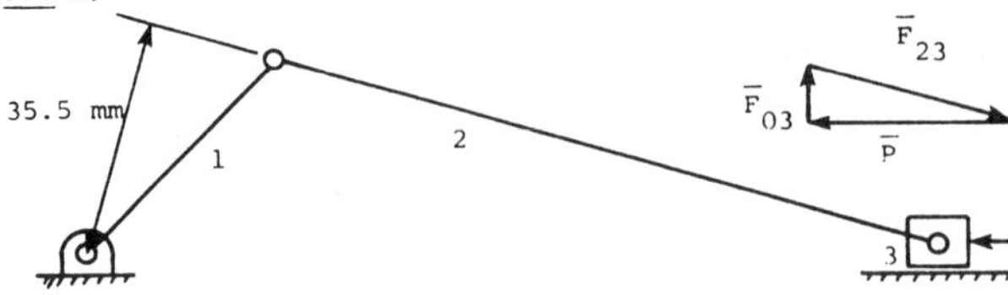

$F_{01} = F_{12} = F_{23} = 520$ N

$F_{03} = 150$ N

$T_1 = 520\ (.0355) = 18.5$ N•m CW

$\overline{P}$

B)

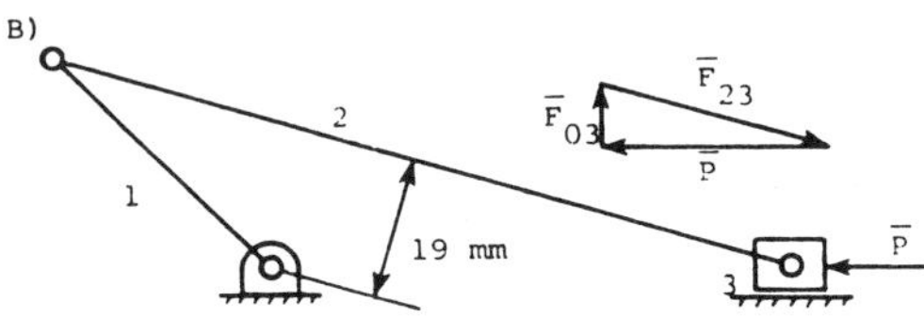

$F_{01} = F_{12} = F_{23} = 520$ N

$F_{03} = 150$ N

$T_1 = 520\ (.019) = 9.9$ N•m CW

C)

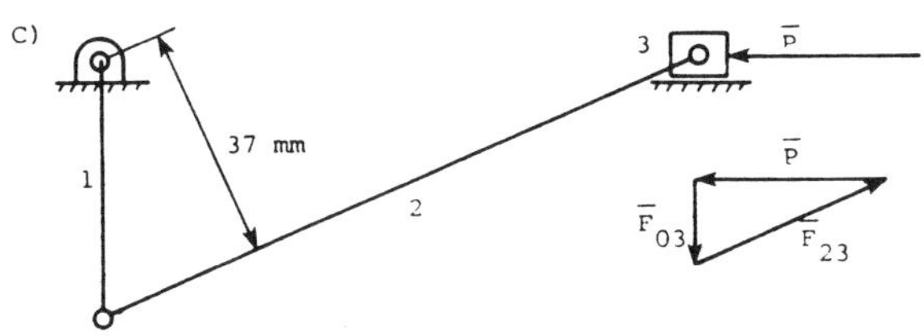

$F_{01} = F_{12} = F_{23} = 550$ N

$F_{03} = 220$ N

$T_1 = 550\ (.037) = 20.4$ N•m CCW

9.2 A)

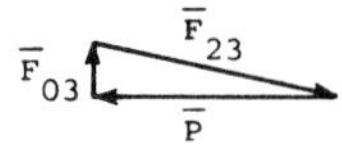

$F_{01} = F_{12} = F_{23} = 105$ lbs

$F_{03} = 20$ lbs

$T_1 = 105\ (1.8) = 189$ lb·in CW

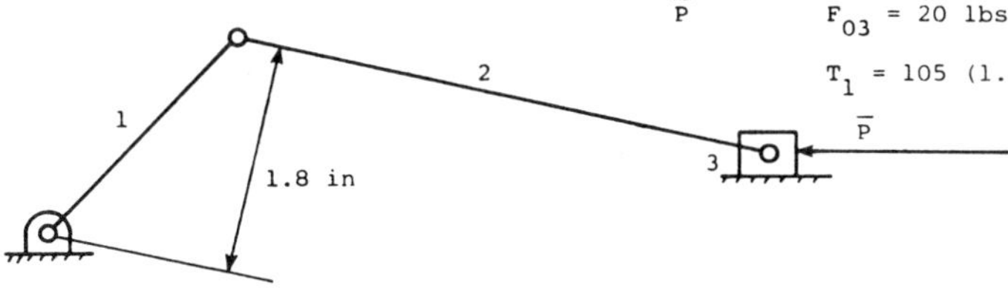

B)

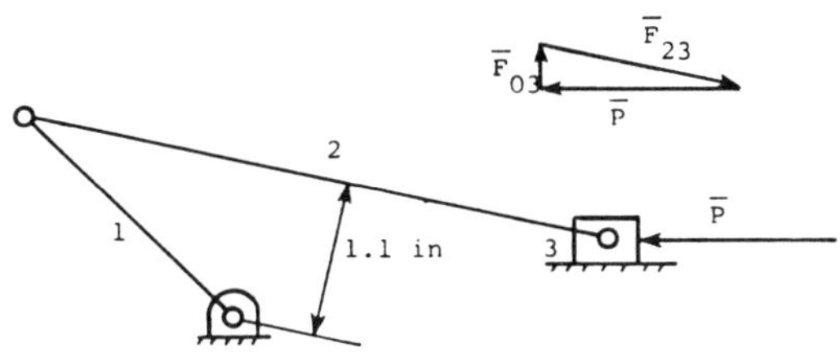

$F_{01} = F_{12} = F_{23} = 105$ lb

$F_{03} = 20$ lbs

$T_1 = 105\ (1.1) = 115.5$ lb·in CW

C)

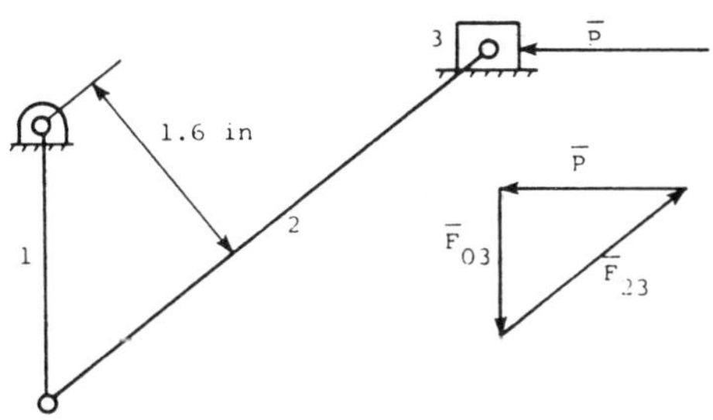

$F_{01} = F_{12} = F_{23} = 125$ lbs

$F_{03} = 80$ lbs

$T_1 = 125\ (1.6) = 200$ lb·in CCW

$F_{01} = F_{12} = F_{23} = 115$ lbs

$F_{03} = 55$ lbs

$T_1 = 115\ (1.9) = 218.5$ lb·in CCW

D)

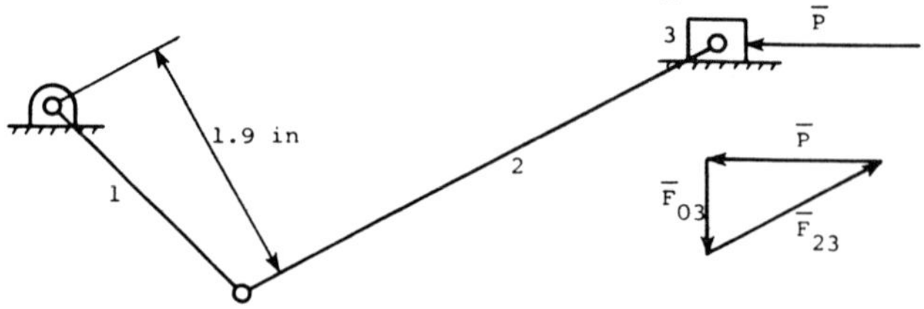

9.3

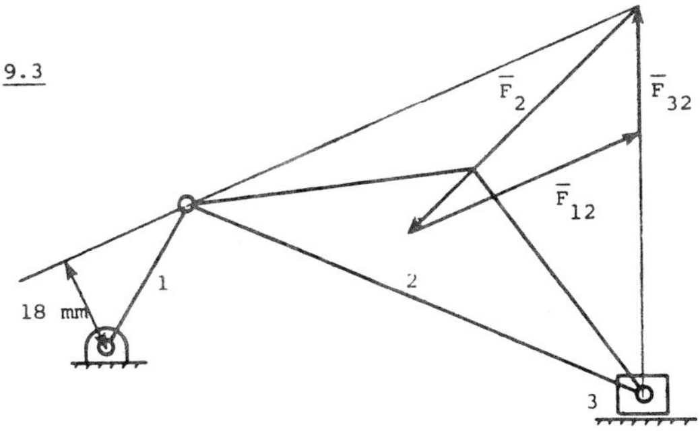

$F_{01} = F_{12} = 156$ N

$F_{23} = F_{03} = 77$ N

$T_1 = 156\ (.018) = 2.8$ N•m CW

9.4

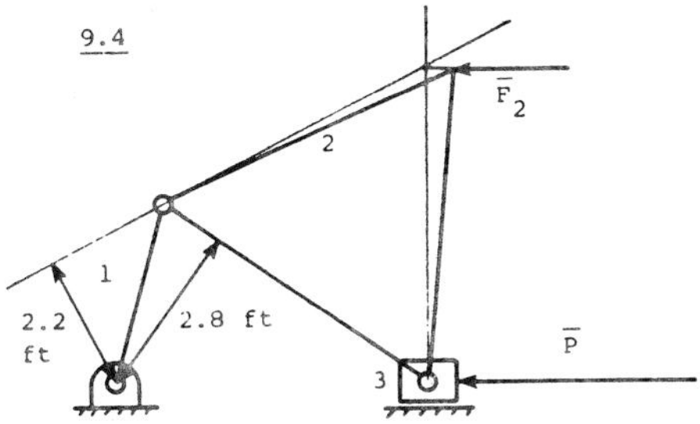

Neglecting $\bar{F}_2$:

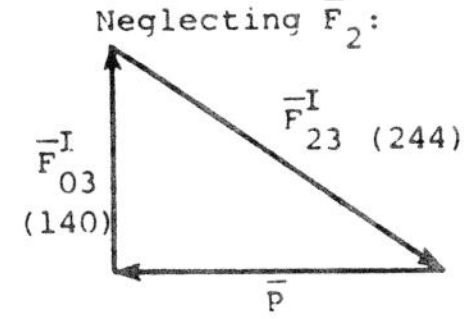

Neglecting $\bar{P}$:

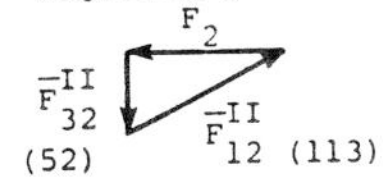

$T^{I}_1 = 244\ (2.8) = 683$ lb•ft CW

$T^{II}_1 = 113\ (2.2) = 249$ lb•ft CW

$T_1 = 683 + 249 = 932$ lb•ft CW

9.5

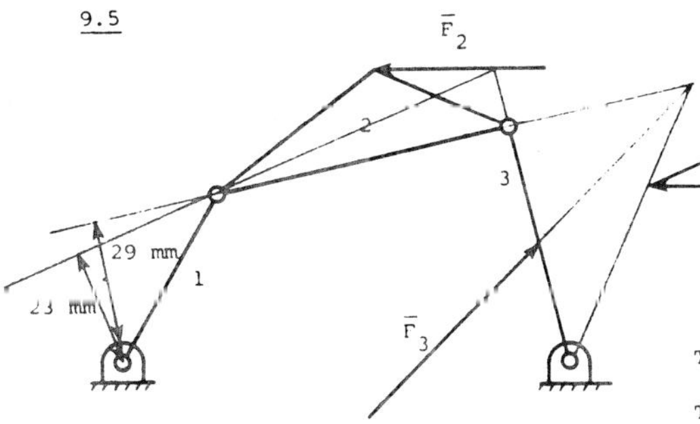

Neglecting $\bar{F}_3$:

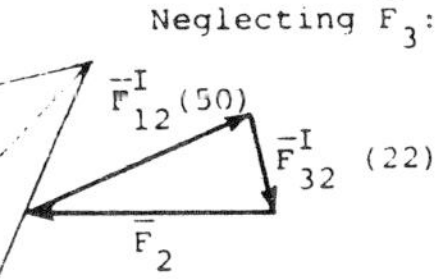

Neglecting $\bar{F}_2$:

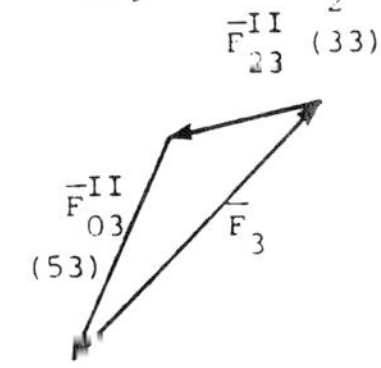

$T^{I}_1 = 50\ (23) = 1150$ N•mm CW

$T^{II}_1 = 33\ (29) = 957$ N•mm CCW

$T_1 = 1150 - 957 = 193$ N•mm CW

(Ans. is low - see Prob 9.22)

9.6

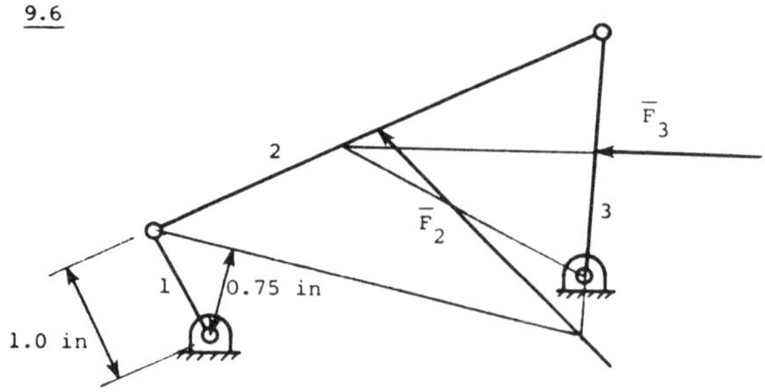

Neglecting $\bar{F}_3$:

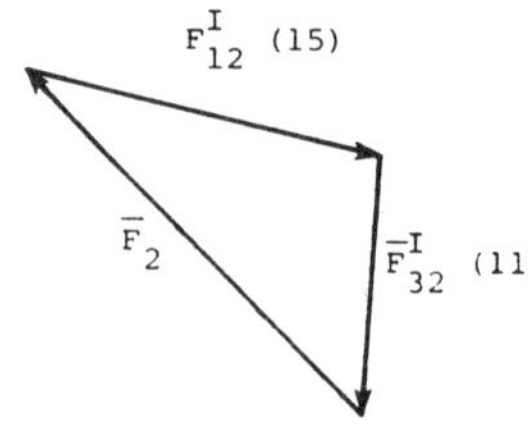

Neglecting $\bar{F}_2$:

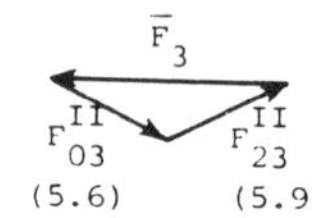

$T_1^{I} = 15\ (0.75) = 11.3$ lb•in CW

$T_1^{II} = 5.9\ (1.0) = 5.9$ lb•in CW

$T_1 = 11.3 + 5.9 = 17.2$ lb•in CW

9.7

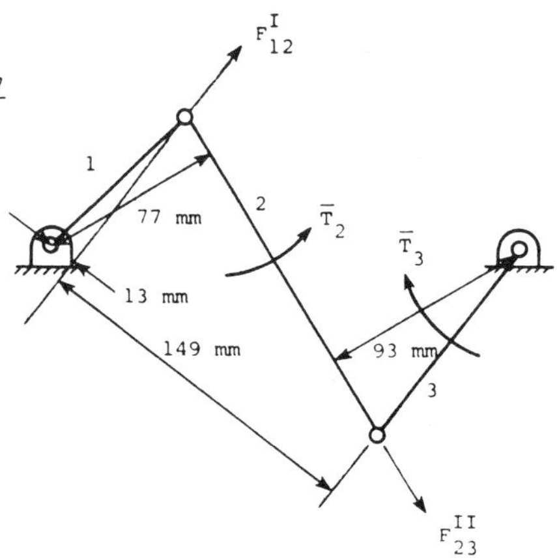

Neglecting $\bar{T}_3$:

$F_{12}^{I} = \frac{10}{.149} = 67.1$ N

$T_1^{I} = 67.1\ (.013) = 0.872$ N•m CCW

Neglecting $\bar{T}_2$:

$F_{23}^{II} = F_{12}^{II} = \frac{7}{.093} = 75.3$ N

$T_1^{II} = 75.3\ (.077) = 5.798$ N•m CW

$T = 5.798 - 0.872 = 4.926$ N•m CW

9.8

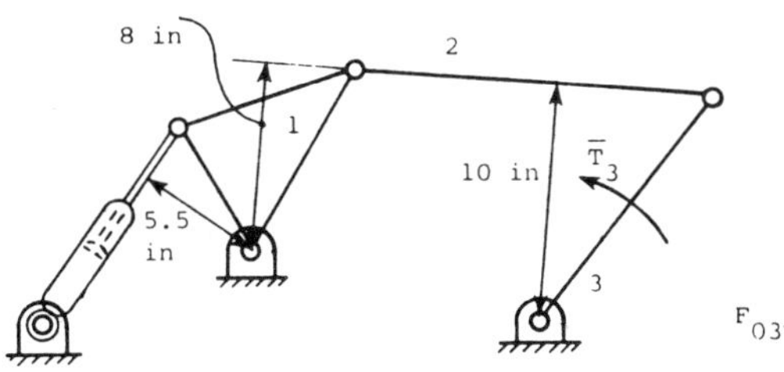

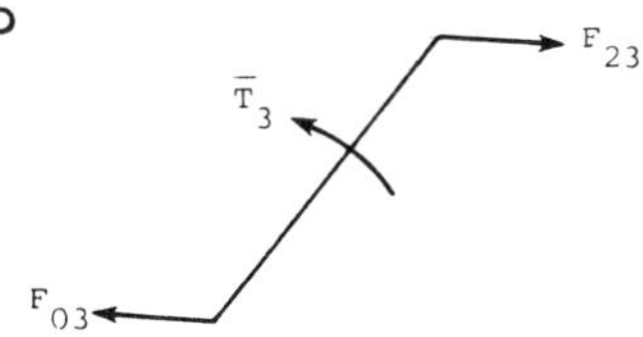

$T_1 = 100\ (8) = 800\ \text{lb}\cdot\text{in CW}$

$F_{23} = F_{12} = \frac{1000}{10} = 100\ \text{lbs}$

$F_p = \frac{800}{5.5} = 145.5\ \text{lbs},\ Ap = \pi(.75)^2 = 1.767\ \text{in}^2$

$\text{pressure} = \frac{145.5}{1.767} = 82.3\ \text{psi}$

9.9

Neglecting $\bar{F}_2$ and $\bar{F}_3$:

$T_1^I = 7\ (9.806)\ (.07) = 4.80\ \text{N}\cdot\text{m CCW}$

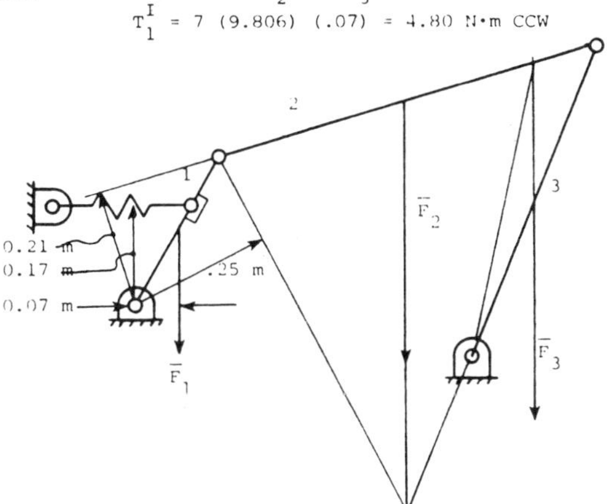

Neglecting $\bar{F}_1$ and $\bar{F}_3$:

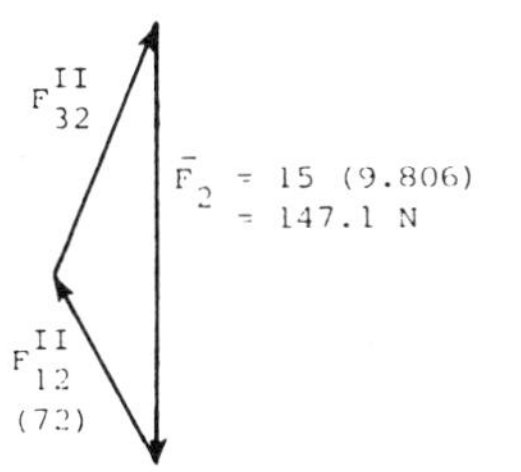

$T_1^{II} = 72\ (.25) = 18.00\ \text{N}\cdot\text{m CCW}$

Neglecting $\bar{F}_1$ and $\bar{F}_2$:

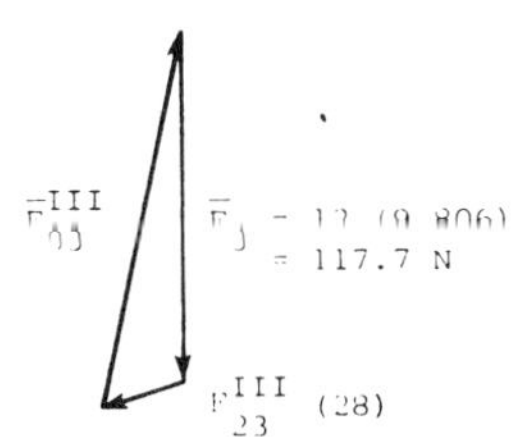

$T_1^{III} = 28\ (.21) = 5.88\ \text{N}\cdot\text{m CCW}$

$T_1 = 4.8 + 18.0 + 5.88 = 28.68\ \text{N}\cdot\text{m CCW}$

$F_s = \frac{\text{[illegible]}}{0.17} = 168.7\ \text{N}$

9.10

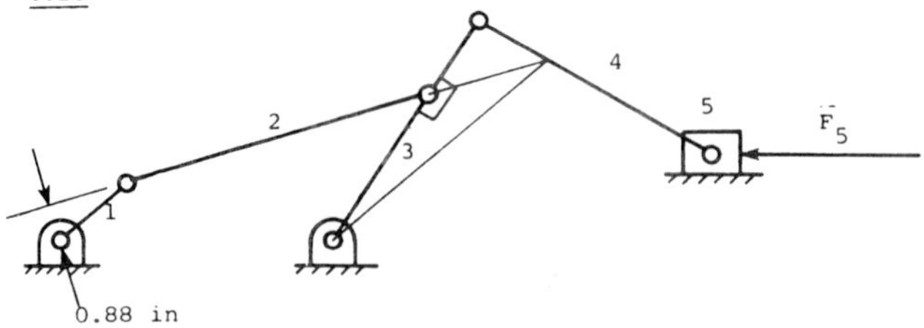

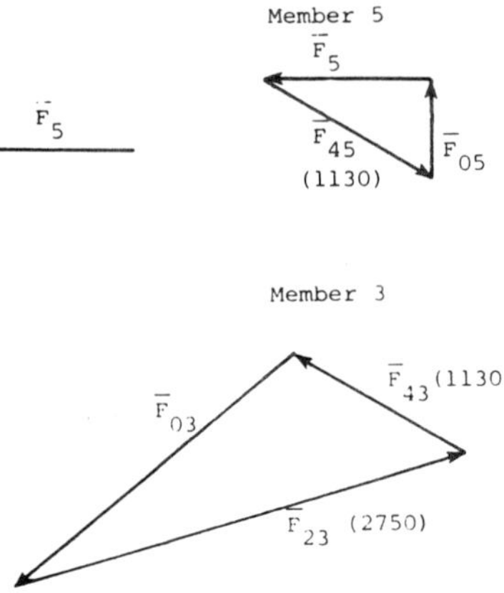

T_1 = 2750 (.88) = 2420 lb·in CW

9.11

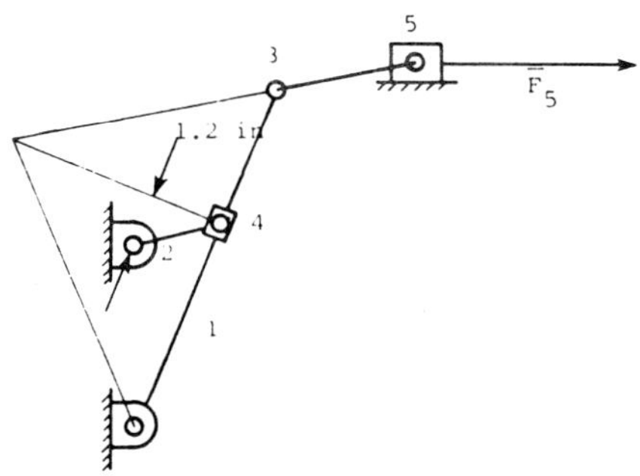

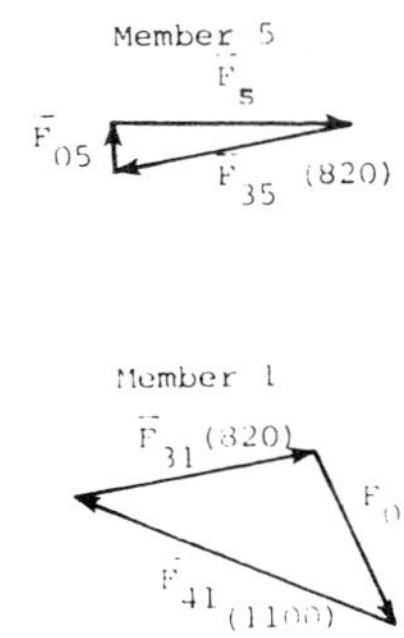

T_2 = 1100 (1.2) = 1320 lb·in CCW

9.12

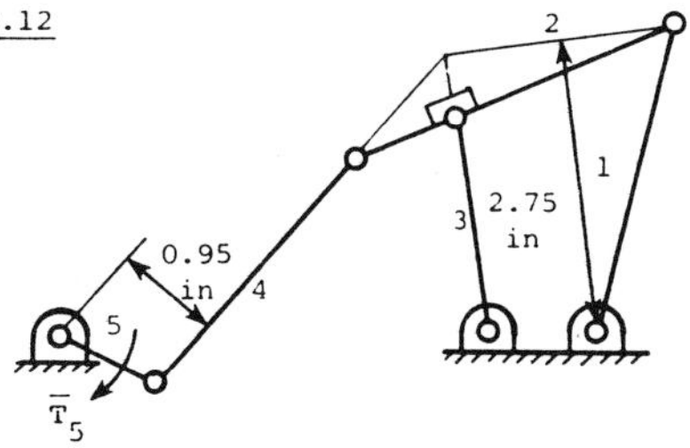

$T_1 = 460\ (2.75) = 1270$ lb·in CW

Member 5

$$F_{45} = \frac{600}{0.95} = 632 \text{ lb}$$

Member 2

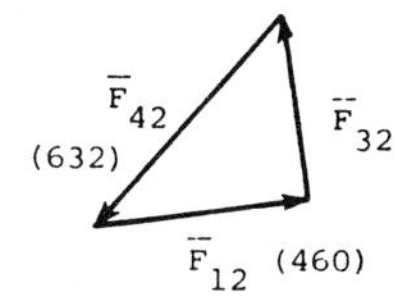

9.13

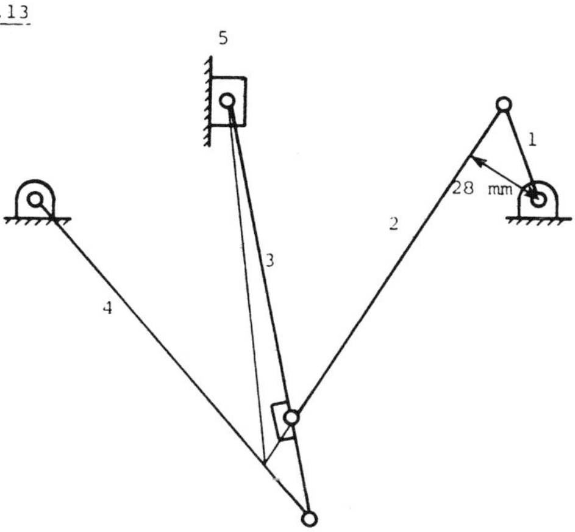

$T_1 = 290\ (.028) = 8.12$ N·m CW

Member 5

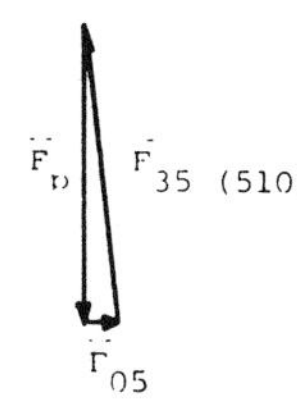

Member 3

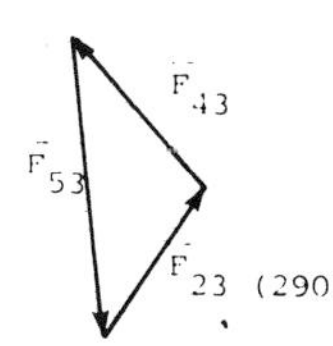

9.14 A) $F_{21} = \frac{100}{0.95} = 105$ lbs

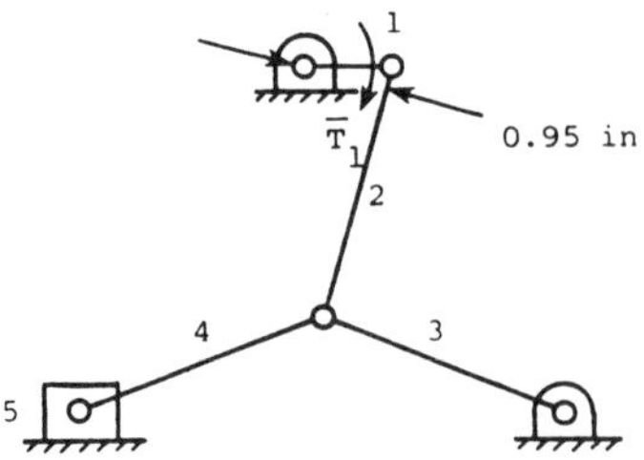

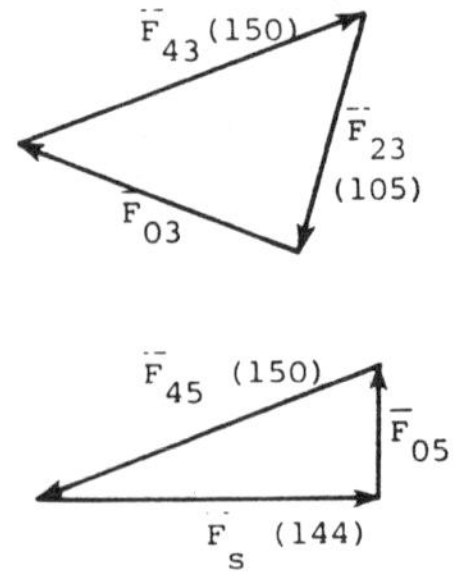

Slider Force = 144 lbs

B) $F_{21} = \frac{100}{0.95} = 105$ lbs

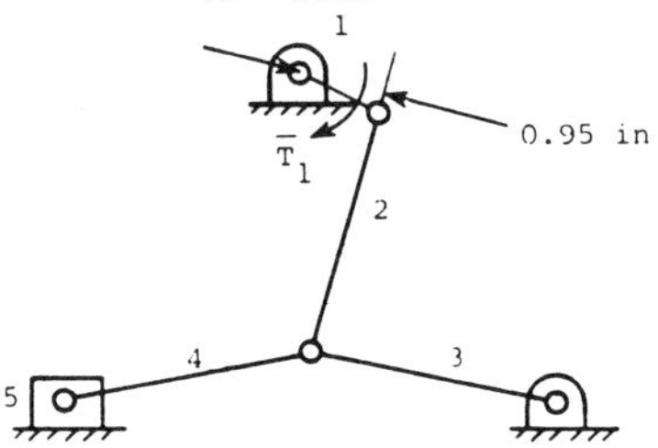

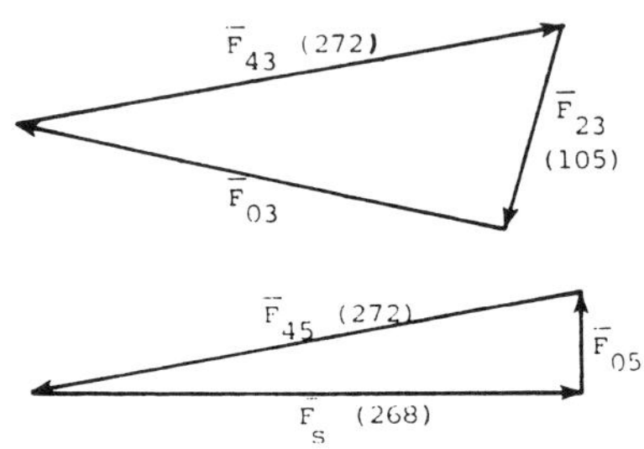

Slider Force = 268 lbs

C) $F_{21} = \frac{100}{0.7} = 143$ lbs

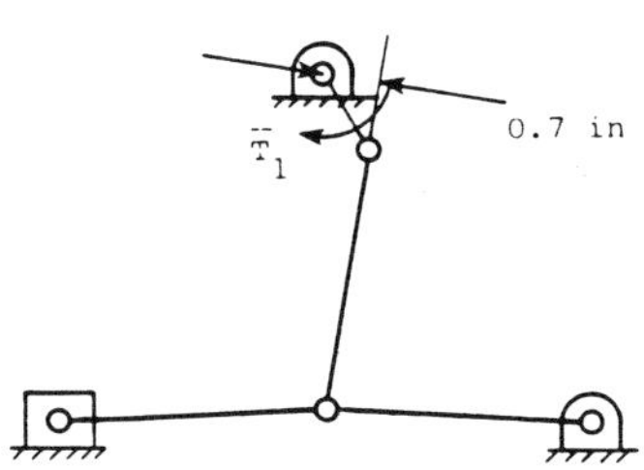

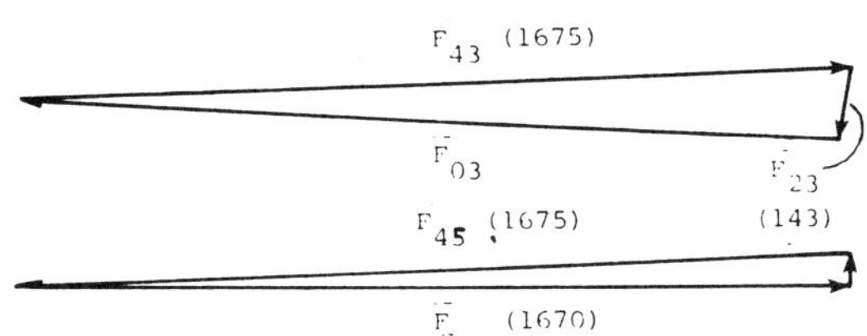

Slider Force = 1670 lbs

9.15

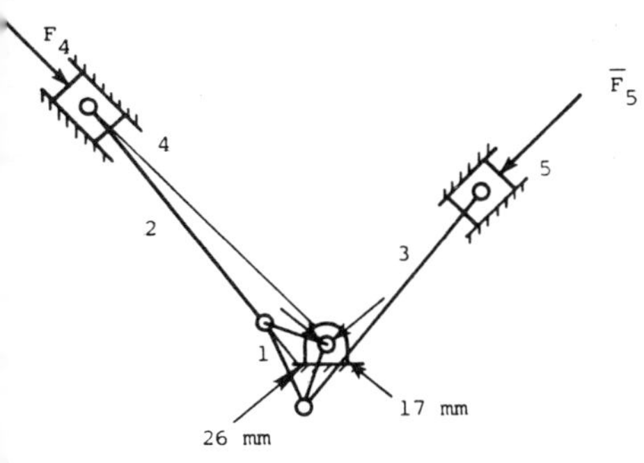

Piston Force $\bar{F}_4$

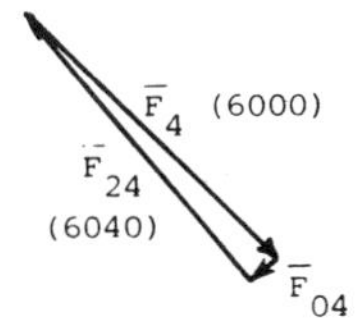

$$T_1^I = 6040\ (.026) = 157.0\ \text{N}\cdot\text{m CW}$$

Piston Force $\bar{F}_5$

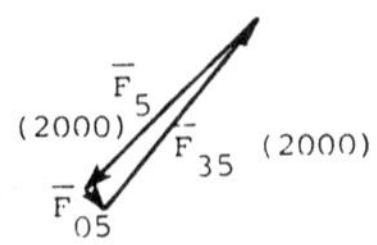

$$T_1^{II} = 2000\ (.017) = 34.0\ \text{N}\cdot\text{m CCW}$$

$$T = 157.0 - 34.0 = 123.0\ \text{N}\cdot\text{m CW}$$

9.16

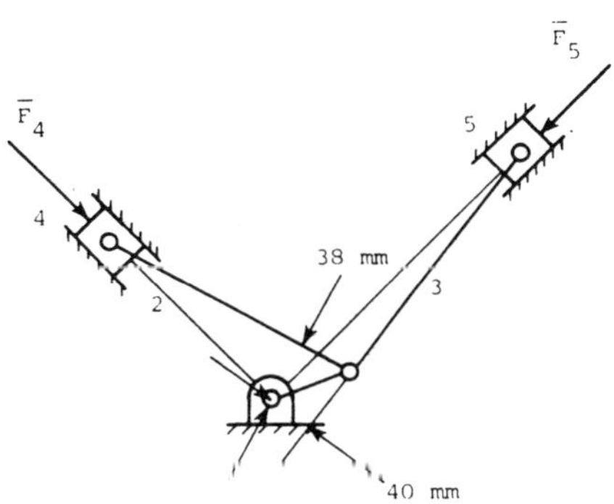

Piston Force $\bar{F}_4$

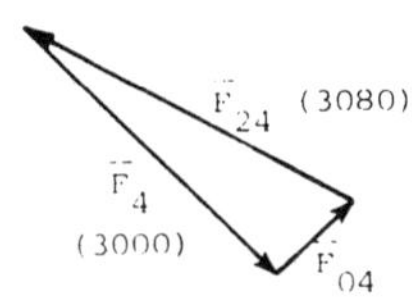

$$T_1^I = 3080\ (.038) = 117.0\ \text{N}\cdot\text{m CCW}$$

Piston Force $\bar{F}_5$

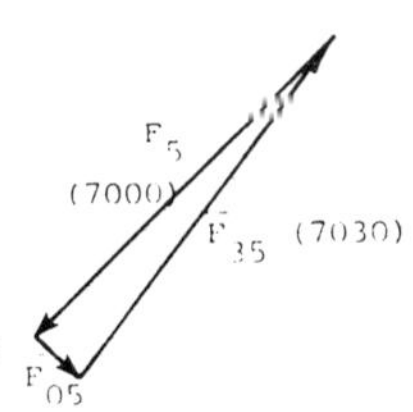

$$T_1^{II} = 7030\ (.040) = 281.2\ \text{N}\cdot\text{m CCW}$$

$$T = 117.0 + 281.2 = 398.2\ \text{N}\cdot\text{m CCW}$$

9.17 Member 1 is the frame.

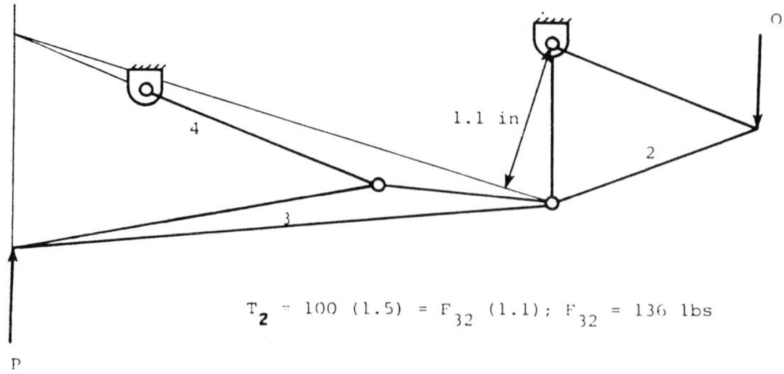

$T_2 = 100\ (1.5) = F_{32}\ (1.1);\ F_{32} = 136$ lbs

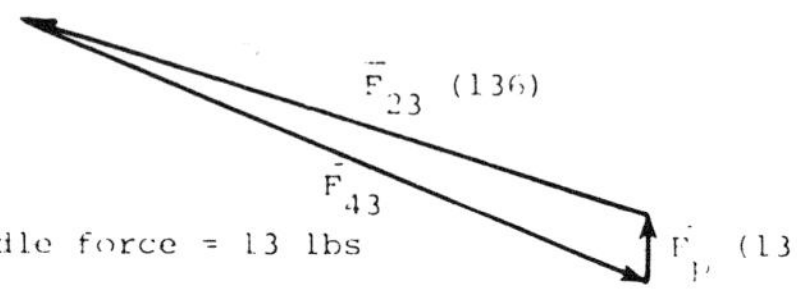

handle force = 13 lbs

.18 $T = -Pr\sin\phi \left(1 + \frac{r\cos\phi}{\sqrt{\ell^2 - r^2\sin^2\phi}}\right)$, $r = 0.04$ m, $\ell = 0.1$ m, $P = 500$ N,

A) $\phi = 45^\circ$, $T_1 = -18.31$ N·m, B) $\phi = 135^\circ$, $T_1 = -9.97$ N·m,

C) $\phi = 270^\circ$, $T_1 = 20.00$ N·m

.19 $\sin\theta = \frac{r\sin\phi - h}{\ell}$, $T = -Pr\left[\sin\phi + \frac{\cos\phi\,(r\sin\phi - h)}{\sqrt{\ell^2 - (r\sin\phi - h)^2}}\right]$, $r = 2$ in,

$\ell = 4$ in, $h = 0.5$ in, $P = 100$ lb, A) $\phi = 45^\circ$, $T_1 = -174.6$ lb·in,

B) $\phi = 135^\circ$, $T_1 = -108.2$ lb·in, C) $\phi = 270^\circ$, $T_1 = 200.0$ lb·in,

D) $\phi = 315^\circ$, $T_1 = 218.5$ lb·in

.20 $F_{o3x} = F_{32x} = 0$, Solve three equilibrium equations for link 2.

(1) $200\cos(225^\circ) + F_{12x} = 0$, (2) $200\sin(225^\circ) + F_{12y} + F_{32y} = 0$,

(3) $F_{32y}\, 90\cos(-24^\circ) - 200\cos(225^\circ)\, 35\sin(-24^\circ) + 200\sin(225^\circ)\, 35\cos(-24^\circ) = 0$,

$F_{32y} = F_{03y} = 79.47$ N, $F_{12x} = F_{o1x} = 141.4$ N, $F_{12y} = F_{o1y} = 61.93$ N,

$T_1 - F_{21x}(30)\sin(60^\circ) + F_{21y}(30)\cos(60^\circ) = 0$; $T_1 = -2745$ N·mm

.21 Problem I - neglect $\overline{P}$, $F_{o3x} = F_{32x} = 0$, Solve three equilibrium equations

for link 2. (1) $-100 + F_{12x} = 0$, (2) $F_{12y} + F_{32y} = 0$, (3) $F_{32y}(5)\cos(-35^\circ)$

$+ 100(-4)\sin(-35^\circ) = 0$, $F_{32y} = F_{o3y} = -56.01$ lbs, $F_{12x} = F_{o1x} = 100$ lbs,

$F_{12y} = F_{o1y} = 56.01$ lbs, $T_1 - F_{21x}(3)\sin(75^\circ) + F_{21y}(3)\cos(75^\circ) = 0$;

$T_1 = -246.3$ lb.ft, Problem II - neglect $\overline{F}$, $F_{o3x} = 0$, $F_{23x} = F_{12x} = F_{o1x}$

$= 200$ lbs, $F_{23y} = -F_{03y} = F_{12y} = F_{o1y} = F_{23x}\tan(-35^\circ) = -140.0$ lbs,

$T_1 - F_{21x}(3)\sin(75^\circ) + F_{21y}(3)\cos(75^\circ) = 0$; $T_1 = -688.3$ lb.ft,

Superposition, $F_{o1x} = F_{12x} = 300$ lbs, $F_{o3x} = 0$, $F_{23x} = 200$ lbs,

$F_{o1y} = F_{12y} = F_{23y} = -F_{o3y} = -83.99$ lbs, $T_1 = -934.6$ lb·ft

9.22 $\ell_1 = 40$ mm, $\ell_2 = 60$ mm, $\ell_3 = 50$ mm, $\phi_1 = 60°$, $\phi_2 = 13°$, $\phi_3 = 103°$,

$F_2 = 50$ N @ 180°, $r_2 = 111$ mm, $F_3 = 75$ N @ 45°, $r_3 = 25$ mm,

$F_1 = r_1 = C_1 = C_2 = C_3 = 0$; $a_{11} = -48.72$, $a_{12} = -11.25$, $a_{21} = 13.50$,

$a_{22} = -58.46$, $b_1 = 1590.1$, $b_2 = -1248.4$, $F_{23} = 38.0$ N @ 160°,

$F_{o3} = 68.4$ N @ -105°, $F_{12} = 19.4$ N @ 42.5°, $F_{01} = 19.4$ N @ 42.5°,

$T_1 = -234$ N·mm

9.23 $\ell_1 = 1$ in, $\ell_2 = 4$ in, $\ell_3 = 2$ in, $\phi_1 = 120°$, $\phi_2 = 24°$, $\phi_3 = 86°$,

$F_2 = 20$ lbs @ 135°, $r_2 = 2$, $F_3 = 10$ lbs @ 180°, $r_3 = 1$,

$F_1 = r_1 = C_1 = C_2 = C_3 = 0$; $a_{11} = -1.995$, $a_{12} = 0.140$, $a_{21} = 1.627$,

$a_{22} = -3.654$, $b_1 = -9.976$, $b_2 = -37.34$, $F_{23} = 14.1$ lbs @ 65°,

$F_{o3} = 13.5$ lbs @ -72°, $F_{12} = 20.1$ lbs @ -3.7°, $F_{01} = 20.1$ lbs @ -3.7°,

$T_1 = -16.71$ lb·in

9.24 $\ell_1 = 0.08$ m, $\ell_2 = 0.16$ m, $\ell_3 = 0.10$ m, $\phi_1 = 45°$, $\phi_2 = -57°$,

$\phi_3 = -127°$, $C_2 = 10$ N·m, $C_3 = -7$ N·m, $F_1 = F_2 = F_3 = r_1 = r_2 = r_3 = C_1 = 0$,

$a_{11} = 0.0799$, $a_{12} = -0.0602$, $a_{21} = -0.1342$, $a_{22} = -0.0871$, $b_1 = 7.00$,

$b_2 = -10.00$, $F_{o3} = 81.1$ N @ 173°, $F_{23} = F_{12} = F_{o1} = 81.1$ N @ -6.6°,

$T_1 = -5.09$ N·m

9.25 Neglecting the hydraulic cylinder, use the method of section 9.4.3 to find the torque on link 1. $\ell_1 = 9$ in, $\ell_2 = 15$ in, $\ell_3 = 12$ in, $\phi_1 = 60^\circ$, $\phi_2 = 355^\circ$, $\phi_3 = 51^\circ$, $C_3 = 1000$ lb in, $F_1 = F_2 = F_3 = r_1 = r_2 = r_3 = C_1 = C_2 = 0$; $a_{11} = -9.33$, $a_{12} = 7.55$, $a_{21} = -1.31$, $a_{22} = -14.94$, $b_1 = -1000.$, $b_2 = 0.0$; $F_{o3} = 100.5$ lbs @ 175°, $F_{23} = F_{12} = 100.5$ lbs @ -5.0°, $T_1 = -819.9$ lb·in, Neglecting links 2 and 3, set the summation of moments about point 0_1 equal to zero to find F_p. $\Sigma M_{o1} = 0 = 819.9 - F_p \cos(58^\circ)$ $6 \sin(120^\circ) + F_p \sin(58^\circ) 6 \cos(120^\circ)$, $F_p = 154.8$ lbs, $A_p = \pi(0.75)^2 = 1.767$ in^2, $P = 154.8 \div 1.767 = 87.6$ psi

9.26 Neglecting the spring, use the method of section 9-4-3 to find the torque on link 1. $\ell_1 = 0.3$ m, $\ell_2 = 0.7$ m, $\ell_3 = 0.6$ m, $\phi_1 = 60^\circ$, $\phi_2 = 16^\circ$, $\phi_3 = 68^\circ$, $F_1 = 9.806(7) = 68.64$ N @ 270°, $r_1 = 0.15$ m, $C_1 = 0$, $F_2 = 9.806(15) = 147.09$ N @ 270°, $r_2 = 0.35$ m, $C_2 = 0$, $F_3 = 9.806(12) = 117.67$ N @ 270°, $r_3 = 0.3$ m, $C_3 = 0$, $a_{11} = -0.5563$, $a_{12} = 0.2248$, $a_{21} = 0.1929$, $a_{22} = -0.6729$, $b_1 = 13.22$, $b_2 = 49.49$, $F_{23} = 109.2$ N @ 123°, $F_{03} = 217.2$ N @ 74°, $F_{12} = 82.6$ N @ 137°,

$T_1 = 29.29$ N·m, Neglecting links 2 and 3, set the summation of moments about point 0_1 equal to zero to find the spring force. $\Sigma M_{o1} = 0 = -29.29 + F_s(0.2)$ $\sin(60^\circ)$; $F_s = 169.1$ N

9.27 $F_{o5x} = 0$, $F_{45x} = 1000$ lbs, $F_{45y} = -F_{o5y} = F_{45x}\tan(-30^\circ) = -577.4$ lbs, $F_{43} = 1154.7$ lbs @ 150°, Use the method of section 9-4-3. $\ell_1 = 2$ in, $\ell_2 = 7$ in, $\ell_3 = 4$ in, $\phi_1 = 45^\circ$, $\phi_2 = 17^\circ$, $\phi_3 = 56^\circ$, $F_3 = 1154.7$ @ 150°, $r_3 = 6$ in, $F_1 = r_1 = F_2 = r_2 = C_1 = C_2 = C_3 = 0$; $a_{11} = -3.316$, $a_{12} = 2.237$, $a_{21} = 2.047$, $a_{22} = -6.694$, $b_1 = -6911.3$, $b_2 = 0$, $F_{23} = 2745.6$ lbs @ 17.0°, $F_{o3} = 2132.4$ lbs @ -139.7°, $F_{12} = 2745.6$ lbs @ 17.0°, $T_1 = -2577.9$ lb·in, $F_{o1} = F_{12}$

9.28 $F_{o5x} = 0$, $F_{35x} = F_{13x} = -800$ lbs, $F_{35y} = F_{13y} = -F_{o5y} = F_{35x}\tan(12^\circ)$ $= -170.0$ lbs. Solve three equilibrium equations for link 1, with the line of action of F_{41} perpendicular to link 1. (1) $F_{o1x} + F_{41}\cos(157^\circ) + F_{31x} = 0$, (2) $F_{o1y} + F_{41}\sin(157^\circ) + F_{31y} = 0$, (3) $-F_{41}\cos(157^\circ)5\sin(67^\circ) + F_{41}\sin(157^\circ)$ $5\cos(67^\circ) - F_{31x}(8)\sin(67^\circ) + F_{31y}(8)\cos(67^\circ) = 0$, $F_{41} = 1071.9$ lbs, $F_{41x} = F_{24x} = F_{o2x} = -986.7$ lbs, $F_{41y} = F_{24y} = F_{o2y} = 418.8$ lbs, $F_{o1x} = 186.7$ lbs, $F_{o1y} = -588.8$ lbs, $T_2 - F_{42x}(2)\sin(15^\circ) + F_{42y}(2)\cos(15^\circ)$ $= 0$, $T_2 = 1320$ lb·in

9.29 $\Sigma M = 0 = -600 - F_{45}\cos(48^\circ)1\sin(-27^\circ) + F_{45}\sin(48^\circ)1\cos(-27^\circ)$, $F_{45} = 621.2$ lbs, $F_{45x} = -F_{o5x} = F_{24x} = 415.6$ lbs, $F_{45y} = -F_{o5y} = F_{24x} = 461.6$ lbs, Use the method of section 9-4-3. $\ell_1 = 3$ in, $\ell_2 = 2.25$ in, $\ell_3 = 2$ in, $\phi_1 = 75^\circ$, $\phi_2 = 203^\circ$, $\phi_3 = 98^\circ$, $F_2 = 621.2$ lbs @ 228°, $r_2 = 3.25$ in,

$F_1 = r_1 = F_3 = r_3 = C_1 = C_2 = C_3 = 0$, $a_{11} = -\ 1.981$, $a_{12} = -0.2[illegible]9$,

$a_{21} = -0.879$, $a_{22} = 2.071$, $b_1 = 0$, $b_2 = -853.2$, $F_{23} = 392.6$ lbs @ -82.0°,

$F_{o3} = 392.6$ lbs @ 98.0°, $F_{12} = 475.9$ lbs @ 8.8°, $F_{01} = 475.9$ lbs @ 8.8°,

$T_1 = -1306.2$ lb• in

9.30 $F_{o5y} = 0$, $F_{35y} = 500N$, Solve three equilibrium equations for link 3.

(1) $F_{53x} + F_{23}\cos(56^\circ) + F_{43}\cos(130^\circ) = 0$, (2) $F_{53y} + F_{23}\sin(56^\circ) + F_{43}\sin(130^\circ) = 0$, (3) $-F_{23}\cos(56^\circ)(110)\sin(-79^\circ) + F_{23}\sin(56^\circ)(110)\cos(-79^\circ) - F_{43}\cos(130^\circ)(145)\sin(-79^\circ) + F_{43}\sin(130^\circ)(145)\cos(-79^\circ) = 0$, $F_{53x} = F_{o5x} = 45.4$ N,

$F_{23} = 298.2$ N, $F_{23x} = F_{12x} = F_{o1x} = 166.8$ N, $F_{23y} = F_{12y} = F_{o1y} = 247.2$ N,

$F_{43} = 330.0$ N, $F_{43x} = F_{o4x} = -212.1$ N, $F_{43y} = F_{o4y} = 252.8$ N,

$T_1 - F_{21x}(35)\sin(110^\circ) + F_{21y}(35)\cos(110^\circ) = 0$, $T_1 = -8445$ N•mm

9.31 $T_1 - F_{21}\cos\phi_2\ell_1\sin\phi_1 + F_{21}\sin\phi_2\ell_1\cos\phi_1 = 0$, $F_{21} = -F_{o1} = -F_{23}$,

$F_{o3}\cos\phi_3 + F_{23}\cos\phi_2 + F_{43}\cos\phi_4 = 0$, $F_{o3}\sin\phi_3 + F_{23}\sin\phi_2 + F_{43}\cos\phi_4 = 0$,

$F_{45} = -F_{43}$, $F_{o5x} = 0$, $F_{o5y} = -F_{45}\sin\phi_4$, $F_s = -F_{45}\cos\phi_4$, A) $\ell_1 = 1$ in,

$T_1 = 100$ lb in, $\phi_1 = 0^\circ$, $\phi_2 = -105^\circ$, $\phi_3 = 159^\circ$, $\phi_4 = 21^\circ$,

$F_{21} = -F_{o1} = -F_{23} = -103.6$ lbs @ -105°, $F_{o3} = 125.2$ lbs @ 159°,

$F_{43} = -F_{45} = 154.0$ lbs @ 21°, $F_{o5y} = 55.2$ lbs, $F_s = 143.8$ lbs,

B) $\ell_1 = 1$ in, $T_1 = 100$ lb in, $\phi_1 =$ [illegible], $\phi_2 = -105^\circ$, $\phi_3 = 168^\circ$,

$\phi_4 = 12^\circ$, $F_{21} = -F_{o1} = -F_{23} = -103.6$ lbs @ -105°, $F_{o3} = 226.9$ lbs @ 168°,

$F_{43} = -F_{45} = 254.4$ lbs @ 12°, $F_{o5y} = 52.9$ lbs, $F_s = 248.8$ lbs,

c) $\ell_1 = 1$ in, $T_1 = 100$ lb·in, $\phi_1 = -60^\circ$, $\phi_2 = -100^\circ$, $\phi_3 = 177^\circ$,

$\phi_4 = 3^\circ$, $F_{21} = -F_{o1} = -F_{23} = -155.6$ lbs @ -100°, $F_{o3} = 1450.4$ lbs @ 177°,

$F_{43} = -F_{45} = 1477.5$ lbs @ 3°, $F_{o5y} = 77.3$ lbs, $F_s = 1475.5$ lbs

9.32 Use equation (9-14) twice. $r = 0.05$ m, $\ell = 0.21$ m, Piston force $\bar{F}_4$:

$\phi = 25^\circ$, $P = 6000$ N, $T_1^{\,I} = -154.3$ N·m, Piston force $\bar{F}_5$: $\phi = 205^\circ$,

$P = 2000$ N, $T_1^{\,II} = 33.1$ N·m, $T_1 = -154.3 + 33.1 = -121.2$ N·m

9.33 Use equation (9-14) twice. $r = 0.06$ m, $\ell = 0.2$ m, Piston force $\bar{F}_4$:

$\phi = 245^\circ$, $P = 3000$ N, $T_1^{\,I} = 141.6$ N·m, Piston force $\bar{F}_5$: $\phi = 335^\circ$,

$P = 7000$ N, $T_1^{\,II} = 226.2$ N·m, $T_1 = 141.6 + 226.2 = 367.8$ N·m

9.34 Solve six equilibrium equations for links 2 and 3. (1) $F_{12x} + F_{32x} = 0$,

(2) $F_{12y} + F_{32y} - 100 = 0$, (3) $F_{32x}(1.2) - 100(1.5) = 0$, (4) $F_{43x} + F_{23x} = 0$, (5) $F_{43y} + F_{23y} + P = 0$, (6) $-F_{23x}1.3\sin(-6^\circ) + F_{23y}1.3\cos(-6^\circ) - P\left[4 - 1.3\cos(-6^\circ)\right] = 0$, From (3), (1) and (4): $F_{32x} = 125$ lbs, $F_{12x} = -125$ lbs, $F_{43x} = 125$ lbs, $F_{43y} = F_{43x}\tan(-22^\circ) = -50.5$ lbs,

Simultaneously solve (2),(5) and (6): $F_{32y} = -38.4$ lbs, $F_{12y} = 138.4$ lbs,

$P = 12.1$ lbs

9.35 For crank speed $\dot{\phi} = 1$ rad/s, $\dot{x} = 40\sin\phi\left[1 + \dfrac{40\cos\phi}{\sqrt{10{,}000 - 1600\sin^2\phi}}\right]$,

where positive velocity is to the left. Then, $P\dot{x} + T_1\dot{\phi} = 0$, from which $T_1 = -P\dot{x}/\dot{\phi}$

9.36 The piston velocity is given by the following expression:

$$\dot{x} = \dot{\phi}\left[r\sin\phi + \frac{(r\sin\phi - e)r\cos\phi}{\sqrt{\ell^2 - (r\sin\phi - e)^2}}\right],$$ where $r = OB$, $\ell = BC$, e is the offset,

and positive velocity is to the left. Then, $P\dot{x} + T_1\dot{\phi} = 0$, from which $T_1 = -P\dot{x}/\dot{\phi}$

9.37 From a velocity analysis with $\dot{\phi} = 1$ rad/s, the velocity of the point of application of force $\overline{F}_2$ is $(-24.9\,\overline{i} + 5.61\,\overline{j})$mm/s. In vector form, force $\overline{F}_2 = (-141\,\overline{i} - 141\,\overline{j})$N. Then, $T_1(1\text{rad/s}) + (-141\,\overline{i} - 141\,\overline{j})\cdot(-24.9\,\overline{i} + 5.61\,\overline{j}) = 0$, from which $T_1 = -2720$ N·mm

9.38 From a velocity analysis with $\dot{\phi} = 1$ rad/s, $\overline{v}_C = -3.45\,\overline{i}$ ft/s and $\overline{v}_D = (-2.50\,\overline{i} - 0.09\,\overline{j})$ft/s. Then, $T_1(1\text{ rad/s}) + (-200\,\overline{i})\cdot(-3.45\,\overline{i}) + (-100\,\overline{i})\cdot(-2.50\,\overline{i} - 0.09\,\overline{j}) = 0$, from which $T_1 = -940$ ft·lb

9.39 For an arbitrarily assigned input angular velocity of $\omega_1 = 1$ rad/s, the velocity at force $\overline{F}_2$ is $(-23.2\,\overline{i} + 5.86\,\overline{j})$ mm/s and the velocity at force $\overline{F}_3$ is $(-14.2\,\overline{i} - 3.29\,\overline{j})$ mm/s. Then, $T_1(1\text{ rad/s}) + (-50\,\overline{i})\cdot(-23.2\,\overline{i} + 5.86\,\overline{j}) + (53\,\overline{i} + 53\,\overline{j})\cdot(-14.2\,\overline{i} - 3.29\,\overline{j}) = 0$, from which $T_1 = -233$ N·mm

9.40 For an arbitrarily assigned input angular velocity of $\omega_1 = 1$ rad/s, the velocity at force $\overline{F}_2$ is $(-0.995\,\overline{i} - 0.211\,\overline{j})$ in/s and the velocity at force $\overline{F}_3$ is $(-0.562\,\overline{i} + 0.039\,\overline{j})$ in/s. Then, $T_1(1\text{ rad/s}) + (-14.1\,\overline{i} + 14.1\,\overline{j})\cdot(-0.995\,\overline{i} - 0.211\,\overline{j}) + (-10\,\overline{i})\cdot(-0.562\,\overline{i} + 0.039\,\overline{j}) = 0$, from which $T_1 =$ [illegible] lb·in

9.41 For an input angular velocity of $\omega_1 = 1$ rad/s, $\omega_2 = -0.074$ rad/s and $\omega_3 = -0.833$ rad/s. Then, $T_1(1) + T_2(-0.074) + T_3(-0.833) = 0$, or $T_1 + (10)(-0.074)+(-7)(-0.833) = 0$, from which $T_1 = -5.09$ N·m

9.42 For an input angular velocity of $\omega_1 = 1$ rad/s, $\omega_3 = 0.820$ rad/s, and $\overline{V}_A = (-5.2\ \overline{i} - 3\ \overline{j})$ in/s. Then, $(F_p \cos 54^\circ\ \overline{i} + F_p \sin 54^\circ\ \overline{j}) \cdot \overline{V}_A + T_3 \omega_3 = 0$, from which $F_p = 150$ lb and $P = 85$ psi

9.43 For an input angular velocity of $\omega_1 = 1$ rad/s, the velocity of the slider is 2.57 in/s to the left. Then, T_1 (1 rad/s) + 1000(2.57) = 0, from which $T_1 = -2570$ in·lb

9.44 For an input angular velocity of $\omega_2 = 1$ rad/s, the velocity of the slider is 1.64 in/s to the left. Then, T_2 (1 rad/s) - 800(1.64) = 0, from which $T_2 = 1312$ in·lb

9.45 For a crank velocity of $\dot{\phi} = 1$ rad/s, the piston velocities are $v_C = 25.7$ mm/s in the same sense as $\overline{F}_4$ and $v_E = 16.5$ mm/s opposite to $\overline{F}_5$. Then, $T_1(1) + 6000(25.7) - 2000(16.5) = 0$, from which $T_1 = -121$ N·m

9.46 For a crank velocity of $\dot{\phi} = 1$ rad/s, the piston velocities are $v_C = 47.2$ mm/s and $v_D = 32.3$ mm/s, both opposite in sense to the applied piston forces. Then, $T_1(1) - 3000(47.2) - 7000(32.3) = 0$, from which $T_1 =$ 368 N·m

9.48 $\phi = 45^\circ$, $r = 40$ mm, $\ell = 100$ mm, $P = 500$ N, $r_{o1} = 20$ mm, $r_{12} = r_{23} = 15$ mm, $\mu_{o1} = 0.1$, $\mu_{12} = \mu_{23} = 0.15$, $\mu_{o3} = 0.2$, sign $(\dot{s}) = +1$, sign $(\omega_3 - \omega_2) = -1$, sign $(\omega_1 - \omega_2) = -1$, sign $(\omega_o - \omega_1) = +1$,

$\theta = 16.4^{\circ}$, $k = -4.5$, $\psi = \arcsin(\frac{4.5}{100}) - 16.4^{\circ} = 161^{\circ}$,

$F_{32} = -500/(\cos 161^{\circ} + 0.2\sin 161^{\circ}) = 568$ N, $F_{21x} = F_{32}\cos\psi = -537$ N,

$F_{21y} = F_{32}\sin\psi = 185$ N, $T_1 = -0.1(20)(568) - 0.15(15)(568) - 185(40)\cos 45^{\circ}$

$- 537(40)\sin 45^{\circ} = -22.8$ N·m, $e = 18.3/22.8 = 0.803$

9.50 $\phi = 45^{\circ}$, $r = 2$ in, $\ell = 4$ in, $e = 0.5$ in, $P = 100$ lb, $r_{o1} = r_{12} = r_{23}$

$= 1$ in, $\mu_{01} = 0.15$, $\mu_{12} = 0.25$, $\mu_{23} = 0.1$, $\mu_{o3} = 0.2$, $\text{sign}(\dot{s}) = +1$,

$\text{sign}(\omega_3 - \omega_2) = -1$, $\text{sign}(\omega_1 - \omega_2) = -1$, $\text{sign}(\omega_o - \omega_1) = +1$, $\theta = 13.2^{\circ}$,

$k = -0.35$, $\psi = \arcsin(\frac{0.35}{4}) - 13.2^{\circ} = 161.8^{\circ}$, $F_{32} = -100/(\cos 161.8^{\circ} + 0.2$

$\sin 161.8^{\circ}) = 113$ lb, $F_{21x} = F_{32}\cos\psi = -107$ lb, $F_{21y} = F_{32}\sin\psi = 35.3$ lb,

$T_1 = -0.15(1)(113) - 0.25(1)(113) - 35.3(2)\cos 45^{\circ} - 107(2)\sin 45^{\circ} = -246$ lb·in,

$e = 175/246 = 0.711$

9.55 Summing moments on the follower about point 0_2, $T_2 - F_{12}\, r_2 \sin 37^{\circ} = 0$,

$F_{12} = 50/(3.5\sin 37^{\circ}) = 23.8$ lb, Summing moments on the cam about point 0_1,

$T_1 = -F_{21}\, 0_1C\sin 37^{\circ} = -(23.8)(0.75)\sin 37^{\circ} = -10.7$ lb·in

9.56 Summing moments on the follower about point 0_2, $T_2 - F_{12}r_2\sin 37^{\circ} + \mu F_{12}$

$(r_2 - r_2\cos 37^{\circ}) = 0$, $50 - F_{12}3.5\sin 37^{\circ} + 0.15F_{12}(3.5 - 3.5\cos 37^{\circ}) = 0$,

$F_{12} = 25.1$ lb, Summing moments on the cam about point 0_1,

$T_1 = -F_{21}0_1C\sin 37^{\circ} - \mu F_{21}(r_1 + 0_1C\cos 37^{\circ}) = -25.1(0.75)\sin 37^{\circ} - 0.15(25.1)$

$(1.5 + 0.75\cos 37^{\circ}) = -19.2$ lb·in

9.57 Summing moments on the cam about point 0_1, $T_1 = F_{21}0_1C \cos 45^\circ = F_2 0_1 C \cos 45^\circ$ $= 20\ (1) \cos 45^\circ = 14.1$ lb·in

9.58 Summing vertical forces on the follower, $F_{o2} = \mu F_{12}$, Summing horizontal forces on the follower, $F_{12} - F_2 - \mu F_{o2} = 0$, or $F_{12} - F_2 - \mu^2 F_{12} = 0$, Then, $F_{12} = F_2/(1 - \mu^2) = 20/(1 - 0.15^2) = 20.5$ lb, Summing moments on the cam about point 0_1, $T_1 = F_{21}0_1C \cos 45^\circ + \mu F_{21}(r - 0_1C \sin 45^\circ) = 20.5(1) \cos 45^\circ$ $+ 0.15(20.5)(2 - \sin 45^\circ) = 18.5$ lb·in

CHAPTER 10

DYNAMIC-FORCE ANALYSIS

10.1 From a kinematic analysis, $\bar{a}_{G_2} = 319\underline{/206^\circ}$ m/s^2, $\alpha_2 = 2690$ rad/s^2 CCW

Therefore, $\bar{F}_2 = -m_2\bar{a}_{G2} = -0.2(319)\underline{/206^\circ} = 63.8\ \underline{/26^\circ}$ N,

$h_2 = \dfrac{I_{G2}\alpha_2}{F_2} = \dfrac{.3(2690)}{63.8} = 12.6$ mm, $\bar{F}_2$ must be positioned so as to produce a CW moment about G_2

10.2

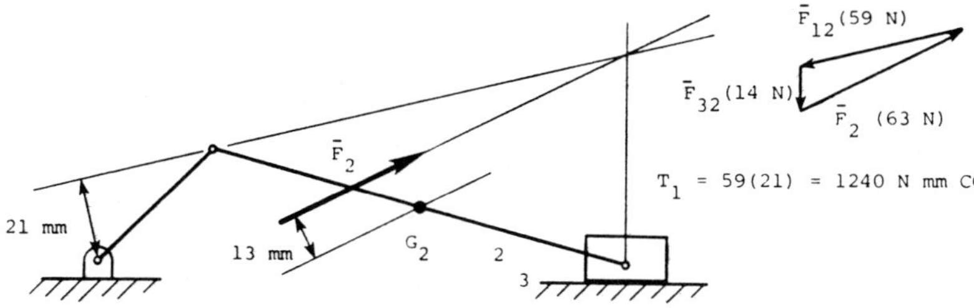

$T_1 = 59(21) = 1240$ N mm CCW

10.3 From a kinematic analysis, $\bar{a}_{G3} = 290\ \underline{/180^\circ}$ m/s^2, Therefore, $\bar{F}_3 = -m_3\bar{a}_{G3} = -0.3(290)\ \underline{/180^\circ} = 87.0\ \underline{/0^\circ}$ N, $T_1 = 87(40)\sin 45^\circ \left[1 + \dfrac{0.4\cos 45^\circ}{\sqrt{1-(0.4\sin 45^\circ)^2}}\right] =$

3190 N·mm CCW, Total torque $T_1 = 1240 + 3190 = 4430$ N·mm CCW

10.4 From a kinematic analysis, $\bar{a}_{G2} = 45.4\ \underline{/234^\circ}$ ft/s^2, $\alpha_2 = 27.1$ rad/s^2 CCW,

$\bar{F}_2 = -m_2\bar{a}_{G2} = -\left(\dfrac{1000}{32.2}\right) 45.4\ \underline{/234^\circ} = 1410\ \underline{/54^\circ}$ lb., $h_2 = \dfrac{I_{G2}\alpha_2}{F_2} = \dfrac{70(27.1)}{1410} =$

1.34 ft.

10.5 From a kinematic analysis, $\bar{a}_{G2} = 167 \underline{/227^\circ}$ m/s^2, $\alpha_2 = 730$ rad/s^2 CCW, $\bar{a}_{G3} = 78.5 \underline{/216^\circ}$ m/s^2, $\alpha_3 = 2900$ rad/s^2 CCW, $\bar{F}_2 = -m_2\bar{a}_{G2} = -0.3\ (167) \underline{/227^\circ} = 50 \underline{/47^\circ}$ N, $h_2 = \dfrac{I_{G2}\alpha_2}{F_2} = \dfrac{(1.0)(730)}{50} = 15$ mm,

$\bar{F}_3 = -m_3\bar{a}_{G3} = -0.2(78.5)\underline{/216^\circ} = 16 \underline{/36^\circ}$ N

$h_3 = \dfrac{I_{G3}\alpha_3}{F_3} = \dfrac{0.15(2900)}{16} = 27$ mm

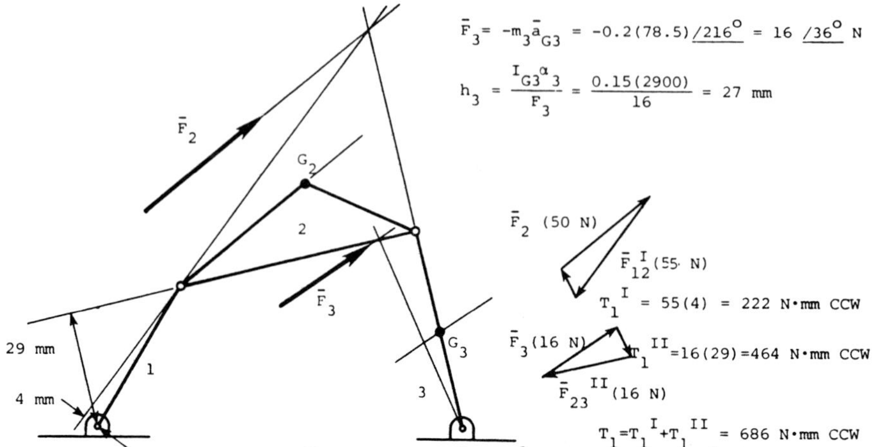

10.6 From a kinematic analysis, $\bar{a}_{G2} = 2170 \underline{/180^\circ}$ in/s^2, $\alpha_2 = 415$ rad/s^2 CW, $\bar{a}_{G3} = 5420 \underline{/180^\circ}$ in/s^2, $\bar{F}_{i2} = -m_2\bar{a}_{G2} = -(\frac{2}{386.4})\ 2170 \underline{/180^\circ} = 11.2 \underline{/0^\circ}$ lb., $C_{i2} = -I_{G2}\alpha_2 = -(0.04)(-415) = 16.6$ in lb, $\bar{F}_{i3} = -m_3\bar{a}_{G3} = -(\frac{1}{386.4})5420 \underline{/180^\circ} = 14.0 \underline{/0^\circ}$ lb, A) Summing horizontal forces and moments about point B for member 2, $F_{12x} + F_{i2} = 0$, $-F_1\ 10 \cos 50^\circ - F_{12x}\ 10 \sin 50^\circ + C_{i2} - F_{i2}\ 6 \sin 50^\circ = 0$, Then $F_1 = \dfrac{C_{i2} + 4F_{i2} \sin 50^\circ}{10 \cos 50^\circ} = \dfrac{16.6 + 4(11.2)(0.766)}{10(0.643)} = 7.92$ lb, B) Assuming that member 2 is massless, and summing forces on slide 3, $F_{23x} = -F_{i3}$, $F_{23y} = -F_{23x} \tan 50^\circ = F_{i3} \tan 50^\circ$, Then $F_1 = -F_{32y} = F_{i3} \tan 50^\circ = 14.0(1.19) = 16.7$ lb, Therefore, the total force required is

$F_1 = 7.92 + 16.7 = 24.6$ lb.

10.7 From the acceleration polygon, $\bar{a}_{G2} = -2140\,\bar{i} - 2140\,\bar{j}$ in/s^2, $\alpha_2 = 1080$ rad/s^2 CCW, $\bar{a}_{G3} = -1780\,\bar{i}$ in/s^2, $-m_2 a_{G2x} = -m_2 a_{G2y} = -\left(\frac{2}{386.4}\right)(-2140) = 11.1$ lb, $-I_{G2}\alpha_2 = -0.009(1080) = -9.72$ in$\cdot$lb., $-m_3 a_{G3} = -\left(\frac{1.5}{386.4}\right)(-1780) = 6.91$ lb, We will utilize Eqs. (10-8), with angle θ determined as follows: $\sin\theta = \frac{2\sin 60^\circ - 0.5}{4} = 0.308$, $\theta = 18^\circ$, Then, $F_{32x} = 6.91$ lb,

$$F_{32y} = \frac{-6.91(4)\sin 18^\circ - 11.1(2)\sin 18^\circ - 11.1(2)\cos 18^\circ + 9.72}{4\cos 18^\circ} = -7.04 \text{ lb.}$$

$F_{21x} = 6.91 + 11.1 = 18.0$ lb, $F_{21y} = -7.04 + 11.1 = 4.06$ lb,

$T_1 = 18.0(2)\sin 60^\circ - 4.06\,(2)\cos 60^\circ = 27.1$ in lb CCW

10.8 From the given information, $C_1 = -I_{G1}\alpha_1 = -4.0(-200) = 800$ N$\cdot$mm,

$\bar{F}_3 = -m_3\bar{a}_{G3} = -8(5.8)\,\underline{/0^\circ} = 46.4\,\underline{/180^\circ}$ N, $\bar{F}_2 = -m_2\bar{a}_{G2} = -15(3.81)\,\underline{/325^\circ} = 57.2\,\underline{/145^\circ}$ N, $h_2 = \frac{I_{G2}\alpha_2}{F_2} = \frac{7.5(66.3)}{57.2} = 8.7$ mm

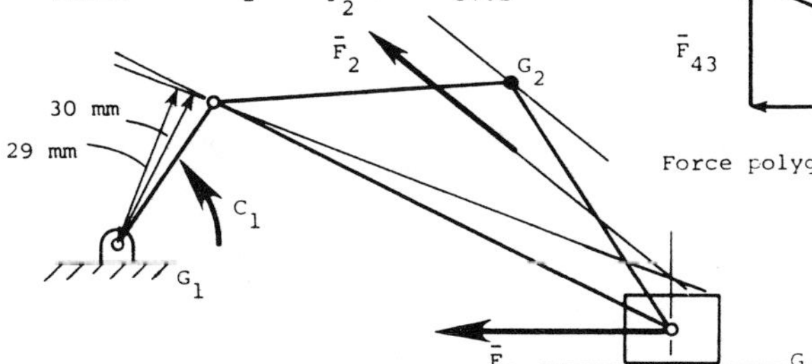

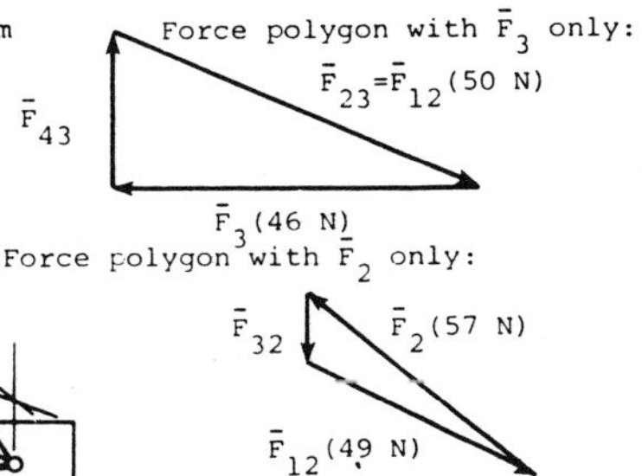

Combining, $T_1 = -800 - 50(30) - 49(29) = -3720$ N$\cdot$mm

10.9 $F_{i2} = -m_2 a_{G2} = -\left(\frac{1}{386.4}\right)$ 34,500 $\underline{/298^\circ} = 107\,\underline{/118^\circ}$ lb, $C_{i2} = -I_{G2}\alpha_2 = -0.03(2670) = -80.1$ lb in, $\bar{F}_{i3} = -m_3\bar{a}_{G3} = -\left(\frac{1.0}{386.4}\right)$ 14,600 $\underline{/294^\circ} = 37.8\,\underline{/114^\circ}$ lb,

$C_{i3} = -I_{G3}\alpha_3 = -0.002(-6940) = 13.9$ lb in, B) $\ell_1 = 1$ in, $\ell_2 = 4$ in, $\ell_3 = 2$ in,

$\phi_1 = 120^\circ$, $\phi_2 = 24^\circ$, $\phi_3 = 86^\circ$, $r_1 = 0$, $r_2 = 2$ in, $r_3 = 1$ in,

$F_{1x} = F_{1y} = C_1 = 0$, $F_{2x} = 107 \cos 118^\circ = -50.2$ lb, $F_{2y} = 107 \sin 118^\circ$

$= 94.5$ lb, $C_2 = -80.1$ lb in, $F_{3x} = 37.8 \cos 114^\circ = -15.4$ lb,

$F_{3y} = 37.8 \sin 114^\circ = 34.5$ lb, $C_3 = 13.9$ lb in, $\overline{F}_{23} = 48.8\ \underline{/67^\circ}$ lb,

$\overline{F}_{o3} = 79.6\ \underline{/267^\circ}$ lb, $\overline{F}_{01} = 85.1\ \underline{/324^\circ}$ lb, $T_1 = -35.2$ in lb

10.10 From the acceleration polygon, $\overline{a}_{G2} = 100\ \underline{/353^\circ}$ m/s^2, $\overline{a}_{G3} = 150\ \underline{/20^\circ}$ m/s^2,

$\alpha_2 = \frac{578,000}{160} = 3610$ rad/s^2 CCW, $\alpha_3 = \frac{168,000}{100} = 1680$ rad/s^2 CCW,

$\overline{F}_{i2} = -m_2\overline{a}_{G2} = -(1.0)100\ \underline{/353^\circ} = 100\ \underline{/173^\circ}$ N, $C_{i2} = -I_{G2}\alpha_2 = -(0.7)(3610)$

$= -2530$ N·mm, $\overline{F}_{i3} = -m_3\overline{a}_{G3} = -(0.6)150\ \underline{/20^\circ} = 90\ \underline{/200^\circ}$ N,

$C_{i3} = -I_{G3}\alpha_3 = -(0.5)(1680) = -840$ N·mm, B) $\ell_1 = 80$ mm, $\ell_2 = 160$ mm,

$\ell_3 = 100$ mm, $\phi_1 = 45^\circ$, $\phi_2 = 302^\circ$, $\phi_3 = 233^\circ$, $r_1 = 0$, $r_2 = 100$ mm,

$r_3 = 50$ mm, $F_{1x} = F_{1y} = C_1 = 0$, $F_{2x} = 100 \cos 173^\circ = -99.2$ N,

$F_{2y} = 100 \sin 173^\circ = 12.2$ N, $C_2 = -2530$ N·mm, $F_{3x} = 90 \cos 200^\circ = -84.6$ N,

$F_{3y} = 90 \sin 200^\circ = -30.8$ N, $C_3 = -840$ N·mm, $\overline{F}_{23} = 88.0\ \underline{/255^\circ}$ N,

$\overline{F}_{03} = 158\ \underline{/47^\circ}$ N, $\overline{F}_{12} = \overline{F}_{01} = 124\ \underline{/308^\circ}$ N, $T = -9820$ N·mm

10.11 For clockwise input rotation, the forces and torques are the same as for Problem 10.10. If the input speed is doubled, the forces and torques will be quadrupled. These results are due to the fact that the accelerations, and therefore the inertia loads, are proportional to the square of the input angular velocity.

10.12 $\overline{F}_{i1} = -m_1\overline{a}_{G1} = -(\frac{3}{386.4})\ 636\ \underline{/195^{\circ}} = 4.94\ \underline{/15^{\circ}}$ lb, $C_{i1} = -I_{G1}\alpha_1 = -(0.1)(100)$

$= -10.0$ in· lb, $\overline{F}_{i2} = -m_2\overline{a}_{G2} = -(\frac{5}{386.4})\ 1240\ \underline{/189^{\circ}} = 16.0\ \underline{/9^{\circ}}$ lb,

$C_{i2} = -I_{G2}\alpha_2 = -(0.3)(19) = -5.70$ in· lb, $\overline{F}_{i3} = -m_3\overline{a}_{G3} = -(\frac{4}{386.4})610\ \underline{/182^{\circ}}$

$= 6.31\ \underline{/2^{\circ}}$ lb, $C_{i3} = -I_{G3}\alpha_3 = -(0.3)(77) = -23.1$ in· lb, B) $\ell_1 = 9$ in,

$\ell_2 = 15$ in, $\ell_3 = 12$ in, $\phi_1 = 60^{\circ}$, $\phi_2 = 355^{\circ}$, $\phi_3 = 51^{\circ}$, $r_1 = 4.5$ in,

$r_2 = 7.5$ in, $r_3 = 6$ in, $F_{1x} = 4.94 \cos 15^{\circ} = 4.77$ lb,

$F_{1y} = 4.94 \sin 15^{\circ} = 1.28$ lb, $C_1 = -10.0$ in· lb, $F_{2x} = 16.0 \cos 9^{\circ} = 15.8$ lb,

$F_{2y} = 16.0 \sin 9^{\circ} = 2.50$ lb, $C_2 = -5.70$ in· lb, $F_{3x} = 6.31 \cos 2^{\circ} = 6.31$ lb,

$F_{3y} = 6.31 \sin 2^{\circ} = 0.22$ lb, $C_3 = -23.1$ in lb, $\overline{F}_{23} = 4.43\ \underline{/154^{\circ}}$ lb,

$\overline{F}_{03} = 3.14\ \underline{/223^{\circ}}$ lb, $\overline{F}_{12} = 19.8\ \underline{/182^{\circ}}$ lb, $\overline{F}_{01} = 24.6\ \underline{/184^{\circ}}$, $T_1 = 177$ in·lb

10.13 $r = 0.1$ m, $\ell = 0.5$ m, $m_3 = 3$ kg, $m_B = \frac{0.35}{0.5}(2) = 1.4$ kg,

$m_c = \frac{0.15}{0.5}(2) = 0.6$ kg, $\omega = (\frac{200}{60})2\pi = 20.9$ rad/s,

$T = 15.8 \sin\phi(\cos\phi + 0.2 \cos 2\phi)(1 + \frac{\cos\phi}{\sqrt{25 - \sin^2\phi}})$ N·m, When $\phi = 30^{\circ}$,

$T = 8.96$ N·m

10.14 $F_{32x} = 131(\cos\phi + 0.2 \cos 2\phi)$, $F_{32y} = F_{03y} = \frac{-157 \sin\phi(\cos\phi + 0.2 \cos 2\phi)}{\sqrt{25 - \sin^2\phi}}$

$F_{12x} = F_{01x} = -61.2 \cos\phi - 157(\cos\phi + 0.2\cos 2\phi)$,

$F_{12y} = F_{01y} = -61.2 \sin\phi + \frac{157 \sin\phi(\cos\phi + 0.2 \cos 2\phi)}{\sqrt{25 - \sin^2\phi}}$

When $\phi = 45^{\circ}$, $F_{32x} = 92.6$ N, $F_{32y} = F_{03y} = -15.8$ N, $F_{12x} = F_{01x} = -154$ N,

$F_{12y} = F_{01y} = -27.4$ N

10.15 $r = 2$ ft, $\ell = 7$ ft, $m_3 = \frac{100}{32.2}$, $m_B = (\frac{5}{7})(\frac{75}{32.2}) = \frac{53.6}{32.2}$, $m_C = (\frac{2}{7})(\frac{75}{32.2})$ $= \frac{21.4}{32.2}$, $\omega = (\frac{60}{60})2\pi = 6.28$ rad/s, $T = 595 \sin\phi(\cos\phi + 0.286 \cos 2\phi)$ $(1 + \frac{2\cos\phi}{\sqrt{49 - 4\sin^2\phi}})$ ft·lb, When $\phi = 30^\circ$, $T = 375$ ft·lb

10.16 $F_{32x} = (\frac{100}{32.2})(2)(6.28)^2 \left[\cos\phi + (\frac{2}{7}) \cos 2\phi\right] = 245 (\cos\phi + 0.286 \cos 2\phi)$,

$F_{32y} = F_{03y} = \frac{-595 \sin\phi (\cos\phi + 0.286 \cos 2\phi)}{\sqrt{49 - 4\sin^2\phi}}$,

$F_{12x} = F_{01x} = -131 \cos\phi - 297 (\cos\phi + 0.286 \cos 2\phi)$,

$F_{12y} = F_{01y} = -131 \sin\phi + \frac{595 \sin\phi(\cos\phi + 0.286 \cos 2\phi)}{\sqrt{49 - 4\sin^2\phi}}$,

When $\phi = 45^\circ$, $F_{32x} = 173$ lb, $F_{32y} = F_{03y} = -43.4$ lb, $F_{12x} = F_{01x}$ $= -303$ lb, $F_{12y} = F_{01y} = -49.2$ lb

10.17 $T = F_{21x} r\sin\phi - F_{21y} r\cos\phi = -F_{12x} r\sin\phi + F_{12y} r\cos\phi$,

$F_{12x} = -F_{32x} + m_C a_C + m_B a_{Bx} = F_{23x} + m_C a_C + m_B a_{Bx} = m_B a_{Bx} + (m_3 + m_C)a_C$,

$F_{12y} = -F_{32y} + m_B a_{By} = F_{23y} + m_B a_{By} = m_B a_{By} - (m_3 + m_C)a_C \tan\theta$,

$a_{Bx} = -r\omega^2 \cos\phi$, $a_{By} = -r\omega^2 \sin\phi$, $T = m_B r^2\omega^2 \sin\phi\cos\phi - (m_3 + m_C)a_C r\sin\phi -$ $m_B r^2\omega^2 \sin\phi\cos\phi - (m_3 + m_C)a_C \tan\theta\, r\cos\phi$, $T = -(m_3 + m_C) r a_C(\sin\phi + \cos\phi\tan\theta)$,

$R_C = r\cos\phi + \ell\cos\theta$ where $r\sin\phi = e + \ell\sin\theta$ or $\sin\theta = \frac{r\sin\phi - e}{\ell}$,

$\cos\theta = \sqrt{1 - \sin^2\theta} \approx 1 - \frac{1}{2}\sin^2\theta = 1 - (\frac{r^2\sin^2\phi - 2er\sin\phi + e^2}{2\ell^2})$,

$R_C = (\ell - \frac{r^2}{4\ell} - \frac{e^2}{2\ell}) + r\cos\phi + \frac{er}{\ell}\sin\phi + \frac{r^2}{4\ell}\cos 2\phi$, $a_C = -r\omega^2(\cos\phi + \frac{e}{\ell}\sin\phi + \frac{r}{\ell}$ $\cos 2\phi)$, $\tan\theta = \frac{r\sin\phi - e}{\sqrt{\ell^2 - (r\sin\phi - e)^2}}$, $T = (m_3 + m_C) r^2\omega^2(\cos\phi + \frac{e}{\ell}\sin\phi + \frac{r}{\ell}\cos 2\phi)$ $(\sin\phi + \cos\phi\tan\theta)$, $e = 0.1$m, $r = 0.1$m, $\ell = 0.5$m, $m_3 = 3$ kg,

$m_c = 0.6$ kg, $\omega = 20.9$ rad/s, $T = 15.7\,(\cos\phi + 0.2\sin\phi + 0.2\cos 2\phi)$

$(\sin\phi + \cos\phi\tan\theta)$, $\tan\theta = \dfrac{0.2\,(\sin\phi - 1)}{\sqrt{1-0.04(\sin\phi-1)^2}}$ when $\phi = 30^\circ$, $\tan\theta = -0.100$,

$T = 6.92$ N·m

10.18 $r = 50$ mm, $\ell = 210$ mm, $m_4 = m_5 = 1$ kg, $m_c = m_E = \frac{1}{2}(0.8) = 0.4$ kg,

$\omega = (\frac{1000}{60})2\pi = 105$ rad/s, For mechanism O_1 BC, the crank angle for use in the torque equation is $160^\circ - 90^\circ - 45^\circ = 25^\circ$, Then, $T_{11} = (1.4)(50)^2(105)^2$

$\sin 25^\circ(\cos 25^\circ + \frac{50}{210}\cos 50^\circ)\left[1 + \dfrac{50\cos 25^\circ}{\sqrt{210^2-(50)^2\sin^2 25^\circ}}\right] = 21.0$ N·m CCW,

For mechanism O_1DE, the crank angle for use in the torque equation is $160^\circ + 90^\circ - 45^\circ = 205^\circ$, Then, $T_{12} = (1.4)(50)^2(105)^2\sin 205^\circ(\cos 205^\circ + \frac{50}{210}\cos 50^\circ)$

$\left[1 + \dfrac{50\cos 205^\circ}{\sqrt{210^2 - 50^2\sin^2 205^\circ}}\right] = 9.62$ N·m CCW, The total torque is $T_1 = T_{11} + T_{12}$

$= 30.6$ N m CCW

10.19 $r = 60$ mm, $\ell = 200$ mm, $m_4 = m_5 = 1$ kg, $m_c = m_D = \frac{1}{2}(0.8) = 0.4$ kg,

$\omega = (\frac{1000}{60})2\pi = 105$ rad/s, For mechanism O_1BC, the crank angle for use in the torque equation is $-90^\circ - (45^\circ - 20^\circ) = -115^\circ = 245^\circ$, Then, $T_{11} = (1.4)(60)^2$

$(105)^2\sin 245^\circ\,(\cos 245^\circ + \frac{60}{200}\cos 130^\circ)\left[1 + \dfrac{60\cos 245^\circ}{\sqrt{200^2 - 60^2\sin^2 245^\circ}}\right] = 26.9$ N·m CCW,

For mechanism O_1BD, the crank angle for use in the torque equation is $-25^\circ = 335^\circ$,

Then, $T_{12} = (1.4)(60)^2(105)^2\sin 335^\circ\,(\cos 335^\circ + \frac{60}{200}\cos 310^\circ)$

$\left[1 + \dfrac{60\cos 335^\circ}{\sqrt{200^2-60^2\sin^2 335^\circ}}\right]$ = [illegible] N·m = [illegible] N·m CW, The net torque is $T_1 = T_{11}$ +

$T_{12} = 6.0$ N·m CW

10.20 B) $m_c r_c = \left[(3\cos 0 + 6\cos 120^\circ + 2\cos 240^\circ)^2 + (3\sin 0 + 6\sin 120^\circ + 2\sin 240^\circ)^2\right]^{1/2} = \left[(-1)^2 + (3.464)^2\right]^{1/2} = 3.60$ kg·mm,

$\theta_c = \arctan\left(\frac{-3.464}{1}\right) = 286.1^\circ$

10.21 B) $s_1 = 6$ in, $s_2 = 12$ in, $s_Q = 18$ in, $w_Q r_Q s_Q = \left\{\left[(4)(2)(6)\cos 90^\circ + (2)(2.5)(12)\cos 0\right]^2 + \left[(4)(2)(6)\sin 90^\circ + (2)(2.5)(12)\sin 0\right]^2\right\}^{1/2}$

$= \left[(60)^2 + (48)^2\right]^{1/2} = 76.84$ oz·in^2, $w_Q r_Q = \frac{76.84}{18} = 4.27$ oz in,

$\theta_Q = \arctan\left(\frac{-48}{-60}\right) = 218.7^\circ$, $w_p r_p = \left[(8\cos 90^\circ + 5\cos 0 + 4.27\cos 218.7^\circ)^2 + (8\sin 90^\circ + 5\sin 0 + 4.27\sin 218.7^\circ)^2\right]^{1/2} = \left[(1.668)^2 + (5.330)^2\right]^{1/2}$

$= 5.58$ oz·in, $\theta_p = \arctan\left(\frac{-5.330}{-1.668}\right) = 252.6^\circ$

10.22 B) $s_1 = 4$ in, $s_2 = 13$ in, $s_R = 20$ in, $w_R r_R s_R = \left\{\left[(6)(4)(4)\cos 90^\circ + (4)(5.25)(13)\cos 210^\circ\right]^2 + \left[(6)(4)(4)\sin 90^\circ + (4)(5.25)(13)\sin 210^\circ\right]^2\right\}^{1/2}$

$= \left[(-236.4)^2 + (-40.5)^2\right]^{1/2} = 239.8$ lb·in^2, $w_R r_R = \frac{239.8}{20} = 12.0$ lb·in,

$\theta_R = \arctan\left(\frac{40.5}{236.4}\right) = 9.7^\circ$, Therefore, remove $\left(\frac{12.0}{3.5}\right) = 3.43$ lb at $\theta = 189.7^\circ$,

$w_L r_L = \left[(24\cos 90^\circ + 21\cos 210^\circ + 12\cos 9.7^\circ)^2 + (24\sin 90^\circ + 21\sin 210^\circ + 12\sin 9.7^\circ)^2\right]^{1/2}$, $w_L r_L = \left[(-6.358)^2 + (15.52)^2\right]^{1/2} = 16.8$ lb·in,

$\theta_L = \arctan\left(\frac{-15.52}{6.358}\right) = 292.3^\circ$, Therefore, add $\left(\frac{16.8}{3}\right) = 5.60$ lb at $\theta = 292.3^\circ$

10.23 B) $s_1 = 3$ in, $s_2 = 1$ in, $s_3 = -2$ in, $s_Q = 3$ in,

$w_Q r_Q s_Q = \left[(18\cos 60^\circ + 9\cos 150^\circ - 8\cos 270^\circ)^2 + (18\sin 60^\circ + 9\sin 150^\circ - 8\sin 270^\circ)^2\right]^{1/2} = \left[(1.206)^2 + (28.09)^2\right]^{1/2} = 28.12$ oz·in^2

$$w_Q r_Q = \frac{28.12}{3} = 9.37 \text{ oz in}, \qquad \theta_Q = \arctan\left(\frac{-28.09}{-1.206}\right) = 267.5^\circ,$$

$$w_p r_p = \Big[(6\cos 60^\circ + 9\cos 150^\circ + 4\cos 270^\circ + 9.37\cos 267.5^\circ)^2 + (6\sin 60^\circ + 9\sin 150^\circ$$

$$+ 4\sin 270^\circ + 9.37\sin 267.5^\circ)^2\Big]^{1/2} = \Big[(-5.20)^2 + (-3.66)^2\Big]^{1/2} = 6.36 \text{ oz in},$$

$$\theta_p = \arctan\left(\frac{3.66}{5.20}\right) = 35.1^\circ$$

10.24 B) $s_1 = 0.5$ m, $s_2 = 1.4$ m, $s_3 = 2.0$ m, $\theta_1 = 315^\circ$, $\theta_2 = 90^\circ$,

$$\theta_3 = 240^\circ, \qquad m_Q r_Q s_Q = \Big[(750\cos 315^\circ + 2800\cos 90^\circ + 3000\cos 240^\circ)^2$$

$$+ (750\sin 315^\circ + 2800\sin 90^\circ + 3000\sin 240^\circ)^2\Big]^{1/2} = \Big[(-969.7)^2 + (-328.3)^2\Big]^{1/2}$$

$$= 1024 \text{ kg·mm·m}, \qquad s_Q = \frac{m_Q r_Q s_Q}{m_Q r_Q} = \frac{1024}{1000} = 1.024 \text{ m}, \qquad \theta_Q = \arctan\left(\frac{328.3}{969.7}\right) = 18.7^\circ,$$

$$m_p r_p = \Big[(1500\cos 315^\circ + 2000\cos 90^\circ + 1500\cos 240^\circ + 1000\cos 18.7^\circ)^2$$

$$+ (1500\sin 315^\circ + 2000\sin 90^\circ + 1500\sin 240^\circ + 1000\sin 18.7^\circ)^2\Big]^{1/2}$$

$$= \Big[(1258)^2 + (-39.05)^2\Big]^{1/2} = 1259 \text{ kg mm}, \qquad \theta_p = \arctan\left(\frac{39.05}{-1258}\right) = 178.2^\circ$$

10.25 Let the y-axis coincide with crank 1. Then $\theta_1 = 0, \theta_2 = 180^\circ$, $\theta_3 = 90^\circ$,

$\theta_4 = 270^\circ$, Measuring from the Q plane, $s_1 = 50$ mm, $s_2 = 150$ mm, $s_3 = 250$ mm

$$s_4 = 350 \text{ mm}, \qquad s_p = 400 \text{ mm}, \qquad \sum_{i=1}^{4} m_i s_i r_i \cos\theta_i = 400\Big[50(1) + 150(-1) + 250\,(0)$$

$$+ 350\,(0)\Big] = -40{,}000 \text{ kg·mm}^2, \qquad \sum_{i=1}^{4} m_i s_i r_i \sin\theta_i = 400\Big[50(0) + 150(0) + 250\,(1)$$

$$+ 350\,(-1)\Big] = -40{,}000 \text{ kg·mm}^2, \qquad m_p r_p s_p = \Big[(-40{,}000)^2 + (-40{,}000)^2\Big]^{1/2} = 56{,}568$$

$$\text{kg·mm}^2, \qquad m_p r_p = \frac{56{,}568}{400} = 141 \text{ kg·mm}, \qquad \theta_p = \arctan\left(\frac{40{,}000}{40{,}000}\right) = 45^\circ,$$

$$\sum_{i=1}^{4} m_i r_i \cos\theta_i + m_p r_p \cos\theta_p = 400(1-1+0+0) + 141(.7071) = 99.7 \text{ kg·mm}$$

$$\sum_{i=1}^{4} m_i r_i \sin\theta_i + m_p r_p \sin\theta_p = 400(0+0+1-1) + 141(.7071) = 99.7 \text{ kg·mm}$$

$$m_Q r_Q = \left[(99.7)^2 + (99.7)^2\right]^{1/2} = 141 \text{ kg mm}, \quad \theta_Q = \arctan\left(\frac{-99.7}{-99.7}\right) = 225^\circ$$

10.26 Let the y-axis coincide with crank 1. Then $\theta_1 = 0, \theta_2 = 90^\circ, \theta_3 = 180^\circ, \theta_4 = 270^\circ$.

Measuring from the Q plane, $s_1 = 50$ mm, $s_2 = 150$ mm, $s_3 = 250$ mm, $s_4 = 350$ mm,

$$s_p = 400 \text{ mm}, \quad \sum_{i=1}^{4} m_i s_i r_i \cos\theta_i = 400\left[50(1)+150(0)+250(-1)+350(0)\right] = -80{,}000 \text{ kg}\cdot\text{mm}^2$$

$$\sum_{i=1}^{4} m_i s_i r_i \sin\theta_i = 400\left[50(0)+150(1)+250(0)+350(-1)\right] = -80{,}000 \text{ kg}\cdot\text{mm}^2$$

$$m_p r_p s_p = \left[(-80{,}000)^2 + (-80{,}000)^2\right]^{1/2} = 113{,}100 \text{ kg}\cdot\text{mm}^2,$$

$$m_p r_p = \frac{113{,}100}{400} = 283 \text{ kg}\cdot\text{mm}, \quad \theta_p = \arctan\left(\frac{80{,}000}{80{,}000}\right) = 45^\circ,$$

$$\sum_{i=1}^{4} m_i r_i \cos\theta_i + m_p r_p \cos\theta_p = 400(1+0-1+0)+283(.7071) = 200 \text{ kg}\cdot\text{mm},$$

$$\sum_{i=1}^{4} m_i r_i \sin\theta_i + m_p r_p \sin\theta_p = 400(0+1+0-1)+283(.7071) = 200 \text{ kg}\cdot\text{mm},$$

$$m_Q r_Q = \left[(200)^2 + (200)^2\right]^{1/2} = 283 \text{ kg}\cdot\text{mm}, \quad \theta_Q = \arctan\left(\frac{-200}{-200}\right) = 225^\circ$$

10.27 The only forces that change from Problem 10.14 are F_{01x} and F_{01y}.

$$F_{01x} = F_{01x}(\text{old}) - m_c r_c \omega^2 \cos(\phi + 180^\circ) = -65.2 \cos\phi - 31.4 \cos 2\phi,$$

$$F_{01y} = F_{01y}(\text{old}) - m_c r_c \omega^2 \sin(\phi + 180^\circ) = 91.8 \sin\phi + \frac{157 \sin\phi(\cos\phi + 0.2\cos 2\phi)}{\sqrt{25 - \sin^2\phi}}$$

When $\phi = 45^\circ$, $F_{01x} = -46.1$N, $F_{01y} = 80.8$ N

10.28 The only forces that change from Problem 10.16 are F_{01x} and F_{01y}.

$$F_{01x} = F_{01x}(\text{old}) - m_c r_c \omega^2 \cos(\phi + 180^\circ) = -122 \cos\phi - 84.9 \cos 2\phi,$$

$$F_{01y} = F_{01y}(\text{old}) - m_c r_c \omega^2 \sin(\phi + 180^\circ) = 175 \sin\phi + \frac{595 \sin\phi(\cos\phi + 0.286\cos 2\phi)}{\sqrt{49 - 4\sin^2\phi}}$$

When $\phi = 45^{\circ}$, $F_{O1x} = -86.3$ lb, $F_{O1y} = 167$ lb

10.29 $\theta_1 = \theta_2 = 0$, $\psi_1 = 0$, $\psi_2 = 90^{\circ}$, $s_1 = 0$, $s_2 = s$,

$F_{s1} = mr\omega^2 \cos\omega t + mr\omega^2 \frac{r}{\ell} \cos 2\omega t$, $F_{s2} = mr\omega^2 \cos(\omega t + 90^{\circ}) + mr\omega^2 \frac{r}{\ell} \cos$

$(2\omega t + 180^{\circ}) = -mr\omega^2 \sin\omega t - mr\omega^2 \frac{r}{\ell} \cos 2\omega t$, $F_s = F_{s1} + F_{s2} = mr\omega^2(\cos\omega t - \sin\omega t)$

$= 2000\ (\cos\omega t - \sin\omega t)$ lb, Primary forces not balanced. Secondary forces

balanced, $M_s = M_{s1} + M_{s2}$, $M_{s1} = F_{s1}s_1 = 0$, $M_{s2} = F_{s2}s_2 = -smr\omega^2 \sin\omega t$

$-smr\omega^2 \frac{r}{\ell} \cos 2\omega t = M_s$, Nonzero primary and secondary shaking moments.

$$M_s = F_s a, \quad a = \frac{M_s}{F_s} = \frac{-s\ \sin\omega t - s\frac{r}{\ell}\cos 2\omega t}{\cos\omega t - \sin\omega t} = \frac{4\ \sin\omega t + \cos 2\omega t}{\sin\omega t - \cos\omega t} \text{ in}$$

10.30 $\theta_1 = \theta_2 = 0$, $\psi_1 = 0$, $\psi_2 = 180^{\circ}$, $F_{s1} = mr\omega^2 \cos\omega t + \frac{mr^2\omega^2}{\ell} \cos 2\omega t$,

$F_{s2} = mr\omega^2 \cos(\omega t + 180^{\circ}) + \frac{mr^2\omega^2}{\ell} \cos(2\omega t + 360^{\circ}) = -mr\omega^2 \cos\omega t + \frac{mr^2\omega^2}{\ell} \cos 2\omega t$,

$F_s = F_{s1} + F_{s2} = \frac{2mr^2\omega^2}{\ell} \cos 2\omega t$, Therefore, the counterweight shafts must rotate

at twice the speed of the crankshaft (i.e., 2ω). The total counterweight

correction is given by $m_c r_c (2\omega)^2 = \frac{2mr^2\omega^2}{\ell}$, $m_c r_c = \frac{mr^2}{2\ell}$ (split evenly between the

two counterweight shafts).

10.31 $\theta_1 = 0, \theta_2 = 180^{\circ}$, $\psi_1 = \psi_2 = 0$. All of Equations 10-63 are satisfied except

10-63A. Therefore, primary forces are not balanced, but secondary forces are.

10.32 $\theta_1 = \theta_2 = \theta_3 = \theta_4 = 90^{\circ}$, $\psi_1 = \psi_4 = 0$, $\psi_2 = \psi_3 = 180^{\circ}$,

$$\left.\begin{aligned} \sum_{n=1}^{4} \sin\psi_n &= 0 + 0 + 0 + 0 = 0 \\ \sum_{n=1}^{4} \cos\psi_n &= 1 - 1 - 1 + 1 = 0 \end{aligned}\right\} \quad \text{Primary forces balanced}$$

$$\sum_{n=1}^{4} \cos 2\psi_n = 1 + 1 + 1 + 1 = 4$$

$$\sum_{n=1}^{4} \sin 2\psi_n = 0 + 0 + 0 + 0 = 0$$

Secondary forces not balanced

$$\sum_{n=1}^{4} (n-1)\sin\psi_n = 0 + 0 + 0 + 0 = 0$$

$$\sum_{n=1}^{4} (n-1)\cos\psi_n = 0 - 1 - 2 + 3 = 0$$

Primary moments balanced

$$\sum_{n=1}^{4} (n-1)\cos 2\psi_n = 0 + 1 + 2 + 3 = 6$$

$$\sum_{n=1}^{4} (n-1)\sin 2\psi_n = 0 + 0 + 0 + 0 = 0$$

Secondary moments not balanced

10.33 $\theta_1 = \theta_2 = \theta_3 = \theta_4 = 90^\circ$, $\psi_1 = 0$, $\psi_2 = 90^\circ$, $\psi_3 = 270^\circ$, $\psi_4 = 180^\circ$

$$\sum_{n=1}^{4} \sin\psi_n = 0 + 1 - 1 + 0 = 0$$

$$\sum_{n=1}^{4} \cos\psi_n = 1 + 0 + 0 - 1 = 0$$

Primary forces balanced

$$\sum_{n=1}^{4} \cos 2\psi_n = 1 - 1 - 1 + 1 = 0$$

$$\sum_{n=1}^{4} \sin 2\psi_n = 0 + 0 + 0 + 0 = 0$$

Secondary forces balanced

$$\sum_{n=1}^{4} (n-1)\sin\psi_n = 0 + 1 - 2 + 0 = -1$$

$$\sum_{n=1}^{4} (n-1)\cos\psi_n = 0 + 0 + 0 - 3 = -3$$

Primary moments not balanced

$$\sum_{n=1}^{4} (n-1)\cos 2\psi_n = 0 - 1 - 2 + 3 = 0$$

$$\sum_{n=1}^{4} (n-1)\sin 2\psi_n = 0 + 0 + 0 + 0 = 0$$ Secondary moments balanced

10.34 $\theta_1 = \theta_2 = \theta_3 = \theta_4 = \theta_5 = \theta_6 = 90^\circ$, $\psi_1 = \psi_6 = 0$, $\psi_2 = \psi_5 = 120^\circ$, $\psi_3 = \psi_4 = 240^\circ$, All shaking force and shaking moment conditions are satisfied.

10.35 $N = 2$, $\theta_1 = 0$, $\theta_2 = 90^\circ$, $\psi_1 = 0$, $\psi_2 = 90^\circ$,

Primary:

$$\sum_{n=1}^{2} \cos(\psi_n - \theta_n)\cos\theta_n = 1 + 0 = 1$$

$$\sum_{n=1}^{2} \sin(\psi_n - \theta_n)\cos\theta_n = 0 + 0 = 0$$

$$\sum_{n=1}^{2} \cos(\psi_n - \theta_n)\sin\theta_n = 0 + 1 = 1$$

$$\sum_{n=1}^{2} \sin(\psi_n - \theta_n)\sin\theta_n = 0 + 0 = 0$$

Secondary:

$$\sum_{n=1}^{2} \cos(2\psi_n - 2\theta_n)\cos\theta_n = 1 + 0 = 1$$

$$\sum_{n=1}^{2} \sin(2\psi_n - 2\theta_n)\cos\theta_n = 0 + 0 = 0$$

$$\sum_{n=1}^{2} \cos(2\psi_n - 2\theta_n)\sin\theta_n = 0 + 1 = 1$$

$$\sum_{n=1}^{2} \sin(2\psi_n - 2\theta_n)\sin\theta_n = 0 + 0 = 0$$

10.36 $N = 2$, $\theta_1 = 0$, $\theta_2 = 90^\circ$, $\psi_1 = \psi_2 = 0$

$$\sum_{n=1}^{2} \cos(\psi_n - \theta_n)\cos\theta_n = 1 + 0 = 1$$

$$\sum_{n=1}^{2} \sin(\psi_n - \theta_n)\cos\theta_n = 0 + 0 = 0$$

$$\sum_{n=1}^{2} \cos(\psi_n-\theta_n)\sin\theta_n = 0 + 0 = 0$$

$$\sum_{n=1}^{2} \sin(\psi_n-\theta_n)\sin\theta_n = 0 - 1 = -1$$

$$\sum_{n=1}^{2} \cos(2\psi_n-2\theta_n)\cos\theta_n = 1 + 0 = 1$$

$$\sum_{n=1}^{2} \sin(2\psi_n-2\theta_n)\cos\theta_n = 0 + 0 = 0$$

$$\sum_{n=1}^{2} \cos(2\psi_n - 2\theta_n)\sin\theta_n = 0 - 1 = -1$$

$$\sum_{n=1}^{2} \sin(2\psi_n-2\theta_n)\sin\theta_n = 0 + 0 = 0$$

10.37 $N = 4, \quad \theta_1 = 0, \quad \theta_2 = 90^\circ, \quad \theta_3 = 180^\circ, \quad \theta_4 = 270^\circ, \quad \psi_1 = \psi_2 = \psi_3 = \psi_4 = 0,$

The horizontal shaking force component is

$$F_{s1} - F_{s3} = mr\omega^2\cos\omega t + \frac{mr^2\omega^2}{\ell}\cos2\omega t + mr\omega^2\cos\omega t - \frac{mr^2\omega^2}{\ell}\cos2\omega t = 2mr\omega^2\cos\omega t,$$

The vertical shaking force component is

$$F_{s2} - F_{s4} = mr\omega^2\sin\omega t - \frac{mr^2\omega^2}{\ell}\cos2\omega t + mr\omega^2\sin\omega t + \frac{mr^2\omega^2}{\ell}\cos2\omega t = 2mr\omega^2\sin\omega t,$$

This is equivalent to a rotating unbalance and can be balanced by a rotating counterweight mounted on the crankshaft at a position 180° from the crankthrow. The counterweight correction is $m_c r_c = 2mr$.

CHAPTER ELEVEN

11.1 The coordinates of the fixed pivot are 125, 344. See figure S11.1.

11.2 The coordinates of the fixed pivot are 8.25, 6.82. See figure S11.2.

11.3 Fixed center O_B must lie along a line through a point with coordinates 4.5, 5.5 at an angle of 135°. Fixed center O_C must lie along a line through a point with coordinates 7.23, 8.23 at an angle of 151°. See figure S11.3.

11.4 See program listing S11.4

11.5 The coordinates of the fixed pivot are 125, 344. See computer output S11.5

11.6 The coordinates of the fixed pivot are 3.75, 3.13. See computer output S11.6

11.7 Fixed center O_B must lie along a line through a point with coordinates 4.5, 5.5 at an angle of 135°. Fixed center O_C must lie along a line through a point with coordinates 7.23, 8.23 at an angle of 151°. See computer output S11.7

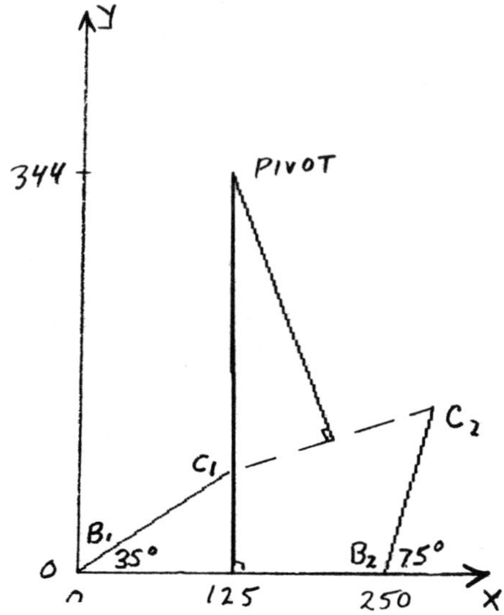
Y
344
PIVOT
C2
C1
B1
35°
B2
75°
0
125
250
X

S11,1

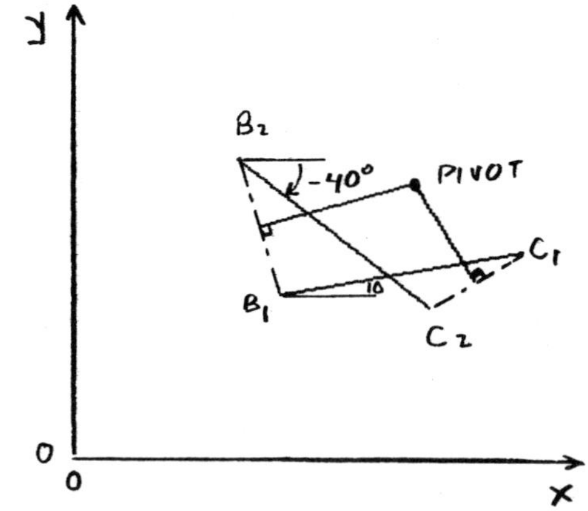
Y
B2
-40°
PIVOT
C1
B1
10
C2
0
0
X

S11,2

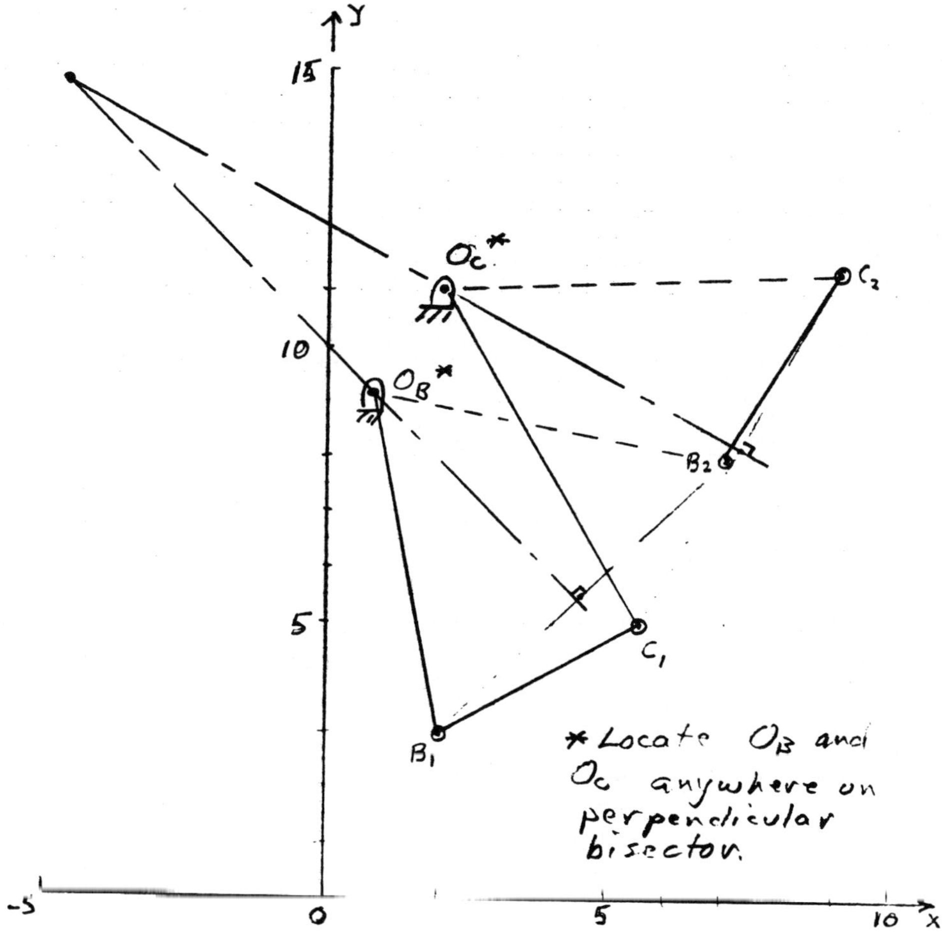
Y
15
10
5
-5
0
5
10
x
O_C*
C_2
O_B*
B_2
C_1
B_1
* Locate O_B and O_C anywhere on perpendicular bisector.

S 11.3

S11.4

```
100 LPRINT"     Plot: 2 position synthesis"
200 INPUT "   link length";L
300 LPRINT"    link length = ";L
500 FOR I = 1 TO 2
550 LPRINT"Position ";I
600 INPUT "Bx, By";BX(I), BY(I)
700 LPRINT "    Coordinates of B: ";BX(I),BY(I)
800 INPUT "Link angle";BETA(I)
900 LPRINT"  Link angle (deg): ";BETA(I)
950 PI=ATN(1)*4
1000 BRAD(I)=BETA(I)*PI/180
1100 CX(I) = BX(I)+L*COS(BRAD(I))
1200 CY(I)=BY(I)+L*SIN(BRAD(I))
1300 LPRINT"  Coordinates of C: ";CX(I),CY(I)
1400 NEXT I
1500 DX=(BX(1)+BX(2))/2
1600 DY=(BY(1)+BY(2))/2
1700 EX=(CX(1)+CX(2))/2: EY=(CY(1)+CY(2))/2
1800 LPRINT"  x and y coordinates of B and C displacement midpoints:
                ";DX,DY,EX,EY
1840 MB=-((BX(2)-BX(1))/(BY(2)-BY(1)))
1860 MC=-((CX(2)-CX(1))/(CY(2)-CY(1)))
1880 LPRINT " Slopes of perpendicular bisectors of B and C displacements:
                "MB, MC
1900 SLB=ATN((BY(2)-BY(1))/(BX(2)-BX(1)))+PI/2
2000 SLC=ATN((CY(2)-CY(1))/(CX(2)-CX(1)))+PI/2
2100 BDEG=(SLB)*180/PI: CDEG=(SLC)*180/PI
2200 LPRINT " Angles of perpendicular bisectors of B and C  displacements (rad
     and deg) "SLB, SLC,BDEG,CDEG
2600 KB=DY-MB*DX
2700 KC=EY-MC*EX
2800 LPRINT" Constants KB and KC for perpendicular bisectors where y = MB*x + KB
  and          y = MC*x + KC :       " KB,KC
2900 X=(KC-KB)/(MB-MC)
3000 Y=MB*(KC-KB)/(MB-MC)+KB
3100 LPRINT "Intersection  x, y  which may be used as a fixed pivot:
           "X,Y
3150 CLS
3200 SCREEN 2
3250 F=1.618   :FX = 15 :  FY=F*FX
3260 GY=100
3300 FOR I = 1 TO 2
3350 XB = BX(I)*FX:  XC = CX(I)*FX
3400 YB=BY(I)*FY:  YC=CY(I)*FY
3450 LINE(GY+YB,XB)-(GY+YC,XC)
3500 NEXT I
3550 LINE(GY+Y*FY,X*FX)-(GY+DY*FY,DX*FX)
3600 LINE(GY+Y*FY,X*FX)-(GY+EY*FY,EX*FX)
4000 END
```

S11.5

```
     Two position syrthesis
   link length =  150
Position  1
   Coordinates of B:  0       0
  Link angle (deg):  35
  Coordinates of C:  122.8728                   86.03648
Position  2
   Coordinates of B:  250     .001
  Link angle (deg):  75
  Coordinates of C:  288.8229                   144.8899
  x and y coordinates of B and C displacement midpoints:
      125      .0005        205.8478      115.4632
 Slopes of perpendicular bisectors of B and C displacements:
    -250000   -2.819718
 Angles of perpendicular bisectors of B and C  displacements (rad       and deg)
  1.570801      1.911634      90.00023      109.5268
 Constants KB and KC for perpendicular bisectors where y = MB*x + KB  and
  y = MC*x + KC :        3.125E+07             695.8961
Intersection  x, y  which may be used as a fixed pivot:
 124.9986      344
```

S11.6

```
     Two position synthesis
   link length =  3.5
Position  1
   Coordinates of B:  2      3
  Link angle (deg):  12
  Coordinates of C:  5.423517              3.727691
Position  2
   Coordinates of B:  5.5    3.2
  Link angle (deg):  190
  Coordinates of C:  2.053173              2.59223
  x and y coordinates of B and C displacement midpoints:
      3.75     3.1          3.738345       3.159961
 Slopes of perpendicular bisectors of B and C displacements:
     -17.5    -2.968261
 Angles of perpendicular bisectors of B and C  displacements (rad and deg)
 1.627877      1.895752      93.27049      108.6186
 Constants KB and KC for perpendicular bisectors where y = MB*x + KB  and
  y = MC*x + KC :        68.72498              14.25634
Intersection  x, y  which may be used as a fixed pivot:
 3.748254      3.130547
```

S11.7

```
     Two position synthesis
   link length =  4
Position  1
   Coordinates of E:  2      3
  Link angle (deg):  30
  Coordinates of C:  5.464102               5
Position  2
   Coordinates of E:  7      8
  Link angle (deg):  60
  Coordinates of C:  9       11.4641
  x and y coordinates of B and C displacement midpoints:              4.5
 5.5         7.232051      8.232051
 Slopes of perpendicular bisectors of B and C displacements:          -1
-.5470054
 Angles of perpendicular bisectors of B and C  displacements (rad and deg)
 2.356195      2.64 052      135           151.3211
 Constants KB and KC for perpendicular bisectors where y = MB*x + KB  and
  y = MC*x + KC :       10   12.18802
Intersection  x, y  which may be used as a fixed pivot
-4.830127      14.83013
```

11.8 Locate fixed pivots O_B at 1.51, - 0.62 and O_C at 5.51, - 2.54 . See figure S11.8.

11.9 Locate fixed pivots O_B at 2.5, -7 and O_C at 1.13, 17.84 . See figure 11.9

11.10 Program listing S11.10 shows one example of such a program.

11.11 Locate fixed pivots O_B at 25.96, 12.02 and O_C at 97.95, 4.59. See computer output S11.11

11.12 Locate fixed pivots O_B at -0.167, 2.833 and O_C at 1.806, 2.956. See computer output S11.12

11.13 Locate fixed pivots O_B at 26.18, - 18.53 and O_C at 125.40, -19.37. See computer output S11.13

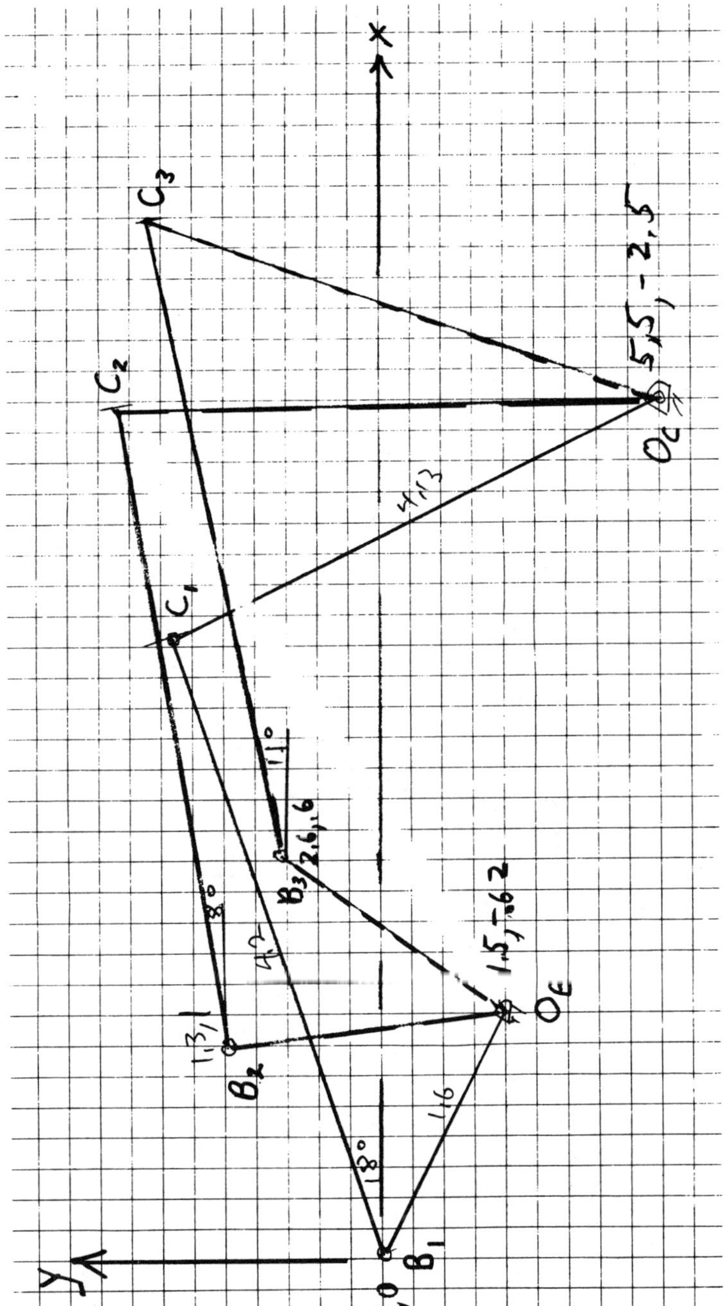
X
Y
C3
C2
C1
B1
B2
B3
OC
OE
5.5, -2.5
4.13
2.6, 1.6
11°
1.5, -.62
8°
4.2
1.3, 1
1.6
18°
0, 0
Fig S11.8

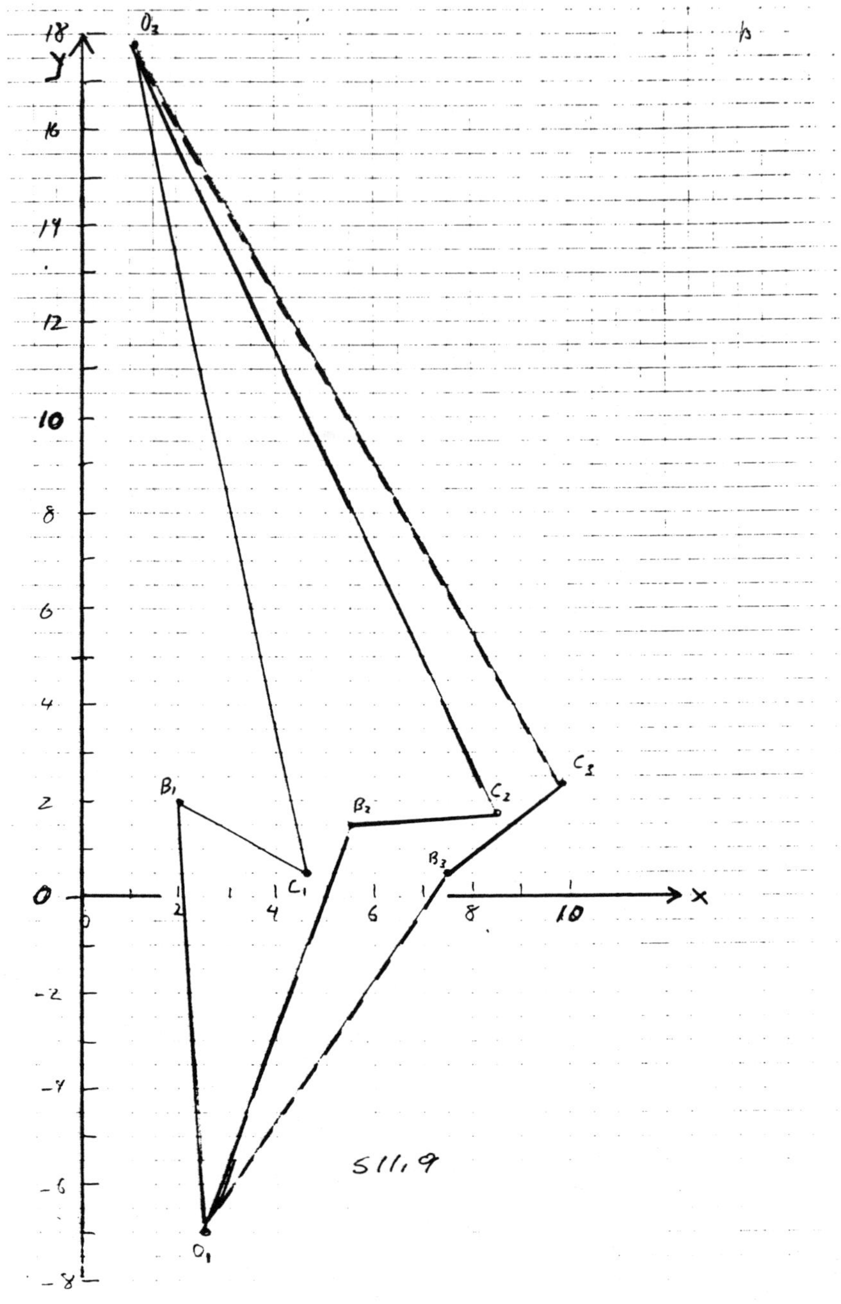

Y
X
O2
O1
B1
B2
B3
C1
C2
C3
18
16
14
12
10
8
6
4
2
0
-2
-4
-6
-8
2
4
6
8
10
S11,9

S11.10

```
100 LPRINT" THREE POSITION SYNTHESIS"
200 INPUT "  link length";L
300 LPRINT"   Link length = ";L
500 FOR I = 1 TO 3
550 LPRINT"Position ";I
600 INPUT "Bx, By";BX(I), BY(I)
700 LPRINT "   Coordinates of B: ";BX(I),BY(I)
800 INPUT "Link angle";BETA(I)
900 LPRINT"  Link angle (deg): ";BETA(I)
950 PI=ATN(1)*4
1000 BRAD(I)=BETA(I)*PI/180
1100 CX(I) = BX(I)+L*COS(BRAD(I))
1200 CY(I)=BY(I)+L*SIN(BRAD(I))
1300 LPRINT"  Coordinates of C: ";CX(I),CY(I)
1400 IF I=1 GOTO 2900
1500 DX(I)=(BX(I-1)+BX(I))/2
1600 DY(I)=(BY(I-1)+BY(I))/2
1700 EX(I)=(CX(I-1)+CX(I))/2: EY(I)=(CY(I-1)+CY(I))/2
1800 LPRINT"  x and y coordinates of B and C displacement midpoints:
                ";DX(I),DY(I), EX(I),EY(I)
1840 MB(I)=-((BX(I)-BX(I-1))/(BY(I)-BY(I-1)))
1860 MC(I)=-((CX(I)-CX(I-1))/(CY(I)-CY(I-1)))
1880 LPRINT " Slopes of perpendicular bisectors of B and C displacements:
                "MB(I),MC(I)
1900 SLB=ATN((BY(I)-BY(I-1))/(BX(I)-BX(I-1)))+PI/2
2000 SLC=ATN((CY(I)-CY(I-1))/(CX(I)-CX(I-1)))+PI/2
2100 BDEG=(SLB)*180/PI: CDEG=(SLC)*180/PI
2200 LPRINT " Angles of perpendicular bisectors of B and C  displacements (rad a
nd deg) "SLB, SLC,BDEG,CDEG
2600 KB(I)=DY(I)-MB(I)*DX(I)
2700 KC(I)=EY(I)-MC(I)*EX(I)
2800 LPRINT" Constants KB and KC for perpendicular bisectors where y = MB*x + KB
  and          y = MC*x + KC :       " KB(I),KC(I)
2900 NEXT I
3000 XB=(KB(3)-KB(2))/(MB(2)-MB(3))
3100 YB=MB(2)*XB+KB(2)
3200 XC=(KC(3)-KC(2))/(MC(2)-MC(3))
3300 YC=MC(2)*XC+KC(2)
3400 LPRINT "Intersections x, y  which may be used as fixed pivots:
        "XB,YB,XC,YC
3500 L1=SQR((BX(1)-XB)^2+(BY(1)-YB)^2)
3600 L3=SQR((CX(1)-XC)^2+(CY(1)-YC)^2)
3700 LPRINT "Crank lengths L1 and L3: "L1,L3
4000 END
```

115/2

5/11,11

```
 THREE POSITION SYNTHESIS
   Link length =  60
Position  1
   Coordinates of B:  0       0
  Link angle (deg):  20
  Coordinates of C:  56.38156              20.52121
Position  2
   Coordinates of B:  20      40
  Link angle (deg):  5
  Coordinates of C:  79.77168              45.22935
  x and y coordinates of B and C displacement midpoints:
      10        20            68.07663      32.87528
 Slopes of perpendicular bisectors of B and C displacements:
     -.5      -.9466566
 Angles of perpendicular bisectors of B and C  displacements (rad and deg)
 2.677945      2.38359       153.4349      136.5696
 Constants KB and KC for perpendicular bisectors where y = MB*x + KB  and
  y = MC*x + KC :         25   97.32046
Position  3
   Coordinates of B: -1.5     4
  Link angle (deg): -15
  Coordinates of C:  56.45555             -11.52914
  x and y coordinates of B and C displacement midpoints:
      9.25      22            68.11362      16.8501
 Slopes of perpendicular bisectors of B and C displacements:
    -.5972222               -.4107956
 Angles of perpendicular bisectors of B and C  displacements (rad and deg)
 2.603218      2.751815      149.1534      157.6674
 Constants KB and KC for perpendicular bisectors where y = MB*x + KB  and
  y = MC*x + KC :         27.52431         44.83087
Intersections x, y  which may be used as fixed pivots:
 25.96429      12.01786      97.95372      4.591927
Crank lengths L1 and L3:  28.61072         44.51951
```

5/11/12

```
 THREE POSITION SYNTHESIS
   Link length =  1.5
Position  1
   Coordinates of B:  1       1
  Link angle (deg):  15
  Coordinates of C:  2.448889                1.388229
Position  2
   Coordinates of B:  2       3
  Link angle (deg):  0
  Coordinates of C:  3.5      3
  x and y coordinates of B and C displacement midpoints:
     1.5       2              2.974445       2.194114
 Slopes of perpendicular bisectors of B and C displacements:
     -.5       -.6521466
 Angles of perpendicular bisectors of B and C  displacements (rad and deg)
 2.677945       2.56371        153.4349       146.8898
 Constants KB and KC for perpendicular bisectors where y = MB*x + KB  and
  y = MC*x + KC :        2.75                    4.133888
Position  3
   Coordinates of B:  0       5
  Link angle (deg): -15
  Coordinates of C:  1.448889                4.611772
  x and y coordinates of B and C displacement midpoints:
     1         4              2.474445       3.805886
 Slopes of perpendicular bisectors of B and C displacements:
     1         1.272582
 Angles of perpendicular bisectors of B and C  displacements (rad and deg)
 .7853983       .9047717       45             51.83959
 Constants KB and KC for perpendicular bisectors where y = MB*x + KB  and
  y = MC*x + KC :        3     .6569529
Intersections x, y  which may be used as fixed pivots:
-.1666667       2.833333       1.806455       2.955815
Crank lengths L1 and L3:  2.173067           1.694121
```

S11.13

```
 THREE POSITION SYNTHESIS
   Link length =  100
Position  1
   Coordinates of B:  0      0
  Link angle (deg): -10
  Coordinates of C:  98.48078              -17.36482
Position  2
   Coordinates of B:  50    -40
  Link angle (deg):  5
  Coordinates of C:  149.6195              -31.28443
  x and y coordinates of B and C displacement midpoints:
      25      -20          124.0501      -24.32462
 Slopes of perpendicular bisectors of B and C displacements:
      1.25      3.673862
 Angles of perpendicular bisectors of B and C  displacements (rad and deg)
 .8960555      1.305042       51.3402        74.77337
 Constants KB and KC for perpendicular bisectors where y = MB*x + KB  and
  y = MC*x + KC :         -51.25                -480.0677
Position  3
   Coordinates of B:  20    -50
  Link angle (deg):  30
  Coordinates of C:  106.6026                  0
  x and y coordinates of B and C displacement midpoints:
      35      -45          128.111       -15.64221
 Slopes of perpendicular bisectors of B and C displacements:
     -3        1.375027
 Angles of perpendicular bisectors of B and C  displacements (rad and deg)
 1.892547      .9420096       108.435        53.97317
 Constants KB and KC for perpendicular bisectors where y = MB*x + KB  and
  y = MC*x + KC :         60  -191.7983
Intersections x, y  which may be used as fixed pivots:
 26.17647     -18.52941      125.398       -19.37262
Crank lengths L1 and L3:  32.07096         26.99203
```

11.14 The solution utilizes Rosenauer's method.

$R_1 = e^{-j1.571}$ $R_2 = 3.606\ e^{j2.159}$ $R_3 = 2.236\ e^{-j0.464}$

See calculations S11.14

11.15 The solution utilizes Rosenauer's method.

$R_1 = j\ 4$ $R_2 = j\ 2$ $R_3 = 6 - j\ 2$

See calculations S11.15 and Fig 11.15

11.16 The solution utilizes Rosenauer's method.

$R_1 = j\ 4$ $R_2 = j\ 2$ $R_3 = 6 - j$

See calculations S11.16 and Fig 11.16

11.17 See computer output S11.17 where vector X represents r1, r2, and r3

11.18 See computer output S11.18 where vector X represents r_1, r_2, and r_3

11.19 See computer output S11.19 where vector X represents r_1, r_2, and r_3

11.20 See computer output S11.20 where vector X represents r_1, r_2, and r_3

S11.14 Link lengths are as follows:

$$\underset{\sim}{R}_1 = \begin{vmatrix} 1 & 1 & 1 \\ 0 & \omega_2 & \omega_3 \\ 0 & j\alpha_2-\omega_2^2 & j\alpha_3-\omega_3^2 \end{vmatrix} = \begin{vmatrix} 1 & 1 & 1 \\ 0 & 1 & 1 \\ 0 & -1 & -1-j \end{vmatrix}$$

$$= -1 - j + 0 + 0 + 1 - 0 - 0 = -j = e^{-j1.571} = 1\angle -90^\circ$$

$$\underset{\sim}{R}_2 = \begin{vmatrix} 1 & 1 & 1 \\ \omega_1 & 0 & \omega_3 \\ j\alpha_1-\omega_1^2 & 0 & j\alpha_3-\omega_3^2 \end{vmatrix} = \begin{vmatrix} 1 & 1 & 1 \\ 2 & 0 & 1 \\ j-4 & 0 & -j-1 \end{vmatrix}$$

$$= 0+j-4+0-0+2j+2-0 = -2+j3 = 3.606e^{j2.159} = 3.606\angle 124^\circ$$

$$\underset{\sim}{R}_3 = \begin{vmatrix} 1 & 1 & 1 \\ \omega_1 & \omega_2 & 0 \\ j\alpha_1-\omega_1^2 & j\alpha_2-\omega_2^2 & 0 \end{vmatrix} = \begin{vmatrix} 1 & 1 & 1 \\ 2 & 1 & 0 \\ j-4 & -1 & 0 \end{vmatrix}$$

$$= 0+0-2-0-0-j+4 = 2-j = 2.236e^{j.464} = 2.236\angle -26.6^\circ$$

$$\underset{\sim}{R}_o = -\underset{\sim}{R}_1 - \underset{\sim}{R}_2 - \underset{\sim}{R}_3 = j + 2 = j3 - 2 + j = -j = e^{-j1.571} = 1\angle -90^\circ$$

See sketch of resulting linkage.

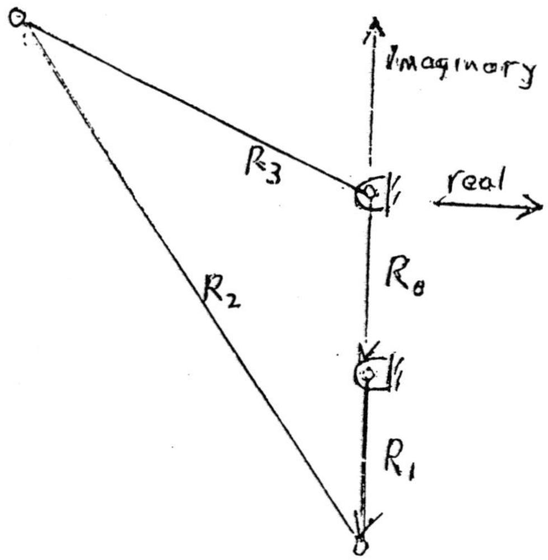

S11.15 a) Link lengths and positions are determined as follows:

$$\begin{bmatrix} 1 & 1 & 1 \\ \omega_1 & \omega_2 & \omega_3 \\ j\alpha_1-\omega_1^2 & j\alpha_2-\omega_2^2 & j\alpha_3-\omega_3^2 \end{bmatrix} \begin{bmatrix} \underset{\sim}{r}_1 \\ \underset{\sim}{r}_2 \\ \underset{\sim}{r}_3 \end{bmatrix} = \begin{bmatrix} -\underset{\sim}{r}_o \\ 0 \\ 0 \end{bmatrix}$$

$$R_1 = \begin{vmatrix} 1 & 1 & 1 \\ 0 & \omega_2 & \omega_3 \\ 0 & j\alpha_2-\omega_2^2 & j\alpha_3-\omega_3^2 \end{vmatrix} \qquad \text{etc}$$

$$\underset{\sim}{R}_1 = \begin{vmatrix} 1 & 1 & 1 \\ 0 & 2 & 0 \\ 0 & -4 & j2 \end{vmatrix} = j4$$

$$\underset{\sim}{R}_2 = \begin{vmatrix} 1 & 1 & 1 \\ -1 & 0 & 0 \\ j-1 & 0 & j2 \end{vmatrix} = j2$$

$$\underset{\sim}{R}_3 = \begin{vmatrix} 1 & 1 & 1 \\ -1 & 2 & 0 \\ j-1 & -4 & 0 \end{vmatrix} = 6-j2$$

$$\underset{\sim}{R}_o = -\underset{\sim}{R}_1 - \underset{\sim}{R}_2 - \underset{\sim}{R}_3 = -6 - j4$$

b) See sketch. $\underset{\sim}{oc} = \underset{\sim}{ob} + \underset{\sim}{bc}$

$ob = j\omega_1 R_1 = j(-1)(j4) = 4 \qquad bc = j\omega_2 R_2 = j \times 2 \times j2 = -4$

$oc = j\omega_3(-R_3) = 0$

$\underset{\sim}{a}^n_C + \underset{\sim}{a}^t_C = \underset{\sim}{a}^n_B + \underset{\sim}{a}^t_B + \underset{\sim}{a}^n_{CB} + \underset{\sim}{a}^t_{CB}$

$a^n_C = -\omega_3^2(-R_3) = 0$

$a^t_C = j\alpha_3(-R_3) = j2(-6+j2) = -4 - j12$

$a^n_B = -\omega_1^2 R_1 = -1(j4) = -j4$

$a^t_B = j\alpha_1 R_1 = j \times 1 \times j4 = -4$

$a^n_{CB} = -\omega_2^2 R_2 = -2^2(j2) = -j8$

$a^t_{CB} = j\alpha_2 R_2 = 0$

The mechanism satisfies the requirements

Fig S11.15

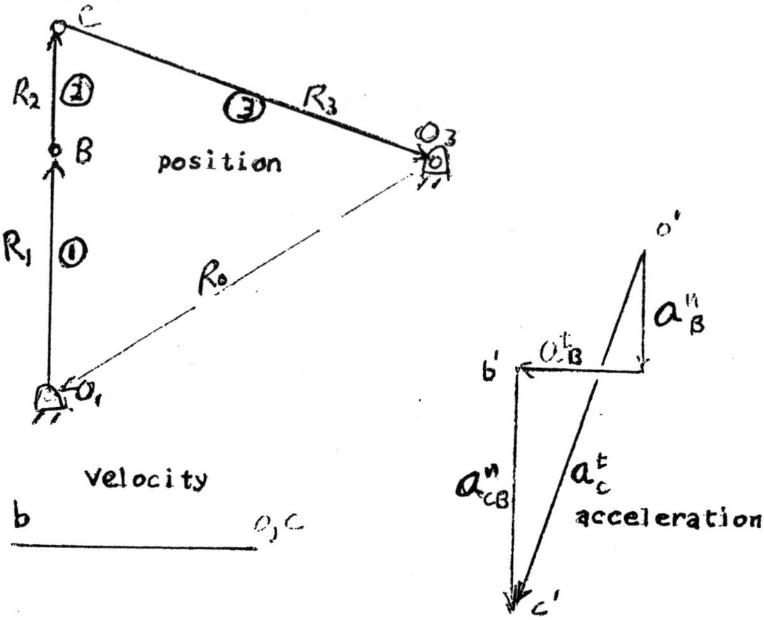

c
R_2
2
3
R_3
O_3
B
position
R_1
1
R_o
O_1
o'
a^n_B
b'
a^t_B
Velocity
b
o, c
a^n_{cB}
a^t_c
acceleration
c'

S 11.16 a. Link lengths and positions are determined as follows:

$$\begin{bmatrix} 1 & 1 & 1 \\ \omega_1 & \omega_2 & \omega_3 \\ j\alpha_1-\omega_1^2 & j\alpha_2-\omega_2^2 & j\alpha_3-\omega_3^2 \end{bmatrix} \begin{bmatrix} \tilde{r}_1 \\ \tilde{r}_2 \\ \tilde{r}_3 \end{bmatrix} = \begin{bmatrix} -\tilde{r}_o \\ 0 \\ 0 \end{bmatrix}$$

$$R_1 = \begin{vmatrix} 1 & 1 & 1 \\ 0 & \omega_2 & \omega_3 \\ 0 & j\alpha_2-\omega_2^2 & j\alpha_3-\omega_3^2 \end{vmatrix} \qquad \text{etc}$$

$$R_1 = \begin{vmatrix} 1 & 1 & 1 \\ 0 & 2 & 0 \\ 0 & j-4 & j2 \end{vmatrix} = j4$$

$$R_2 = \begin{vmatrix} 1 & 1 & 1 \\ -1 & 0 & 0 \\ -1 & 0 & j2 \end{vmatrix} = j2$$

$$R_3 \quad \begin{vmatrix} 1 & 1 & 1 \\ -1 & 2 & 0 \\ -1 & j-4 & 0 \end{vmatrix} = 6-j$$

$$\tilde{R}_o = -\tilde{R}_1 - \tilde{R}_2 - \tilde{R}_3 = -6 - j5$$

b) See sketch

$ob = j\omega_1 R_1 = j(-1)(j4) = 4$

$bc = j\omega_2 R_2 = j2(j2) = -4$

$oc = j\omega_3(-R_3) = 0$

$a^n{}_B = -\omega_1{}^2 R_1 = -(-1)^2 j4 = -j4$

$a^t{}_B = j\alpha_1 R_1 = 0$

$a^n{}_{CB} = -\omega_2{}^2 R_2 = -2^2(j2) = -j8$

$a^t{}_{CB} = j\alpha_2 R_2 = j(1)(j2) = -2$

$a^n{}_C = -\omega_3{}^2(-R_3) = 0$

$a^t{}_C = j\alpha_3(-R_3) = j2(-6 + j) = -2 - j12$

The mechanism satisfies the requirements.

S 11.16

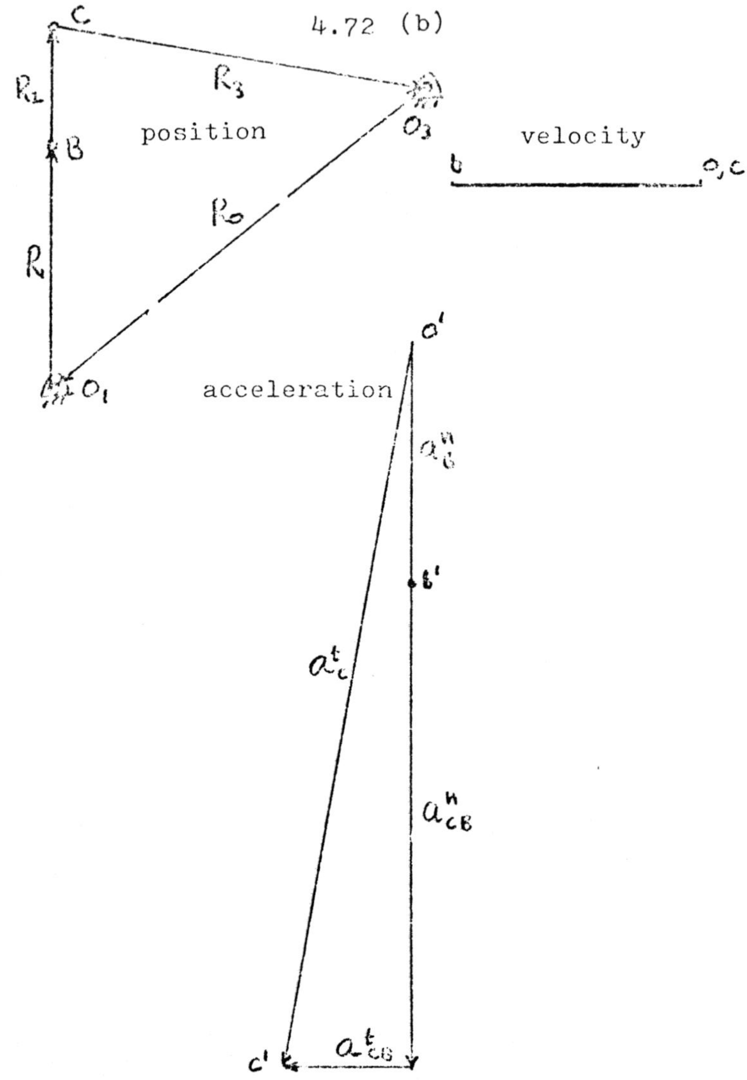

S11.17

$$A := \begin{bmatrix} \omega_1^{\,1} & \omega_2^{\,1} & \omega_3^{\,1} \\ (j\cdot\alpha_1 - \omega_1)^2 & (j\cdot\alpha_2 - \omega_2)^2 & (j\cdot\alpha_3 - \omega_3)^2 \end{bmatrix} \qquad B := \begin{bmatrix} -r_0 \\ 0 \\ 0 \\ 0 \end{bmatrix}$$

$A\,X = B$ $\qquad X := A^{-1}\cdot B \qquad X = \begin{bmatrix} -9.483 - 1.679j \\ -168.099 + 10.075j \\ 77.583 - 8.396j \end{bmatrix} \qquad \begin{bmatrix} r_1 \\ r_2 \\ r_3 \end{bmatrix} := X$

$$r_0 + r_1 + r_2 + r_3 = 4.263\cdot 10^{-14} \qquad \text{(zero)}$$

S11.18

Letting $\quad r_0 := 1$

Given:

$\omega_0 := 0 \qquad \omega_1 := 2 \qquad \omega_2 := 1 \qquad \omega_3 := 1$

$\alpha_0 := 0 \qquad \alpha_1 := 1 \qquad \alpha_2 := 0 \qquad \alpha_3 := -1$

a)

$$A := \begin{bmatrix} \omega_1^{\,1} & \omega_2^{\,1} & \omega_3^{\,1} \\ (j\cdot\alpha_1 - \omega_1)^2 & (j\cdot\alpha_2 - \omega_2)^2 & (j\cdot\alpha_3 - \omega_3)^2 \end{bmatrix} \qquad B := \begin{bmatrix} -1 \\ 0 \\ 0 \end{bmatrix}$$

$A\,X = B$ $\qquad X := A^{-1}\cdot B \qquad X = \begin{bmatrix} 1 \\ -3 - 2j \\ 1 + 2j \end{bmatrix} \qquad \begin{bmatrix} r_1 \\ r_2 \\ r_3 \end{bmatrix} := X$

b)

If $\quad R_0 := -j$

$Y := -j\cdot X \qquad Y = \begin{bmatrix} -j \\ -2 + 3j \\ 2 - j \end{bmatrix}$

5.11.19

Letting $r_0 := 1$

Given:

$\omega_0 := 0 \quad \omega_1 := -1 \quad \omega_2 := 2 \quad \omega_3 := 0$

$\alpha_0 := 0 \quad \alpha_1 := 1 \quad \alpha_2 := 0 \quad \alpha_3 := 2$

a)

$$A := \begin{bmatrix} 1 & 1 & 1 \\ \omega_1 & \omega_2 & \omega_3 \\ j\cdot\alpha_1 - \omega_1^{\,2} & j\cdot\alpha_2 - \omega_2^{\,2} & j\cdot\alpha_3 - \omega_3^{\,2} \end{bmatrix} \qquad B := \begin{bmatrix} -1 \\ 0 \\ 0 \end{bmatrix}$$

$A\,X = B \qquad X := A^{-1}\cdot B \qquad X = \begin{bmatrix} -0.308 - 0.462j \\ -0.154 - 0.231j \\ -0.538 + 0.692j \end{bmatrix} \qquad \begin{bmatrix} r_1 \\ r_2 \\ r_3 \end{bmatrix} := X$

b)

If $R_0 := -6 - 4j$

$Y := (-6 - 4j)\,X \qquad Y = \begin{bmatrix} 4j \\ 2j \\ 6 - 2j \end{bmatrix}$

S11.20

Letting $r_0 := 1$

Given:

$\omega_0 := 0 \quad \omega_1 := -1 \quad \omega_2 := 2 \quad \omega_3 := 0$

$\alpha_0 := 0 \quad \alpha_1 := 0 \quad \alpha_2 := 1 \quad \alpha_3 := 2$

a)

$$A := \begin{bmatrix} 1 & 1 & 1 \\ \omega_1 & \omega_2 & \omega_3 \\ j\cdot\alpha_1 - \omega_1^{\,2} & j\cdot\alpha_2 - \omega_2^{\,2} & j\cdot\alpha_3 - \omega_3^{\,2} \end{bmatrix} \qquad B := \begin{bmatrix} -1 \\ 0 \\ 0 \end{bmatrix}$$

$A\,X = B \qquad X := A^{-1}\cdot B \qquad X = \begin{bmatrix} -0.328 - 0.393j \\ -0.164 - 0.197j \\ -0.508 + 0.59j \end{bmatrix} \qquad \begin{bmatrix} r_1 \\ r_2 \\ r_3 \end{bmatrix} := X$

b)

If $R_0 := -6 - 5j$

$Y := (-6 - 5j)\cdot X \qquad Y = \begin{bmatrix} 4j \\ 2j \\ 6 - j \end{bmatrix}$

11.21 The equations of section 11.6 and the displacement and transmission angle equations for a four-bar linkage were incorporated into a spreadsheet to aid in solving this problem.

a. Link ratios are $r_1/r_0 = 2.32$, $r_2/r_0 = 0.62$ and $r_3/r_0 = 2.43$

See table S11.21a

b. See table S11.21b

c. See figure S11.21

11.22 The equations of section 11.6 and the displacement and transmission angle equations for a four-bar linkage were incorporated into a spreadsheet to aid in solving this problem.

a. Link ratios are $r_1/r_0 = 2.83$, $r_2/r_0 = 1.22$ and $r_3/r_0 = 2.27$

See table S11.22a

b. See table S11.22b

c. See figure S1.22

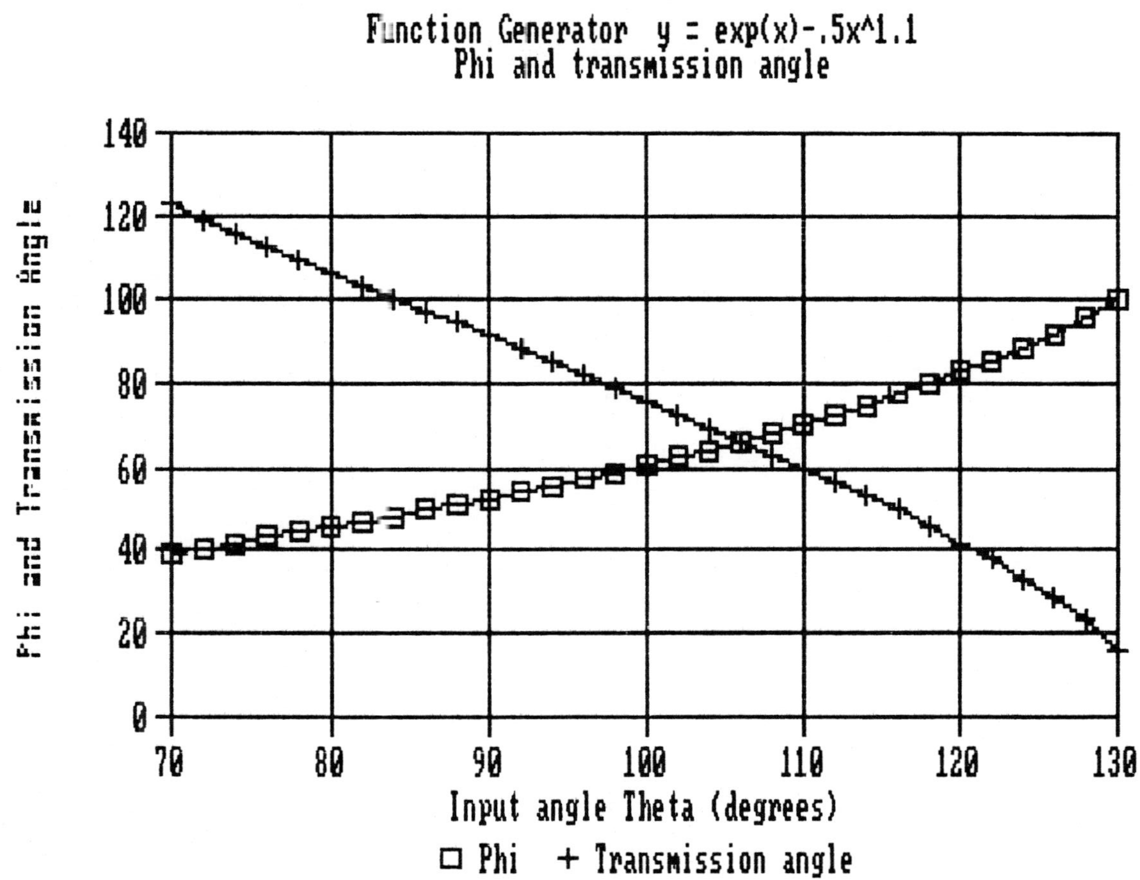
Function Generator y = exp(x)-.5x^1.1
Phi and transmission angle
Phi and Transmission Angle
140
120
100
80
60
40
20
0
70
80
90
100
110
120
130
Input angle Theta (degrees)
□ Phi + Transmission angle

Fig S11.21

Design of a function generator. y = exp(x) - 0.5 x^1.1

x0=	.50	xrange=	1.00	theta0=	70.00	phi0=	40.00
Point	x	theta	y	phi	thetarad	phirad	
1.00	.57	74.02	1.50	42.03	1.29	.73	
2.00	1.00	100.00	2.22	60.50	1.75	1.06	
3.00	1.43	125.98	3.45	91.92	2.20	1.60	
A=	.45	E=	.08	r1/r0=	2.32		
B=	.86	F=	.02	r3/r0=	2.43		
C=	.25	r2x/r0=	-.17	r2/r0=	.62		
D=	.78	r2y/r0=	-.60				

rhteta=	60.00	rphi=	40.00 degrees	y0=	1.42
xmean=	1.00	ymax=	2.98	yrange=	1.57

Table 5 11.21a

Theta deg	Phi deg	x	y	exp(x) -.5x^1.1	Error %
70	39.879	.500	1.411	1.415	-.34
75	42.587	.583	1.517	1.516	.07
80	45.570	.667	1.634	1.628	.36
85	48.831	.750	1.761	1.753	.49
90	52.386	.833	1.900	1.892	.46
95	56.263	.917	2.052	2.047	.28
100	60.503	1.000	2.218	2.218	.00
105	65.161	1.083	2.401	2.408	-.32
110	70.313	1.167	2.602	2.619	-.63
115	76.066	1.250	2.828	2.851	-.83
120	82.593	1.333	3.083	3.108	-.78
125	90.233	1.417	3.382	3.390	-.22
130	100.138	1.500	3.770	3.701	1.88

Table SII.2/b

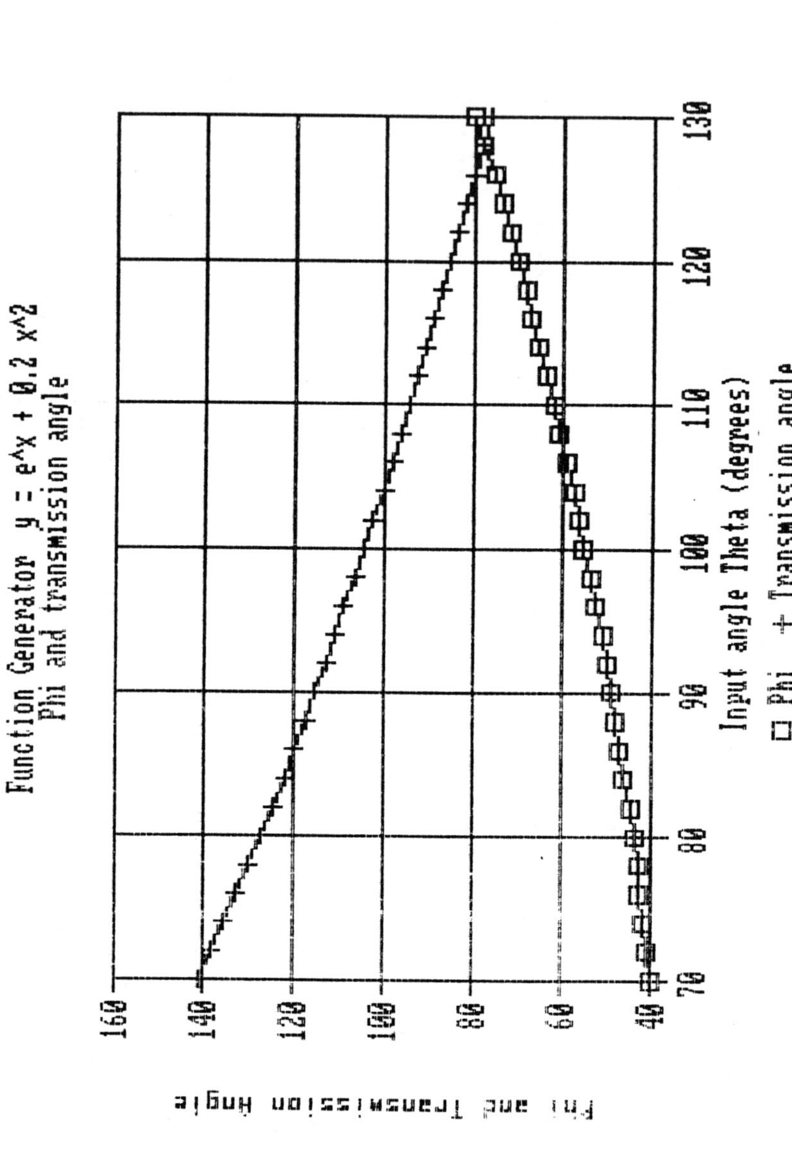

Function Generator y = e^x + 0.2 x^2
Phi and transmission angle
Phi and Transmission Angle
160
140
120
100
80
60
40
70
80
90
100
110
120
130
Input angle Theta (degrees)
□ Phi + Transmission angle

FIG 5.11.22

Design of a function generator. y = exp(x) + 0.2 x^2

x0=	.50	xrange=	1.00	theta0=	70.00	phi0=	40.00
Point	x	theta	y	phi	thetarad	phirad	
1.00	.57	74.02	1.83	41.59	1.29	.73	
2.00	1.00	100.00	2.92	55.09	1.75	.96	
3.00	1.43	125.98	4.60	75.92	2.20	1.33	
0.00	.50	70.00	1.70	40.00	1.22	.70	
4.00	1.50	130.00	4.93	80.00	2.27	1.40	
A=	.45	E=	.14	r1/r0=	2.83		
B=	.86	F=	.20	r3/r0=	2.27		
C=	.18	r2x/r0=	.08	r2/r0=	1.22		
D=	.50	r2y/r0=	-1.22				

rhteta=	60.00	rphi=	40.00 degrees	y0=	1.70
xmean=	1.00	ymax=	4.93	yrange=	3.23

Table S11.22a

Theta deg	Phi deg	x	y	e^x+ .2x^2	Error %
70	40.217	.500	1.716	1.699	1.03
75	41.963	.583	1.857	1.860	-.15
80	44.054	.667	2.026	2.037	-.50
85	46.431	.750	2.218	2.230	-.49
90	49.066	.833	2.431	2.440	-.34
95	51.952	.917	2.665	2.669	-.16
100	55.089	1.000	2.918	2.918	-.00
105	58.487	1.083	3.193	3.189	.11
110	62.157	1.167	3.490	3.483	.17
115	66.117	1.250	3.810	3.803	.18
120	70.385	1.333	4.155	4.149	.13
125	74.980	1.417	4.526	4.525	.03
130	79.920	1.500	4.925	4.932	-.13

Table S11.22b
Performance of a function generator.

CHAPTER 12

INTRODUCTION TO ROBOTIC MANIPULATORS

12.1 (a) DF = 6(4 - 4 - 1) + 1 + 2 + 3 + 1 = 1

(b) DF = 6(4 - 4 - 1) + 1 + 2 + 2 + 2 = 1

(c) DF = 6(4 - 4 - 1) + 1 + 3 + 2 + 1 = 1

(d) DF = 6(4 - 4 - 1) + 1 + 3 + 3 + 1 = 2

12.2 (a) DF = 6(5 - 4 - 1) + 1 + 1 + 2 + 1 = 5

(b) DF = 6(5 - 4 - 1) + 1 + 1 + 1 + 1 = 4

(c) DF = 6(5 - 4 - 1) + 2 + 2 + 2 + 1 = 7

(d) DF = 6(5 - 4 - 1) + 1 + 2 + 2 + 1 = 6

12.3 $^{i}_{j}[R] = [R(x_i,\alpha)][R(z_i,\gamma)][R(y_i,\beta)]$

$$= \begin{bmatrix} 1 & 0 & 0 \\ 0 & \cos\alpha & -\sin\alpha \\ 0 & \sin\alpha & \cos\alpha \end{bmatrix} \begin{bmatrix} \cos\gamma & -\sin\gamma & 0 \\ \sin\gamma & \cos\gamma & 0 \\ 0 & 0 & 1 \end{bmatrix} \begin{bmatrix} \cos\beta & 0 & \sin\beta \\ 0 & 1 & 0 \\ -\sin\beta & 0 & \cos\beta \end{bmatrix}$$

$$= \begin{bmatrix} (\cos\beta\cos\gamma) & (-\sin\gamma) & (\sin\beta\cos\gamma) \\ (\cos\alpha\cos\beta\sin\gamma + \sin\alpha\sin\beta) & (\cos\alpha\cos\gamma) & (\cos\alpha\sin\beta\sin\gamma - \sin\alpha\cos\beta) \\ (\sin\alpha\cos\beta\sin\gamma - \cos\alpha\sin\beta) & (\sin\alpha\cos\gamma) & (\sin\alpha\sin\beta\sin\gamma + \cos\alpha\cos\beta) \end{bmatrix}$$

12.4 $$[R(y,\beta)][R(x,\alpha)] = \begin{bmatrix} \cos\beta & 0 & \sin\beta \\ 0 & 1 & 0 \\ -\sin\beta & 0 & \cos\beta \end{bmatrix} \begin{bmatrix} 1 & 0 & 0 \\ 0 & \cos\alpha & -\sin\alpha \\ 0 & \sin\alpha & \cos\alpha \end{bmatrix}$$

$$= \begin{bmatrix} \cos\beta & \sin\alpha\sin\beta & \cos\alpha\sin\beta \\ 0 & \cos\alpha & -\sin\alpha \\ -\sin\beta & \sin\alpha\cos\beta & \cos\alpha\cos\beta \end{bmatrix}$$

$$[R(x,\alpha)][R(y,\beta)] = \begin{bmatrix} 1 & 0 & 0 \\ 0 & \cos\alpha & -\sin\alpha \\ 0 & \sin\alpha & \cos\alpha \end{bmatrix} \begin{bmatrix} \cos\beta & 0 & \sin\beta \\ 0 & 1 & 0 \\ -\sin\beta & 0 & \cos\beta \end{bmatrix}$$

$$= \begin{bmatrix} \cos\beta & 0 & \sin\beta \\ \sin\alpha\sin\beta & \cos\alpha & -\sin\alpha\cos\beta \\ -\cos\alpha\sin\beta & \sin\alpha & \cos\alpha\cos\beta \end{bmatrix}$$

$$\therefore [R(y,\beta)][R(x,\alpha)] \neq [R(x,\alpha)][R(y,\beta)]$$

12.5 $${}^{i}_{j}[R] = \begin{bmatrix} \cos(-45^\circ) & -\sin(-45^\circ) & 0 \\ \sin(-45^\circ) & \cos(-45^\circ) & 0 \\ 0 & 0 & 1 \end{bmatrix} \begin{bmatrix} \cos 45^\circ & 0 & \sin 45^\circ \\ 0 & 1 & 0 \\ -\sin 45^\circ & 0 & \cos 45^\circ \end{bmatrix} \begin{bmatrix} 1 & 0 & 0 \\ 0 & \cos(-90^\circ) & -\sin(-90^\circ) \\ 0 & \sin(-90^\circ) & \cos(-90^\circ) \end{bmatrix}$$

$$= \begin{bmatrix} 0.707 & 0.707 & 0 \\ -0.707 & 0.707 & 0 \\ 0 & 0 & 1 \end{bmatrix} \begin{bmatrix} 0.707 & 0 & 0.707 \\ 0 & 1 & 0 \\ -0.707 & 0 & 0.707 \end{bmatrix} \begin{bmatrix} 1 & 0 & 0 \\ 0 & 0 & 1 \\ 0 & -1 & 0 \end{bmatrix} = \begin{bmatrix} 0.5 & -0.5 & 0.707 \\ -0.5 & 0.5 & 0.707 \\ -0.707 & -0.707 & 0 \end{bmatrix}$$

$${}^{i}\{P\} = \begin{bmatrix} 0.5 & -0.5 & 0.707 \\ -0.5 & 0.5 & 0.707 \\ -0.707 & -0.707 & 0 \end{bmatrix} \begin{Bmatrix} 1 \\ 1 \\ 1 \end{Bmatrix} = \begin{Bmatrix} 0.707 \\ 0.707 \\ -1.414 \end{Bmatrix}$$

12.6 $${}^{i}_{j}[R] = [R(x_i,90^o)][R(z_i,90^o)][R(y_i,90^o)]$$

$$= \begin{bmatrix} 1 & 0 & 0 \\ 0 & 0 & -1 \\ 0 & 1 & 0 \end{bmatrix} \begin{bmatrix} 0 & -1 & 0 \\ 1 & 0 & 0 \\ 0 & 0 & 1 \end{bmatrix} \begin{bmatrix} 0 & 0 & 1 \\ 0 & 1 & 0 \\ -1 & 0 & 0 \end{bmatrix} = \begin{bmatrix} 0 & -1 & 0 \\ 1 & 0 & 0 \\ 0 & 0 & 1 \end{bmatrix}$$

$${}^{j}_{i}[R] = {}^{i}_{j}[R]^{-1} = [R(y_i,90^o)]^{-1}[R(z_i,90^o)]^{-1}[R(x_i,90^o)]^{-1}$$

$$= [R(y_i,90^o)]^{T}[R(z_i,90^o)]^{T}[R(x_i,90^o)]^{T}$$

$$= \begin{bmatrix} 0 & 0 & -1 \\ 0 & 1 & 0 \\ 1 & 0 & 0 \end{bmatrix} \begin{bmatrix} 0 & 1 & 0 \\ -1 & 0 & 0 \\ 0 & 0 & 1 \end{bmatrix} \begin{bmatrix} 1 & 0 & 0 \\ 0 & 0 & 1 \\ 0 & -1 & 0 \end{bmatrix} = \begin{bmatrix} 0 & 1 & 0 \\ -1 & 0 & 0 \\ 0 & 0 & 1 \end{bmatrix}$$

12.7 $\phi = -90^o$, $u_{x_i} = 0.707$, $u_{y_i} = 0.707$, $u_{z_i} = 0$

$\cos\phi = 0$, $\sin\phi = -1$, $\mathrm{ver}\phi = 1$

$${}^{i}_{j}[R] = \begin{bmatrix} 0.5 & 0.5 & -0.707 \\ 0.5 & 0.5 & 0.707 \\ 0.707 & -0.707 & 0 \end{bmatrix}$$

$${}^{i}\{P\} = {}^{i}_{j}[R]\,{}^{j}\{P\} = \begin{bmatrix} 0.5 & 0.5 & -0.707 \\ 0.5 & 0.5 & 0.707 \\ 0.707 & -0.707 & 0 \end{bmatrix} \begin{Bmatrix} 1 \\ 1 \\ 1 \end{Bmatrix} = \begin{Bmatrix} 0.293 \\ 1.707 \\ 0 \end{Bmatrix}$$

12.8 $${}^{j}_{i}[R] = {}^{i}_{j}[R]^{-1} = [R(x_i,\alpha)]^{-1}[R(y_i,\beta)]^{-1}[R(z_i,\gamma)]^{-1}$$

$$= [R(x_i,\alpha)]^{T}[R(y_i,\beta)]^{T}[R(z_i,\gamma)]^{T} = {}^{i}_{j}[R]^{T}$$

$$= \begin{bmatrix} \cos\beta\cos\gamma & \cos\beta\sin\gamma & -\sin\beta \\ (\sin\alpha\sin\beta\cos\gamma - \cos\alpha\sin\gamma) & (\sin\alpha\sin\beta\sin\gamma + \cos\alpha\cos\gamma) & \sin\alpha\cos\beta \\ (\cos\alpha\sin\beta\cos\gamma + \sin\alpha\sin\gamma) & (\cos\alpha\sin\beta\sin\gamma - \sin\alpha\cos\gamma) & \cos\alpha\cos\beta \end{bmatrix}$$

12.9 $\phi = \alpha$, $u_{x_1} = 1$, $u_{y_1} = 0$, $u_{z_1} = 0$,

$\cos\phi = \cos\alpha$, $\sin\phi = \sin\alpha$, $ver\phi = 1 - \cos\alpha$

$$
{}^{i}_{j}[R] = \begin{bmatrix} 1 & 0 & 0 \\ 0 & \cos\alpha & -\sin\alpha \\ 0 & \sin\alpha & \cos\alpha \end{bmatrix}
$$

12.10 ${}^{i}_{j}[R] = [R(z_i,90^\circ)]\,[R(y_i,90^\circ)]\,[R(x_i,90^\circ)]$

$$
= \begin{bmatrix} 0 & -1 & 0 \\ 1 & 0 & 0 \\ 0 & 0 & 1 \end{bmatrix} \begin{bmatrix} 0 & 0 & 1 \\ 0 & 1 & 0 \\ -1 & 0 & 0 \end{bmatrix} \begin{bmatrix} 1 & 0 & 0 \\ 0 & 0 & -1 \\ 0 & 1 & 0 \end{bmatrix} = \begin{bmatrix} 0 & 0 & 1 \\ 0 & 1 & 0 \\ -1 & 0 & 0 \end{bmatrix}
$$

From Eq. 12.20, $R_{12} + R_{21} = 0 \Rightarrow u_{x_1}u_{y_1}ver\phi = 0$

$R_{13} + R_{31} = 0 \Rightarrow u_{x_1}u_{z_1}ver\phi = 0$

$R_{23} + R_{32} = 0 \Rightarrow u_{y_1}u_{z_1}ver\phi = 0$

Then, from $R_{12} = 0 \Rightarrow u_{z_1}\sin\phi = 0$

and $R_{23} = 0 \Rightarrow u_{x_1}\sin\phi = 0$

One possibility is $\sin\phi = ver\phi = 0 \Rightarrow \phi = 0$

However, this does not satisfy R_{13} *and* R_{31}

$\therefore\ u_{x_1} = u_{z_1} = 0 \Rightarrow u_{y_1} = 1$, $\phi = 90^\circ$

12.11
$$
{}^{1}_{2}[R] = \begin{bmatrix} 1 & 0 & 0 \\ 0 & \cos 90^\circ & -\sin 90^\circ \\ 0 & \sin 90^\circ & \cos 90^\circ \end{bmatrix} \begin{bmatrix} \cos 90^\circ & 0 & \sin 90^\circ \\ 0 & 1 & 0 \\ -\sin 90^\circ & 0 & \cos 90^\circ \end{bmatrix} = \begin{bmatrix} 0 & 0 & 1 \\ 1 & 0 & 0 \\ 0 & 1 & 0 \end{bmatrix}
$$

$$
{}^{1}\{P\} = {}^{1}_{2}[T]\,{}^{2}\{P\} = \begin{bmatrix} 0 & 0 & 1 & 2 \\ 1 & 0 & 0 & 3 \\ 0 & 1 & 0 & 0 \\ 0 & 0 & 0 & 1 \end{bmatrix} \begin{Bmatrix} 3 \\ 4 \\ 1 \\ 1 \end{Bmatrix} = \begin{Bmatrix} 3 \\ 6 \\ 4 \\ 1 \end{Bmatrix}
$$

12.12 $${}^{1}_{2}[R] = \begin{bmatrix} 0 & 1 & 0 \\ -1 & 0 & 0 \\ 0 & 0 & 1 \end{bmatrix} \begin{bmatrix} .707 & 0 & .707 \\ 0 & 1 & 0 \\ -.707 & 0 & .707 \end{bmatrix} \begin{bmatrix} 1 & 0 & 0 \\ 0 & .707 & .707 \\ 0 & -.707 & .707 \end{bmatrix} = \begin{bmatrix} 0 & .707 & .707 \\ -.707 & 0.5 & -0.5 \\ -.707 & -0.5 & 0.5 \end{bmatrix}$$

$${}^{1}\{P\} = {}^{1}_{2}[T]\,{}^{2}\{P\} = \begin{bmatrix} 0 & 0.707 & 0.707 & 1 \\ -0.707 & 0.5 & -0.5 & 0 \\ -0.707 & -0.5 & 0.5 & -2 \\ 0 & 0 & 0 & 1 \end{bmatrix} \begin{Bmatrix} 1 \\ 2 \\ -1 \\ 1 \end{Bmatrix} = \begin{Bmatrix} 1.707 \\ 0.793 \\ -4.207 \\ 1 \end{Bmatrix}$$

12.13 $${}^{0}_{1}[T] = \begin{bmatrix} \cos\theta_A & -\sin\theta_A & 0 & 0 \\ \sin\theta_A & \cos\theta_A & 0 & 0 \\ 0 & 0 & 1 & 0 \\ 0 & 0 & 0 & 1 \end{bmatrix} = \begin{bmatrix} 0.866 & -0.5 & 0 & 0 \\ 0.5 & 0.866 & 0 & 0 \\ 0 & 0 & 1 & 0 \\ 0 & 0 & 0 & 1 \end{bmatrix}$$

$${}^{1}_{2}[T] = \begin{bmatrix} \cos\theta_B & -\sin\theta_B & 0 & \ell_1 \\ \sin\theta_B & \cos\theta_B & 0 & 0 \\ 0 & 0 & 1 & 0 \\ 0 & 0 & 0 & 1 \end{bmatrix} = \begin{bmatrix} 0.866 & -0.5 & 0 & 10 \\ 0.5 & 0.866 & 0 & 0 \\ 0 & 0 & 1 & 0 \\ 0 & 0 & 0 & 1 \end{bmatrix}$$

$${}^{2}_{3}[T] = \begin{bmatrix} \cos\theta_C & -\sin\theta_C & 0 & \ell_2 \\ \sin\theta_C & \cos\theta_C & 0 & 0 \\ 0 & 0 & 1 & 0 \\ 0 & 0 & 0 & 1 \end{bmatrix} = \begin{bmatrix} 0.866 & -0.5 & 0 & 10 \\ 0.5 & 0.866 & 0 & 0 \\ 0 & 0 & 1 & 0 \\ 0 & 0 & 0 & 1 \end{bmatrix}$$

$${}^{0}_{3}[T] - {}^{0}_{1}[T]\,{}^{1}_{2}[T]\,{}^{2}_{3}[T]$$

$$= \begin{bmatrix} \cos(\theta_A+\theta_B+\theta_C) & -\sin(\theta_A+\theta_B+\theta_C) & 0 & \ell_1\cos\theta_A+\ell_2\cos(\theta_A+\theta_B) \\ \sin(\theta_A+\theta_B+\theta_C) & \cos(\theta_A+\theta_B+\theta_C) & 0 & \ell_1\sin\theta_A+\ell_2\sin(\theta_A+\theta_B) \\ 0 & 0 & 1 & 0 \\ 0 & 0 & 0 & 1 \end{bmatrix}$$

$${}^{0}\{P\}_D = {}^{0}_{3}[T]\,{}^{3}\{P\}_D = \begin{bmatrix} 0 & -1 & 0 & 13.66 \\ 1 & 0 & 0 & 13.66 \\ 0 & 0 & 1 & 0 \\ 0 & 0 & 0 & 1 \end{bmatrix} \begin{Bmatrix} 10 \\ 0 \\ 0 \\ 1 \end{Bmatrix} = \begin{Bmatrix} 13.66 \\ 23.66 \\ 0 \\ 1 \end{Bmatrix} \text{ in.}$$

12.14 *Define* x_1, y_1, z_1 *to be coincident with* x_0, y_0, z_0 *at the position shown. Then* $\ell_i = \tau_i = s_n = 0$, $\theta_n = \theta_A$, *and*

$$ {}^0_1[T] = \begin{bmatrix} \cos\theta_A & -\sin\theta_A & 0 & 0 \\ \sin\theta_A & \cos\theta_A & 0 & 0 \\ 0 & 0 & 1 & 0 \\ 0 & 0 & 0 & 1 \end{bmatrix} = \begin{bmatrix} \cos 30^\circ & -\sin 30^\circ & 0 & 0 \\ \sin 30^\circ & \cos 30^\circ & 0 & 0 \\ 0 & 0 & 1 & 0 \\ 0 & 0 & 0 & 1 \end{bmatrix} $$

Locate x_2, y_2, z_2 *with* z_2 *along axis B and* x_2 *along link 2. Then* $\ell_i = \ell_1$, $\tau_i = 90^\circ$, $s_n = 0$, $\theta_n = \theta_B$, *and*

$$ {}^1_2[T] = \begin{bmatrix} \cos\theta_B & -\sin\theta_B & 0 & \ell_1 \\ 0 & 0 & -1 & 0 \\ \sin\theta_B & \cos\theta_B & 0 & 0 \\ 0 & 0 & 0 & 1 \end{bmatrix} = \begin{bmatrix} \cos 30^\circ & -\sin 30^\circ & 0 & 10 \\ 0 & 0 & -1 & 0 \\ \sin 30^\circ & \cos 30^\circ & 0 & 0 \\ 0 & 0 & 0 & 1 \end{bmatrix} $$

Then, for ${}^2_3[T]$, $\ell_i = \ell_2$, $\tau_i = s_n = 0$, $\theta_n = \theta_C$, *and*

$$ {}^2_3[T] = \begin{bmatrix} \cos\theta_C & -\sin\theta_C & 0 & \ell_2 \\ \sin\theta_C & \cos\theta_C & 0 & 0 \\ 0 & 0 & 1 & 0 \\ 0 & 0 & 0 & 1 \end{bmatrix} = \begin{bmatrix} \cos 30^\circ & -\sin 30^\circ & 0 & 10 \\ \sin 30^\circ & \cos 30^\circ & 0 & 0 \\ 0 & 0 & 1 & 0 \\ 0 & 0 & 0 & 1 \end{bmatrix} $$

Therefore,

$$ {}^0_3[T] = \begin{bmatrix} 0.866 & -0.5 & 0 & 0 \\ 0.5 & 0.866 & 0 & 0 \\ 0 & 0 & 1 & 0 \\ 0 & 0 & 0 & 1 \end{bmatrix} \begin{bmatrix} 0.866 & -0.5 & 0 & 10 \\ 0 & 0 & -1 & 0 \\ 0.5 & 0.866 & 0 & 0 \\ 0 & 0 & 0 & 1 \end{bmatrix} \begin{bmatrix} 0.866 & -0.5 & 0 & 10 \\ 0.5 & 0.866 & 0 & 0 \\ 0 & 0 & 1 & 0 \\ 0 & 0 & 0 & 1 \end{bmatrix} $$

$$ = \begin{bmatrix} 0.433 & -0.750 & 0.500 & 16.2 \\ 0.250 & -0.433 & -0.866 & 9.33 \\ 0.866 & 0.500 & 0 & 5 \\ 0 & 0 & 0 & 1 \end{bmatrix} $$

$$ {}^0\{P\}_D = {}^0_3[T]\,{}^3\{P\}_D = {}^0_3[T] \begin{Bmatrix} 10 \\ 0 \\ 0 \\ 1 \end{Bmatrix} = \begin{Bmatrix} 20.5 \\ 11.8 \\ 13.7 \\ 1 \end{Bmatrix} \text{ in.} $$

12.15 *Locate axes* x_1, y_1, z_1 *coincident with axes* x_0, y_0, z_0 *at the position shown. Then, for* ${}^{0}_{1}[T]$, $\ell_i = \tau_i = s_n = 0$, $\theta_n = \theta_A$, *and*

$$
{}^{0}_{1}[T] = \begin{bmatrix} \cos\theta_A & -\sin\theta_A & 0 & 0 \\ \sin\theta_A & \cos\theta_A & 0 & 0 \\ 0 & 0 & 1 & 0 \\ 0 & 0 & 0 & 1 \end{bmatrix} = \begin{bmatrix} \cos45° & -\sin45° & 0 & 0 \\ \sin45° & \cos45° & 0 & 0 \\ 0 & 0 & 1 & 0 \\ 0 & 0 & 0 & 1 \end{bmatrix}
$$

For ${}^{1}_{2}[T]$, $\ell_i = \ell_1$, $\tau_i = 90°$, $s_n = b_1$, $\theta_n = \theta_B$, *and*

$$
{}^{1}_{2}[T] = \begin{bmatrix} \cos\theta_B & -\sin\theta_B & 0 & \ell_1 \\ 0 & 0 & -1 & -b_1 \\ \sin\theta_B & \cos\theta_B & 0 & 0 \\ 0 & 0 & 0 & 1 \end{bmatrix} = \begin{bmatrix} \cos45° & -\sin45° & 0 & 10 \\ 0 & 0 & -1 & -5 \\ \sin45° & \cos45° & 0 & 0 \\ 0 & 0 & 0 & 1 \end{bmatrix}
$$

Then,

$$
{}^{0}_{2}[T] = {}^{0}_{1}[T]\,{}^{1}_{2}[T] = \begin{bmatrix} 0.5 & -0.5 & 0.707 & 10.6 \\ 0.5 & -0.5 & -0.707 & 3.54 \\ 0.707 & 0.707 & 0 & 0 \\ 0 & 0 & 0 & 1 \end{bmatrix}
$$

$$
{}^{0}\{P\}_C = {}^{0}_{2}[T]\,{}^{2}\{P\}_C = \begin{bmatrix} 0.5 & -0.5 & 0.707 & 10.6 \\ 0.5 & -0.5 & -0.707 & 3.54 \\ 0.707 & 0.707 & 0 & 0 \\ 0 & 0 & 0 & 1 \end{bmatrix} \begin{Bmatrix} 10 \\ 5 \\ 0 \\ 1 \end{Bmatrix} = \begin{Bmatrix} 13.1 \\ 6.04 \\ 10.6 \\ 1 \end{Bmatrix} \text{ in.}
$$

12.16 For ${}^0_1[T]$, $\ell_i = \tau_i = \theta_n = 0$, $s_n = s_A$, and

$$ {}^0_1[T] = \begin{bmatrix} 1 & 0 & 0 & 0 \\ 0 & 1 & 0 & 0 \\ 0 & 0 & 1 & s_A \\ 0 & 0 & 0 & 1 \end{bmatrix} = \begin{bmatrix} 1 & 0 & 0 & 0 \\ 0 & 1 & 0 & 0 \\ 0 & 0 & 1 & 5 \\ 0 & 0 & 0 & 1 \end{bmatrix} $$

For ${}^1_2[T]$, $\ell_i = \theta_n = 0$, $\tau_i = -90°$, $s_n = s_B$, and

$$ {}^1_2[T] = \begin{bmatrix} 1 & 0 & 0 & 0 \\ 0 & 0 & 1 & s_B \\ 0 & -1 & 0 & 0 \\ 0 & 0 & 0 & 1 \end{bmatrix} = \begin{bmatrix} 1 & 0 & 0 & 0 \\ 0 & 0 & 1 & 5 \\ 0 & -1 & 0 & 0 \\ 0 & 0 & 0 & 1 \end{bmatrix} $$

$$ {}^0_2[T] = {}^0_1[T]\,{}^1_2[T] = \begin{bmatrix} 1 & 0 & 0 & 0 \\ 0 & 0 & 1 & s_B \\ 0 & -1 & 0 & s_A \\ 0 & 0 & 0 & 1 \end{bmatrix} = \begin{bmatrix} 1 & 0 & 0 & 0 \\ 0 & 0 & 1 & 5 \\ 0 & -1 & 0 & 5 \\ 0 & 0 & 0 & 1 \end{bmatrix} $$

$$ {}^0\{P\}_C = {}^0_2[T]\,{}^2\{P\}_C = \begin{bmatrix} 1 & 0 & 0 & 0 \\ 0 & 0 & 1 & 5 \\ 0 & -1 & 0 & 5 \\ 0 & 0 & 0 & 1 \end{bmatrix} \begin{Bmatrix} 10 \\ 0 \\ 10 \\ 1 \end{Bmatrix} = \begin{Bmatrix} 10 \\ 15 \\ 5 \\ 1 \end{Bmatrix} \text{ in.} $$

12.17 $x = -6$, $y = 12$, $\ell_1 = \ell_2 = 10$

$$ \alpha = \tan^{-1}\left(\frac{y}{x}\right) = \tan^{-1}\left(\frac{12}{-6}\right) = 116.6° $$

$$ \beta = \cos^{-1}\left(\frac{36 + 144 + 100 - 100}{2(10)\sqrt{36 + 144}}\right) = \cos^{-1}\left(\frac{9}{\sqrt{180}}\right) = 47.9° $$

$$ \theta_A = \alpha \pm \beta = 68.7°,\ 164.5° $$

$$ \theta_B = \cos^{-1}\left(\frac{36 + 144 - 100 - 100}{2(10)(10)}\right) = \cos^{-1}(-0.1) = \pm 95.7° $$

$$ \therefore\ \theta_A^{(1)} = 68.7°,\ \theta_B^{(1)} = 95.7°,\ \text{and } \theta_A^{(2)} = 164.5°,\ \theta_B^{(2)} = -95.7° $$

12.18 $x = 12\,,\quad y = 6\,,\quad z = 18\,,\quad s_A = \ell_2 = \ell_3 = 10$

$$\theta_A^{(1)} = \theta_A^{(2)} = \tan^{-1}\left(\frac{6}{12}\right) = 26.6^{\circ}$$

$$\cos\theta_C = \left(\frac{144+36+64-100-100}{2(10)(10)}\right) = 0.220,\ \theta_C = \pm 77.3^{\circ}$$

$$\sin\theta_C = \pm\sqrt{1-(0.22)^2} = \pm 0.975$$

For $\theta_C^{(1)} = -77.3^{\circ}$,

$$\theta_B^{(1)} = \tan^{-1}\left\{\frac{8[10+10(0.220)](0.894)-12(10)(-0.975)}{12[10+10(0.220)]+8(10)(-0.975)(0.894)}\right\} = 69.4^{o}$$

For $\theta_C^{(2)} = 77.3^{\circ}$,

$$\theta_B^{(2)} = \tan^{-1}\left\{\frac{8[10+10(0.220)](0.894)-12(10)(0.975)}{12[10+10(0.220)]+8(10)(0.975)(0.894)}\right\} = -7.8^{\circ}$$

12.19 $x = 12\,,\quad y = 6\,,\quad z = 8\,,\quad \ell_2 = 10$

$s_A = 8\ in.\,,\qquad s_B = \sqrt{144+36} - 10 = 3.42\ in.$

$$\theta_A = \tan^{-1}\left(\frac{-12}{6}\right) = -63.4^{\circ}$$

12.20 $x = [\ell_1 + \ell_2\cos\theta_B + \ell_3\cos(\theta_B+\theta_C)]\cos\theta_A$

$y = [\ell_1 + \ell_2\cos\theta_B + \ell_3\cos(\theta_B+\theta_C)]\sin\theta_A$

$z = \ell_2\sin\theta_B + \ell_3\sin(\theta_B+\theta_C)$

Solving, $\theta_A = \tan^{-1}\left(\frac{y}{x}\right)$

$$\left(\frac{x}{\cos\theta_A} - \ell_1\right)^2 + z^2 = \ell_2^2 + \ell_3^2 + 2\ell_2\ell_3\cos\theta_C$$

or $$\theta_C = \cos^{-1}\left[\frac{\left(\frac{x}{\cos\theta_A} - \ell_1\right)^2 + z^2 - \ell_2^2 - \ell_3^2}{2\ell_2\ell_3}\right]$$

$$(\ell_2 + \ell_3\cos\theta_C)\cos\theta_B - \ell_3\sin\theta_C\sin\theta_B = \frac{x}{\cos\theta_A} - \ell_1$$

$$\ell_3\sin\theta_C\cos\theta_B + (\ell_2 + \ell_3\cos\theta_C)\sin\theta_B = z$$

and $$\theta_B = \tan^{-1}\left[\frac{z(\ell_2+\ell_3\cos\theta_C) - \ell_3\sin\theta_C\left(\frac{x}{\cos\theta_A} - \ell_1\right)}{(\ell_2+\ell_3\cos\theta_C)\left(\frac{x}{\cos\theta_A} - \ell_1\right) + z\ell_3\sin\theta_C}\right]$$